PROTEIN FOLDING AND MISFOLDING: NEURODEGENERATIVE DISEASES

FOCUS ON STRUCTURAL BIOLOGY

Volume 7

Series Editor
ROB KAPTEIN
*Bijvoet Center for Biomolecular Research,
Utrecht University, The Netherlands*

Protein Folding and Misfolding: Neurodegenerative Diseases

Edited by

JUDIT OVÁDI

Hungarian Academy of Sciences, Biological Research Center,
Institute of Enzymology, Hungary

and

FERENC OROSZ

Hungarian Academy of Sciences, Biological Research Center,
Institute of Enzymology, Hungary

Editors
Judit Ovádi
Hungarian Academy of
Sciences
Biological Research Center
Institute of Enzymology
Karolina út 29
Budapest 1113
Hungary
ovadi@enzim.hu

Ferenc Orosz
Hungarian Academy of
Sciences
Biological Research Center
Institute of Enzymology
Karolina út 29
Budapest 1113
Hungary
orosz@enzim.hu

ISBN: 978-1-4020-9433-0   e-ISBN: 978-1-4020-9434-7

DOI 10.1007/978-1-4020-9434-7

Library of Congress Control Number: 2008938904

© Springer Science+Business Media B.V. 2009
No part of this work may be reproduced, stored in a retrieval system, or transmitted
in any form or by any means, electronic, mechanical, photocopying, microfilming, recording
or otherwise, without written permission from the Publisher, with the exception
of any material supplied specifically for the purpose of being entered
and executed on a computer system, for exclusive use by the purchaser of the work.

Printed on acid-free paper

9 8 7 6 5 4 3 2 1

springer.com

Foreword

It was twenty five years ago this year that for the first time a protein underlying a form of human cerebral amyloidosis, the Icelandic-type hereditary cerebral haemorrhage was identified. This, together with the recognition that an amino acid substitution can transform the wild type cystatin C into a disease-associated amyloid-forming protein in this condition, was only a prelude to a series of important discoveries that followed. As a result, pathologically altered proteins have been brought into the centre stage of research into the pathomechanism of a number of neurodegenerative diseases, which include epidemiologically such important conditions as Alzheimer's disease or Parkinson's disease and, among others, also the transmissible spongiform encephalopathies, Huntington's chorea, spinocerebellar ataxias, frontotemporal lobar degenerations and amyotrophic lateral sclerosis. Despite the diversity in the amino acid sequence of the different proteins involved in these neurological diseases, one of the common themes underlying the pathomechanisms of all these conditions is protein misfolding, aggregation – hence the term protein folding disorders –, which can trigger cascades of events ultimately resulting in synapse loss and neuron death with devastating clinical consequences in many of the most precious spheres of human existence including personality, cognition, memory, skilled movements and affection.

It is always a challenging task to unite the different topics of the individual chapters into a common theme in a multi-author volume, but the current book edited by Judit Ovadi and Ferenc Orosz fits this task admirably. The contributors of the chapters are very well-chosen to cover a good number of topical areas of neurodegenerative research. Without exception the chapters set forth clearly the current understanding of their chosen topics, which will allow both the specialist reader and the novice entering into the field to acquire the information they require to find. I have no hesitation in expecting that this wisely edited book will shortly become a well-thumbed text on the bookshelves of many research libraries and offices.

London July 2008

Tamas Revesz FRCPath
Professor of Neuropathology
Institute of Neurology
University College London
UK

Preface

The worldwide ageing of populations has brought the neurodegenerative diseases into the focus of interest. These diseases constitute large variety of pathological conditions originating from the slow, irreversible and systemic loss of cells in different regions of the brain resulting in degenerative problems with distinct clinical symptoms. The pathological behaviors are frequently associated with "proteinopathies", the non-physiological behavior of a specific protein, affecting its processing, functioning, and/or folding. These proteins do not have stable tertiary and/or secondary structures in vivo; they enter into aberrant interactions affecting their folding state and function. A number of the diverse human neurodegenerative diseases are now recognized as conformational diseases because these are caused by aggregations of unfolded or misfolded proteins. Knowledge on the intrinsically unstructured proteins, a relatively newly recognized family of gene products as well as on the misfolded proteins produced by genetic mutations or environmental effects has been extensively accumulated in the past years. These proteins frequently cause proteolytic stress and/or enter into aberrant, non-physiological protein–protein interactions leading to sequestration of protein aggregates which are assemblies of many not-yet-identified components in addition to the deposition of well-characterized misfolded peptides and proteins. Such fate is known in the cases of Aβ peptide and tau protein in Alzheimer's disease, α-synuclein in Parkinson's disease, the extended polyglutamine stretch of mutant huntingtin in Huntington's disease and the prion protein in prion diseases. These protein assemblies display diverse ultrastructures such aggresomes, fibers, oligomers or amorphous structures, however, the nature of these species concerning their cytoprotective or cytotoxic effects has not been clarified yet. The understanding of the course and pathomechanism of the diseases arising from interactions of the so called malfolded proteins is crucial for finding effective therapeutic interventions. The identification of aberrant protein-protein interaction(s) playing constitutive role in aggregate formation contributes to the development of new pharmacofors that could prevent or circumvent the development of neurodegenerative disorders in human.

The main focus of this issue is to review the molecular events initiated by unfolded or misfolded proteins leading to cell death via the development of pathological inclusions, with special emphasis on the macromolecular associations of the malfolded proteins into characteristic ultrastructures found in the cases of neurological disorders, some of them are shown in this issue. There are papers which uncover in details the intriguing interconnections between intrinsic disorder and human neurodegenerative diseases; the characterization of the diseases in relation to their hallmark proteins and ultrastructures. Other papers provide conceptual background of the molecular mechanism of the tendency of disordered proteins for aggregation in vitro and in vivo connected with misfolding diseases. Due to the fundamental biological importance of protein aggregates, and our poor knowledge about the molecular basis or specificity of the general phenomenon of protein aggregation, this problem will be specifically discussed. In the light of the protein based neuropathology the classification of the human neurodegenerative diseases is presented. This book also reviews the structural knowledge accumulated for well-studied and for newly discovered proteins involved in paradigmatic conformational disorders with the aim to broaden our understanding of the pathomechanisms of neurodegeneration, which is crucial for finding effective therapeutic interventions that could prevent or circumvent the development of neurodegenerative disorders in humans.

Acknowledgments We are grateful to the Hungarian National Scientific Research Fund (OTKA) and the European Union FP6-2003-LIFESCIHEALTH-I Biosim Fund for providing many years of valuable support to our research, which has also enabled us to edit this volume.

Budapest
July 2008

Judit Ovádi
Ferenc Orosz
Institute of Enzymology, BRC
Hungarian Academy of Sciences
Budapest, Hungary

Contents

1 **Structural Disorder and Its Connection with Misfolding Diseases** 1
Veronika Csizmók and Peter Tompa

2 **Intrinsic Disorder in Proteins Associated with Neurodegenerative Diseases** 21
Vladimir N. Uversky

3 **Dynamic Role of Ubiquitination in the Management of Misfolded Proteins Associated with Neurodegenerative Diseases** 77
Esther S.P. Wong, Jeanne M.M. Tan and Kah-Leong Lim

4 **Protein Misfolding and Axonal Protection in Neurodegenerative Diseases** 97
Haruhisa Inoue, Takayuki Kondo, Ling Lin, Sha Mi, Ole Isacson and Ryosuke Takahashi

5 **Endoplasmic Reticulum Stress in Neurodegeneration** 111
Jeroen J.M. Hoozemans and Wiep Scheper

6 **Involvement of Alpha-2 Domain in Prion Protein Conformationally-Induced Diseases** 133
Luisa Ronga, Pasquale Palladino, Ettore Benedetti, Raffaele Ragone and Filomena Rossi

7 **Synuclein Structure and Function in Parkinson's Disease** 159
David Eliezer

8 **Inhibition of α-Synuclein Aggregation by Antioxidants and Chaperones in Parkinson's Disease** 175
Jean-Christophe Rochet and Fang Liu

9   **Novel Proteins in α-Synucleinopathies** 207
    Christine Lund Kragh and Poul Henning Jensen

10 **TPPP/p25: A New Unstructured Protein Hallmarking Synucleinopathies** 225
    Ferenc Orosz, Attila Lehotzky, Judit Oláh and Judit Ovádi

11 **Protein-Based Neuropathology and Molecular Classification of Human Neurodegenerative Diseases** 251
    Gabor G. Kovacs and Herbert Budka

Index 273

Contributors

Ettore Benedetti Dipartimento delle Scienze Biologiche & C.I.R.Pe.B.,Università degli Studi Federico II di Napoli, Napoli, Italy

Herbert Budka Institute of Neurology, Medical University of Vienna, AKH 4J, POB 48, A-1097 Wien, Austria

Veronika Csizmók Institute of Enzymology, Biological Research Center, Hungarian Academy of Sciences, Karolina út 29, Budapest, H-1113 Hungary, e-mail: csv@enzim.hu

David Eliezer Weill Cornell Medical College, 1300 York Avenue, New York, NY 10065, USA, e-mail: eliezer@med.cornell.edu

Jeroen J.M. Hoozemans Department of Pathology, VU University Medical Center, P.O. Box 7057, 1007 MB Amsterdam, The Netherlands, e-mail: jjm.hoozemans@vumc.nl

Haruhisa Inoue Department of Neurology, Kyoto University Graduate School of Medicine, 54 Kawahara-cho, Shogoin, Sakyo-ku, Kyoto 606-8507, Japan, e-mail: haruhisa@kuhp.kyoto-u.ac.jp

Ole Isacson Neuroregeneration Laboratories and Center for Neuroregeneration Research at McLean Hospital/Harvard Medical School, NINDS Morris K. Udall Parkinson's Disease Research Center of Excellence, 115 Mill Street, Belmont, MA 02478, USA, e-mail: isacson@mclean.harvard.edu

Poul Henning Jensen Institute of Medical Biochemistry, University of Aarhus, Ole Worms Allé 1.170, DK-8000 Aarhus C, Denmark, e-mail: phj@biokemi.au.dk

Takayuki Kondo Department of Neurology, Kyoto University Graduate School of Medicine, 54 Kawahara-cho, Shogoin, Sakyo-ku, Kyoto 606-8507, Japan

Gabor G. Kovacs Institute of Neurology, Medical University of Vienna, AKH 4J, POB 48, A-1097 Wien, Austria, e-mail: gabor.kovacs@meduniwien.ac.at

Christine Lund Kragh Institute of Medical Biochemistry, University of Aarhus, Ole Worms Allé 1.170, DK-8000 Aarhus C, Denmark, e-mail: clp@biokemi.au.dk

Attila Lehotzky Institute of Enzymology, Biological Research Center, Hungarian Academy of Sciences, Karolina út 29, Budapest, H-1113 Hungary, e-mail: lehotzky@enzim.hu

Kah-Leong Lim Neurodegeneration Research Laboratory, National Neuroscience Institute, Singapore and Duke-NUS Graduate Medical School, Singapore; 11 Jalan Tan Tock Seng, 308433, Singapore, e-mail: Kah_Leong_Lim@nni.com.sg

Ling Lin Neuroregeneration Laboratories and Center for Neuroregeneration Research at McLean Hospital/Harvard Medical School, NINDS Morris K. Udall Parkinson's Disease Research Center of Excellence 115 Mill Street, Belmont, MA 02478, USA, e-mail: llin@mclean.harvard.edu

Fang Liu Department of Medicinal Chemistry and Molecular Pharmacology, Purdue University, West Lafayette, IN 47907-2091, USA

Sha Mi Department of DiscoveryNeurobiology, Biogen Idec Inc., 14 Cambridge Center, Cambridge, MA 02142, United States, e-mail:sha.mi@biogenidec.com

Judit Oláh Institute of Enzymology, Biological Research Center, Hungarian Academy of Sciences, Karolina út 29, Budapest, H-1113 Hungary, e-mail: olju@enzim.hu

Ferenc Orosz Institute of Enzymology, Biological Research Center, Hungarian Academy of Sciences, Karolina út 29, Budapest, H-1113 Hungary, e-mail: orosz@enzim.hu

Judit Ovádi Institute of Enzymology, Biological Research Center, Hungarian Academy of Sciences, Karolina út 29, Budapest, H-1113 Hungary, e-mail: ovadi@enzim.hu

Pasquale Palladino Dipartimento delle Scienze Biologiche & C.I.R.Pe.B., Università degli Studi Federico II di Napoli, Napoli, Italy

Raffaele Ragone Dipartimento di Biochimica e Biofisica & CRISCEB, Seconda Università di Napoli, Napoli, Italy

Tamas Revesz Queen Square Brain Bank, Department of Molecular Neuroscience, UCL Institute of Neurology, Queen Square, London, WC1N 3BG, UK, e-mail: T.Revesz@ion.ucl.ac.uk

Jean-Christophe Rochet Department of Medicinal Chemistry and Molecular Pharmacology, Purdue University, West Lafayette, IN 47907-2091, USA, e-mail: rochet@pharmacy.purdue.edu.

Luisa Ronga Dipartimento delle Scienze Biologiche & C.I.R.Pe.B., Università degli Studi Federico II di Napoli, Napoli, Italy

Filomena Rossi Dipartimento delle Scienze Biologiche-Sez. Biostrutture, Università degli Studi "Federico II" di Napoli, Via Mezzocannone, 16, I-80134 Napoli, Italy, e-mail: filomena.rossi@unina.it

Wiep Scheper Neurogenetics Laboratory and Department of Neurology, Academic Medical Center, University of Amsterdam, Amsterdam, The Netherlands

Ryosuke Takahashi Department of Neurology, Kyoto University Graduate School of Medicine, 54 Kawahara-cho, Shogoin, Sakyo-ku, Kyoto 606-8507, Japan, e-mail: ryosuket@kuhp.kyoto-u.ac.jp

Jeanne M.M. Tan Neurodegeneration Research Laboratory, National Neuroscience Institute, Singapore, Singapore

Peter Tompa Institute of Enzymology, Biological Research Center, Hungarian Academy of Sciences, Karolina út 29, Budapest, H-1113 Hungary, e-mail: tompa@enzim.hu

Esther S.P. Wong Neurodegeneration Research Lab, National Neuroscience Institute, Singapore and Department of Anatomy and Structural Biology, Marion Bessin Liver Research Center, Albert Einstein College of Medicine, USA

Vladimir N. Uversky Center for Computational Biology and Bioinformatics, Department of Biochemistry and Molecular Biology, Institute for Intrinsically Disordered Protein Research, Indiana University School of Medicine, Indianapolis, Indiana 46202, USA and Institute for Biological Instrumentation, Russian Academy of Sciences 142290 Pushchino, Moscow Region, Russia And Molecular Kinetics, Inc., Indianapolis, IN 46268, USA, e-mail: vuversky@iupui.edu

Chapter 1
Structural Disorder and Its Connection with Misfolding Diseases

Veronika Csizmók and Peter Tompa

Abstract Intrinsically disordered proteins or regions of proteins lack a well-defined structure, yet they carry out important functions often associated with the regulation of cell cycle and transcription. Due to these central roles in key cellular processes, their mutations are frequently involved in neurodegenerative diseases. These diseases are usually caused by the structural transition of disordered proteins to insoluble, highly ordered deposits termed amyloids, such a fate has been described in the case of Aβ peptide and tau protein in Alzheimer's disease, α-synuclein in Parkinson's disease or the polyglutamin stretch of huntingtin in Huntington's disease and the prion protein in prion diseases. Due to the involvement of critical conformational change, these diseases are often denoted as "protein misfolding" diseases. Here we provide a brief overview of the rapidly expanding field of protein disorder to provide a conceptual background for the discussion of the essence of molecular mechanisms of these diseases. We will provide a brief overview of the field in general, directing focus on the tendency of disordered proteins for aggregation *in vitro* and also *in vivo*. We will provide some details on neurodegenerative diseases and the proteins involved. It will be shown that the underlying phenomenon of "misfolding" may also result in altering the normal function of proteins (physiological prions). We will wrap up the story by showing that the conformational transition occurs via partially ordered intermediates, which lead to a highly structured cross-β state in amyloids.

1.1 The Concept of Protein Disorder

The classical structure-function paradigm that appeared unshakable for decades rested on the correspondence between function and a well-folded 3D structure. The basic insight provided by this notion into the function of enzymes, receptors

---

V. Csizmók (✉)
Institute of Enzymology, Biological Research Center, Hungarian Academy of Sciences
Karolina út 29, Budapest, H-1113 Hungary
e-mail: csv@enzim.hu

and structural proteins has precluded alternative views. The spectacular advance of structural biology crowned recently by the success of structural genomics programs has made it the central dogma of molecular biology that a unique structure encoded by sequence is the prerequisite of function. More than 50.000 structures deposited in the protein data bank (PDB, www.pdb.org) bear witness to the power of this paradigm. Its generality, however, does not infer universality, as indicated by the recent recognition that many proteins or regions of proteins lack a well-defined three-dimensional structure under native, physiological conditions [1–4]. The recognition that intrinsic disorder is the native, functional state of these proteins, has brought about the demand of re-assessing the structure-function paradigm [5].

The polypeptide chain of intrinsically disordered, or unstructured, proteins (IDPs/IUPs) assumes a fluctuating ensemble of alternative conformations, which is the prerequisite of their functions. In a structural sense, IDPs occupy a continuum of states from a fully disordered state devoid of either short- or long-range intrachain interactions (*random coil*) to a compact state of significant secondary and tertiary contacts (*molten globule*) [6, 7]. These states in many aspects resemble those attained by globular proteins under highly denaturing conditions. Unlike globular proteins, however, which most often carry out their function as enzymes, small-ligand binding receptors or structural proteins, IDP functions stem from the unfolded states, and are mostly involved in regulating processes of signal transduction and transcription regulation [8–10]. Functional classification of IDPs into six categories is based on that in one category (*entropic chains*) their function directly stems from disorder, whereas in the other five categories their function stems from transient (*display sites, chaperones*) or permanent (*effectors, assemblers, scavengers*) binding to partner molecules [2, 3, 11].

The prevalence of structural disorder in regulatory functions results from the functional advantages structural disorder provides. Among many advantages, most often mentioned and discussed are the separation of specificity from binding strength [5], adaptability to various partners [12] and frequent involvement in post-translational modifications [13]. These and other advantages explain the advance of protein disorder in evolution, i.e. its much higher frequency in eukaryotes than prokaryotes [8–10], and its high proportion/dominance in functionally important proteins also noted in disease, such as tau protein [14], p53 [15], α-synuclein [16], prion protein [17], or BRCA1 [18]. The current most complete collection of IDPs is in the DisProt database (www.disprot.org), which contains about 500 proteins, in which biophysical evidence points to the structural disorder of about 1100 regions [19]. DisProt, and previous less-complete collections of disordered proteins enabled the development of about 25 bioinformatics predictors [20, 21]. The application of such predictors to whole genomes and/or proteomes has suggested that about 5–15% of proteins are fully disordered, and 30–50% of proteins contain at least one long disordered region in higher organisms [8–10].

1.2 Biophysical and Bioinformatics Characterization of Disorder

1.2.1 Biophysical Techniques

The primary observation of the unusual behavior of proteins of heat-stability and anomalous SDS-PAGE mobility, circular dichroism (CD) and NMR spectra suggesting a "denatured" state, as well as the frequent observation of missing coordinates from X-ray structures, have led to the formulation of the concept of protein disorder. The first collection of disorder datasets then led to the creation of bioinformatics tools which brought about the recognition of the generality of protein disorder. To respect this historical order of events, we first survey the most important biophysical methods used for recognition and characterization of disorder, followed by a brief overview of the bioinformatics methods. The physical characteristics of IDPs contrasting globular proteins is apparent with all possible approaches, which explains the multiplicity of methods that can be applied for studying protein disorder [2, 7, 20, 22].

Observations by indirect techniques may provide the first line of evidence for the disorder of a protein. IDPs are resistant to heat and low pH, which form the basis of enrichment strategies employed for their proteomic identification [23, 24]. Their aberrant mobility on SDS-PAGE, suggestive of an unusual amino acid composition, has also been frequently noted in the literature [2]. The open and exposed structural character of their unfolded polypeptide chain is also signaled by an extreme proteolytic sensitivity, which also manifests itself in their ubiquitination-independent degradation by the 20S proteasome, termed "degradation by default" [25]. Proteolytic sensitivity can not only provide a binary classification in terms of order/disorder, but the application of proteases at very low concentrations can also provide low-resolution structural information on the topology of IDPs [26]. Another indirect technique, differential scanning calorimetry provides information on the lack of a globularity, i.e., the absence of compact, cooperative structure of IDPs [27].

Hydrodynamic approaches constitute the most coherent group of techniques for the structural characterization of IDPs. The primary observables are the radius of gyration (R_G) or Stokes radius (R_S), which translate into a large apparent molecular weight (M_W). Such behavior is apparent by size-exclusion chromatography (gel-filtration), dynamic light scattering, and analytical ultracentrifugation. More thorough characterization of hydrodynamic behavior can be attained by small-angle X-ray scattering, which not only enables to determine overall dimensions of the protein, but by careful analysis of scattering intensities it provides low-resolution structural topology-model of the molecule [28]. Thus, hydrodynamic techniques not only provide evidence for disorder, but they also enable description of its type and the overall structural topology of an IDP.

Description of disorder in most detail can be achieved by spectroscopic techniques. UV fluorescence, sensitive to the exposure of Trp residues, enables the rapid identification of IDPs. The application of a quencher, such as iodine or acrylamide, provides further evidence for the exposure of aromatic residues. CD spectroscopy is sensitive to repetitive secondary structural elements (α-helix and β-strand), or coil conformation, the latter being abundant in IDPs. Even more structural detail can be obtained by a less well-known technique, Raman optical activity measurement, which provides information on details of structure and dynamics of IDPs [29]. The most powerful spectroscopic technique for studying IDPs is NMR, which enables their atomic-level structural characterization [22]. A range of NMR observables, such as secondary chemical shifts, relaxation rates and residual dipolar coupling enable detailed description of equilibrium structural features and also dynamic characteristics of IDPs. To mention just a few examples, NMR has been used to provide evidence for the overall disorder of proteins [30–32], it enabled characterizing residual structure within IDPs [33–35], and also detailed analysis of the mechanism of binding of an IDP to its partner [36]. Recently, NMR even made possible the *in vivo* characterization of IDPs by the application of in-cell NMR techniques [37].

1.2.2 Bioinformatics Techniques

Followed by the recognition of protein disorder, several bioinformatics algorithms have been developed in rapid succession, which can be used to approach disorder at the residue level [20, 21]. The application of predictors to studying single proteins and/or entire proteomes has contributed basically to the development of the field. Although the predictors are based on different principles, they all rely on common attributes of IDPs, namely, that they are depleted in order-promoting amino acids (WCFIYVL) and are enriched in disorder promoting amino acids (KEPSQRA) [38]. There are more than 20 predictors of disorder, and they can be roughly classified into three groups, such as (i) predictors relying on simple statistics, (ii) predictors applying machine-learning algorithms, and (iii) predictors applying some structural considerations.

The most straightforward approach relies on simple statistics of amino acid propensities, as implemented in the charge-hydropathy plot, by plotting net charge of proteins vs. their net hydrophobicity [6]. IDPs are found in the high net charge – low mean hydrophobicity half of this 2D plane, which suggests a clear interpretation of the physical factors underlying disorder. By calculating the distribution of these features for a pre-defined sequence window, this approach can be made sequence-specific [39].

Arguably the most advanced predictors are those which rely on machine learning approaches, i.e. neural networks and support vector machines. These are trained on datasets of disorder and order, and capture the inherent differences in implicit ways. The classical neural network predictor, PONDR [40], has recently been developed to be able to distinguish between short and long disorder (VSL2) [41]. In other cases, the input data can be generated by sequence alignment, as in the case of DISO-PRED2 [9], which relies on a support vector machine algorithm. The power of these methods also comes from that they can readily accommodate other factors, such as predicted secondary structure or solvent accessibility of the polypeptide chain. Although these algorithms usually perform the best when performance of predictors is compared in the community-wide experiment "critical assessment of structure prediction algorithms" (CASP) [42], their limitation may come from uncertainties inherent in the underlying databases.

A completely different principle has been exploited in the construction of the IUPred algorithm [43, 44]. This approach uses low-resolution force-fields to estimate the total pairwise interaction energy of a (segment of a) protein. The underlying idea is that IDPs lack stable structure because their amino acid composition is not compatible with the formation of interresidue interactions in numbers sufficiently large to overcome the large unfavorable decrease in conformational entropy that accompanies folding. Because IUPred and other similar algorithms, such as Fold Unfold [45] and Ucon [46] do not rely on actual data on protein disorder, their assessment of the structural status of a protein as disordered may be considered as an independent evidence for disorder, and actually for the very existence of intrinsically disordered proteins.

1.3 Disorder *In Vivo*, the Effect of Crowding?

The structural ensemble of IDPs is very sensitive to variations in environmental conditions, which makes it rather difficult to appreciate the actual structural state of these proteins. Among the variety of factors, *crowding* caused by extremely high macromolecular concentrations is of special interest, because it may basically influence the structural state of IDPs [47, 48]. Typical concentrations of macromolecules in the cell are on the order of 300–400 mg/ml, which gives rise very large excluded volume effect that favors reactions accompanied by reduction of volume, such as folding and aggregation. In the case of denatured globular proteins crowding does promote them to assume their native-like compact states and to regain at least partial activity [49, 50]. This issue of promoting native structure is of particular importance in the case of IDPs, because it would be logical to assume that crowding may enforce them to fold, and behave as globular proteins, *in vivo*.

Studies addressing this issue either apply high concentrations of macromolecular crowding agents, such as Dextran or Ficoll 70 (occasionally a small molecular osmolyte, TMAO), or actually follow the behavior of the IDP within a living cell. The results are rather mixed, and they overall suggest that crowding makes IDPs to locally fold or assume more compact structural states, but never to transformation to a unique ordered state. For example, crowding had no effect on two IDPs, the KID domain of p27^{Kip1}, and the trans-activator domain of c-Fos [51], but leads to some compaction of α-synuclein [52]. Under real *in vivo* conditions, i.e. within a living cell, some IDPs, such as FlgM [53] or tau protein [54] undergo partial ordering, whereas others, such as α-synuclein [37], retain their fully disordered character.

In principle, aggregation, being a second- or higher-order reaction, is particularly sensitive to the effect of crowding. The formation of aggregates is sensitive to the concentration – in fact the chemical activity – of interacting chains, which is basically influenced by the excluded volume effect [47, 55]. As shown by experiment and also theoretical considerations, crowding may increase the rate of aggregation orders of magnitude. For example, the formation α–synuclein fibrils has a lag time of 80–90 days for a concentration of 300 μM, but addition of polyethylene glycol, Dextran or Ficoll 70 reduces this lag time to 8–10 days [56].

1.4 Disorder and Aggregation

Early on after the recognition of protein disorder, it has been realized that the open and extended conformation of IDPs may be particularly adapted to interactions leading to aggregation, making them, in principle, particularly prone to aggregation [2]. Although several of the proteins involved in amyloid diseases are IDPs, most IDPs are not known for their involvement in aggregation, which suggests that these proteins use some countermeasures against aggregation [57]. Studies of sequences of proteins involved in amyloid diseases unveil that certain features are directly related with disorder. Because the key structural feature of amyloids is an extended H-bonding network of backbone amides in a cross-β scaffold, the exposure of these moieties is key to the misfolding reaction leading to the amyloid state. In accord, deficient local shielding (under-wrapping) of backbone H-bonds is a critical factor in the amyloidogeneicity of proteins [58]. Due to their total structural exposure, IDPs are inherently more prone to form amyloids than globular proteins.

In studies of protein aggregation of a range of mutants under conditions favoring the unfolded states of globular proteins, it was found that amyloidogeneicity shows a significant positive correlation with hydrophobicity and β-sheet forming potential, and negative correlation with total charge [59]. These results are entirely relevant with respect to how IDPs remain soluble despite their exposed polypeptide chain. As suggested above, IDPs possess high mean net charge and low mean hydrophobicity [6], they are depleted in order-promoting amino acids (WCFIYVL) and are enriched in disorder promoting amino acids (KEPSQRA) [38]. These biases in composition act strongly against amyloid formation. By applying TANGO, the algorithm developed to asses β-aggregation propensity of proteins [60] it was found that globular proteins contain almost three times as much aggregation nucleating regions as IDPs, and formation of the ordered structure of globular proteins can only be achieved at the expense of a higher β-aggregation propensity [61]. In general, formation of structure and aggregates rely on very similar physico-chemical characteristics.

Thus, amino acid sequences of IDPs appear to significantly counteract the inherent propensity of their open structure for aggregation and amyloid formation. Limiting the occurrence of amino acids of significant β-sheet forming potential [62] is probably also of significant inhibitory potential. This may also rationalize the presence of conspicuous conserved Pro or Gly residues in proteins [63], which are inhibitory to the formation of extended β-structures, either due to their restricted

(Pro) or unrestricted (Gly) conformational freedom, serving as "guardians" against aggregation [64]. These considerations can also explain the presence of residues in IDPs known for significant β-breaking potential, such as Pro, Gln and Ser [2]. In a related study, it was observed [65] that the positions flanking aggregating stretches are enriched with residues such as Pro, Lys, Arg, Glu and Asp. These residues are either β-breakers, or are located in the bottom of the aggregation propensity scales. In the *E. coli* proteome, at least one of these five residues occur at the first position on either side of an aggregation-prone segment, and thus appear to act as "gatekeepers" against aggregation [65].

1.5 Disorder in Neurodegenerative Diseases

Despite these effective countermeasures, disordered proteins do show significant association with aggregation involved in neurodegenerative diseases. Because they are caused by formation of insoluble aggregates, they belong to the family of amyloidoses. Amyloids are highly ordered deposits of misfolded protein, which often originate from full-length proteins, but sometimes from processed segments. Since the diseases and the proteins involved are discussed in detail in the next chapter (See Sect. 1.4), here we concentrate on the underlying structural principles. Amyloidoses are caused by the deposition of insoluble, highly ordered fibrillar aggregates of proteins, which are not related in any aspect [57]. Amyloid diseases are classified by the causative protein that forms the amyloid (Table 1.1), the manner of deposition of the aggregate (systemic vs. tissue-specific cases, the latter primarily meaning neurodegeneration), and the cause of aggregation (sporadic, inherited, or infectious, the latter meaning prion diseases). As seen, the proteins involved might be globular (lysozyme, transthyretin or immunoglobulin), but often they are intrinsically disordered. The diseases are usually intractable and progressive, i.e. they cannot be cured, and either are caused by organ failure (primarily in systemic cases) or impairment in higher-order brain function (cognitive disorder, psychological problems, impairment of movements).

Due to their involvement in diseases, proteins of neurodegenerative diseases have been studied in great detail, and were among the first for which structural disorder

Table 1.1 Amyloid diseases

Disease	Protein	Region	Structural status
Alzheimer's	APP	Aβ peptide	Disordered
Huntington's	Huntingtin	polyQ region	Disordered
Parkinson's	α-Synuclein	Whole protein	Disordered
Prion diseases	Prion	Whole protein	Half disordered
Lysozyme amyloidosis	Lysozyme	Whole protein	Ordered
Senile systemic amyloidoses	Transthyretin	Whole protein	Ordered
AL amyloidosis	Ig light chain	Whole protein	Ordered

The table enlists some of the best known amyloid diseases. The major point is that amyloid deposits may be formed from either ordered or disordered proteins, but in neurodegenerative disorders mostly IDPs are involved.

has been established. In Alzheimer's disease (AD), extracellular protein deposits (senile plaques) are formed from the 40–42 amino-acid long fragment of amyloid precursor protein (APP), termed amyloid-β peptide (Aβ), which is disordered [66]. Intracellular inclusions also form in AD, from the microtubule-associated protein tau, which was among the first proteins described as disordered [14]. The causative agent of Parkinson's disease is α-synuclein, also termed "non-A beta component of Alzheimer's disease amyloid plaque (NACP)", because it is a minor peptide component of the insoluble fibrillar core of AD plaque. The protein was shown to be disordered by a battery of techniques, such as heat-stability, sedimentation, CD, Fourier-transform infrared spectroscopy (FTIR) and UV spectroscopy [16]. The protein has been at the focus of intense interest ever since, which has resulted in ample detail on the structural ensemble of α-synuclein structure *in vitro* [67] and also in vivo [37]. Huntingon's disease, and a range of other diseases are caused by the pathologic expansion of glutamin-repeats in proteins. In Huntingon's disease, the CAG-repeat region in exon1 of huntingtin encodes for a run of Gln residues less than 40 in healthy individuals, which undergoes expansion to above 40 residues under pathologic conditions (thus the diseases are also termed CAG-repeat diseases). The repeat region is intrinsically disordered [68], and can undergo transition to the amyloid state.

A special case of amyloidoses is prion diseases, in which the transmission of the amyloid can elicit infectious propagation of the amyloid state. Prions have been first noted as non-conventional infectious entities in mammals, which were shown later to be proteins which may exist in two different structural states, a cellular state and a prion state. The prion state is contagious, and is implicated in a variety of diseases, such as kuru and Creutzfeldt-Jakob disease of humans, bovine spongiform encephalopathy, and scrapie of sheep [69]. Transmission of prions results from that the scrapie state can convert the cellular form to the scrapie form in a self-sustaining, autocatalytic reaction. The two forms are identical at the level of sequence or post-translational modifications [70], and thus the only information that distinguishes them is protein conformation. The structure of the cellular form solved by NMR has an N-terminal disordered and a C-terminal ordered half [17, 71]. Since the cellular form is dominated by disorder and α-helical regions, whereas the scrapie state is largely β-strand, the prion disease constitutes a special class of transmissible protein misfolding diseases [72].

1.6 Physiological Prions

As suggested in the previous section, prions have the dreadful connotation of causing lethal and somewhat mysterious diseases. It is generally held that the propagation of prion diseases results from the conversion of the cellular form to the scrapie state in an autocatalytic reaction [57]. In this section we will discuss that the above structural principle of prion propagation, i.e., the autocatalytic structural conversion from a soluble form to the highly ordered amyloid state, may also serve the physiological function of proteins. In the case of about 10 proteins it has been shown that their normal cellular function results from their capacity to undergo prion-like structural conversion. Unlike their pathological counterparts, these physiological amyloids/prions do no harm to their host cells, but may confer adaptive advantages under certain conditions [73]. The variety of cases can be exemplified by the *curli* protein of bacteria, involved in biofilm formation and host invasion, *URE2p* of yeast involved in the regulation of nitrogen catabolism, or *Pmel17* of humans, which functions in scaffolding and sequestration of toxic intermediates during melanin synthesis. Often these proteins are noted for the presence of Q/N-rich, disordered, portable prion domains [74]. Their function and action can be best illustrated by two interesting well-characterized examples, *Sup35p* of yeast and *cytoplasmic polyadenylation element-binding protein (CPEB)* of *Drosophila melanogaster*.

Sup35p in yeast is a protein component of the translational termination complex. Intriguingly, it has been discovered as a non-Mendelian genetic element, [PSI^+], which causes translational read-through in yeast cells [75]. Later, it has been recognized that the genetic element corresponds to the altered structural state of the cellular protein, Sup35p, which is composed of a Q/N-rich disordered amino-terminal domain and a globular carboxy-terminal domain. When the amino-terminal domain attains an amyloid-like prion conformation [76], it prevents the globular domain from taking part in the translation termination complex. The physiological readout of this change is the inability of the cell to terminate translation at stop codons, and the resulting read-through might be functionally advantageous under some circumstances [77].

A completely different example is the CPEB protein of *D. melanogaster*. This neuronal protein regulates mRNA translation by promoting polyadenylation and activation of mRNA localized in the cytoplasm [78, 79]. Its amino-terminal Q/N-rich domain has the capacity to undergo a transition to a prion state, as demonstrated by fusion constructs in yeast. In the activated synapses of fruit-fly it converts to the prion state and provides a molecular marker of the synapses. Its activated prion-like form stimulates translation of CPEB-regulated mRNA, and promotes synaptic growth associated with the maintenance of long-term facilitation. In all, this prion/amyloid functions in synaptic communication and memory formation [78, 79].

1.7 Structural Transition to Amyloid: Partially Folded Intermediates

Although the proteins involved in amyloid formation have practically nothing in common [57], their structural transitions to the amyloid state share common characteristics, both in terms of their kinetics, the mechanism of the structural transition and the final structure attained.

The kinetics of amyloid formation shows two characteristic features, i.e., i) the process involves a lag-phase, i.e. the rate-limiting formation of a critical seed, followed by an exponentially accelerating growth phase, and ii) the lag-phase can be abolished by the addition of small pieces of amyloid, i.e., pre-formed seeds. These features are reminiscent of the process of crystallization, and amyloid formation can be considered as one-dimensional crystal growth. To account for these observations, two models have been developed. The model of "nucleation-polymerization" and "template-assistance" [80, 81] differ in the thermodynamic nature of the critical step. In nucleation-polymerization, it is assumed that the structurally altered molecule is less stable than the original protein species, and it only becomes stable when incorporated into an oligomeric (amyloid) form. Thus, the rate-limiting step is the assembly of a seed of critical size, followed by the practically uninhibited transformation of further molecules upon interaction with the seed. The key assumption of the other model, template-assistance, is that the transformed state is inherently more stable than the solution state, but it is kinetically inaccessible due to a high energy barrier. Molecules already transformed can lower the energy barrier, and bring about conversion in an autocatalytic conversion. In this case, the rate-limiting step is the formation of an effective catalytic molecule. Both models adequately describe the kinetic course of the reaction, and they actually mechanistically converge if the seed size in the template-assistance model is thought to be a monomer.

There are also mechanistic parallels in the misfolding process that leads to the formation of amyloids [82], which appear to apply to both globular and disordered proteins. In the case of globular proteins, fibrillation occurs when the native structure is partially destabilized, because to arrive at the common cross-β structure profound conformational rearrangements have to occur, which cannot take place within the structural confines of the native globular state. In accord, most mutations associated with accelerated fibrillation of globular proteins destabilize the native structure, as demonstrated in the case of lysozyme [83], transthyretin [84], and immunoglobulin light chains [85]. Destabilization of structure by non-native conditions, such as low- or high pH, high temperatures, or the presence of denaturants, also lead to an increased fibrillation, as shown in the case of the SH3 domain of PI3K [86] and the Fn III module of murine fibronectin [87]. The generality of this relation and the importance of an increase in the concentration of partially folded conformers [82, 88] is also underscored by that amyloidogenicity of proteins can be significantly reduced by stabilization of the native structure by ligand binding, for example [89]. The critical involvement of partially structured intermediates is also apparent in the case of IDPs, where the primary step of fibrillogenesis is the stabilization of a partially folded conformation. It has been shown in the case of α-synuclein [90], or islet amyloid polypeptide (IAPP) [91] that the presence of amyloidogenic intermediates is strongly correlated with the enhanced formation of fibrils (Fig. 1.1). In all, it appears that the structural prerequisite of amyloid formation is the transformation of a polypeptide chain into a partially folded conformation.

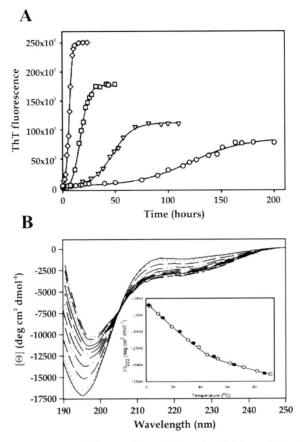

Fig. 1.1 Transition to the amyloid state via partially ordered intermediate. The experiment shows aggregation of α-synuclein at increasing temperatures from 3.0°C up to 92.0°C in equal increments (11 temperatures). (**A**) The kinetics of fibril formation monitored by the enhancement of thioflavin T fluorescence, at four temperatures, 27°C (circles), 37°C (inverted triangles), 47°C (squares), and 57°C (diamonds). (**B**) CD spectra at all temperatures, which show that upon increasing the temperature, in parallel with shortening the lag phase, there is an increase in residual structure of the protein. Adapted with permission from Ref. [90] (Uversky et al. 2001, J Biol Chem 276:10737–10744)

1.8 The Structure of Amyloid: Cross-Beta Models and Flexibility

The third unifying feature of amyloids involved in misfolding diseases is that in spite of the great variety of protein precursor, the resulting amyloid is structurally very similar in most of the cases. Amyloid fibrils visualized by transmission electron microscopy or atomic force microscopy usually consist of 2–6 protofilaments, each about 2–5 nm in diameter, twisted together to form rope-like fibrils typically

7–13 nm wide [92]. As shown by X-ray fiber diffraction, the polypeptide chain runs perpendicular to the axis of the fiber in a β-strand conformation, thus forming an extended β-sheet along the fiber. The structure is highly ordered, and in this sense clearly differs from general protein aggregates [93]. Structural uniformity of amyloids is also signaled by their common tinctorial properties, because they all can be stained by specific dyes such as thioflavin T and Congo red.

Whereas gross similarities of amyloid obtained from different proteins have been apparent for long, establishing its structure at high-resolution remained intractable for many years. Recently, the combined application of solid-state NMR, X-ray crystallography, electron microscopy and electron paramagnetic resonance (EPR) spectroscopy, have begun to provide atomic-level structural information on the structure of amyloids fibers [57, 93]. For example, the X-ray crystal structure of a model amyloid, the heptapeptide Gly-Asn-Asn-Gln-Gln-Asn-Tyr of yeast prion Sup35p [76], suggested a pair of parallel β-sheets composed of β-strands contributed by individual peptide molecules. The strands are stacked, parallel and are located in register in both sheets. The side-chains of the two sheets interdigitate so tightly that water is excluded from the interface, which lead to suggesting the model "steric zipper" (Fig. 1.2), to extend on the previous model "polar zipper" of the extended

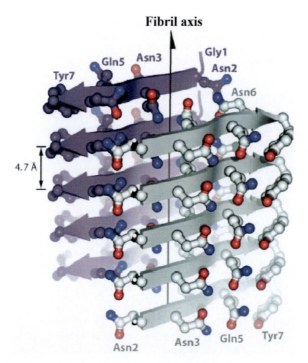

Fig. 1.2 Steric-zipper structural model of amyloids. The fibril formed by a heptapeptide segment (GNNQQNY) from the N-terminus of the yeast prion protein Sup35p. Reproduced with permission from Ref. [76] (Nelson et al. 2005, Nature 435:773–778)

H-bond network supporting the β-sheet structure of polyQ amyloids [94]. Similar results were obtained in the case of the fast-folding WW domain FBP28 [95].

Solid-state NMR results were combined with computational energy minimization procedures to obtain a detailed picture of the amyloid fibrils formed from the Aβ(1–40) peptide of AD [96]. The molecule makes up two β-strands, connected by a short loop, stacked upon each other, parallel and are in register. The two strands participate in the formation of two distinct sheets within the same protofilament. EPR spectroscopy, in which spectra from a series of labeled molecules have been obtained, also lent support to the highly structured, parallel and in-register arrangement of the strands [97]. A similar arrangement, i.e., single-molecule layers that stack on top of one another with parallel, in-register alignment of β-strands has been observed by the same technique in the case of fibrils formed from α-synuclein [98] and human prion protein [99].

These and many other studies have corroborated that amyloid fibrils have a tightly-packed cross-β core region, which lends stability to the structure. Outside of the core the structure is much less defined, the polypeptide chain is exposed and rather flexible, and is often explicitly stated as disordered. For example, in the case of the Aβ(1–42) molecule in AD, residues 13–21 and 30–39 are highly structured in the fibrils by EPR spectroscopy, whereas high flexibility and exposure to the solvent within the N-terminal region is apparent [97], also corroborated by hydrogen-deuterium (H/D) exchange, limited proteolysis and Pro-scanning mutagenesis [57]. In the case of α-synuclein, restricted motility is apparent in the segment 35–97, which roughly corresponds to the NAC region, whereas outside this region the chain is rather flexible in the fibril by EPR [98]. H/D exchange has also shown that in β2-microglobulin fibrils most residues in the middle segment form a rigid beta-sheet core, whereas the N- and C-termini are excluded from this core [100]. In the case of amylin, residues 12–17, 22–27, and 31–37 form stacked β-sandwiches, whereas the N-terminal tail is disordered with a disulfide bridge between Cys2 and Cys7 [101]. The fibril formed by the NM region of Sup35p is largely stabilized by interactions between residues that belong to region N, whereas regions 1–20 and 158–250 (the latter being M) remain largely disordered, as shown by fluorescence proximity analysis [102]. Similar conclusions can be drawn in the case of the human prion, in which the protease-resistant core corresponds to about 140 amino acids encompassing region 90–230, whereas the N-terminal 90 amino acids are largely disordered, as shown by limited proteolysis [69] and electron crystallography [103].

1.9 Conclusions

Neurodegenerative diseases are a major health problem in developed countries, resulting primarily from the deposition of neuronal proteins in the form of insoluble, highly ordered protein aggregates in the brain. Often, the protein or segment of protein involved in the disease is intrinsically disordered, and undergoes a major structural transmission toward the cross-β structure characteristic of the amyloid fibrils. Understanding the cause of transition and the structural details of both the soluble and aggregated states is a long way ahead, but inevitably will provide the insight required for designing successful intervention strategies against these debilitating diseases.

Acknowledgments This work was supported by grants K60694 and NK71582 from the Hungarian Scientific Research Fund (OTKA), and ETT 245/2006 from the Hungarian Ministry of Health.

Abbreviations

Aβ   amyloid β-peptide
AD   Alzheimer's disease
APP   amyloid precursor protein
CD   circular dichroism
CPEB   cytoplasmic polyadenylation element-binding protein
EPR   electron paramagnetic resonance
IDP   intrinsically disordered protein

References

1. Dyson HJ, Wright PE (2005) Intrinsically unstructured proteins and their functions. Nat Rev Mol Cell Biol 6:197–208
2. Tompa P (2002) Intrinsically unstructured proteins. Trends Biochem Sci 27:527–533
3. Tompa P (2005) The interplay between structure and function in intrinsically unstructured proteins. FEBS Lett 579:3346–3354
4. Uversky VN, Oldfield CJ, Dunker AK (2005) Showing your ID: intrinsic disorder as an ID for recognition, regulation and cell signaling. J Mol Recognit 18:343–384
5. Wright PE, Dyson HJ (1999) Intrinsically unstructured proteins: re-assessing the protein structure-function paradigm. J Mol Biol 293:321–331
6. Uversky VN, Gillespie JR, Fink AL (2000) Why are "natively unfolded" proteins unstructured under physiologic conditions? Proteins 41:415–427
7. Uversky VN (2002) Natively unfolded proteins: A point where biology waits for physics. Protein Sci 11:739–756
8. Iakoucheva L, Brown C, Lawson J, Obradovic Z, Dunker A (2002) Intrinsic disorder in cell-signaling and cancer-associated. Proteins J Mol Biol 323:573–584
9. Ward JJ, Sodhi JS, McGuffin LJ, Buxton BF, Jones DT (2004) Prediction and functional analysis of native disorder in proteins from the three kingdoms of life. J Mol Biol 337: 635–645
10. Tompa P, Dosztányi Z, Simon I (2006) Prevalent structural disorder in E coli and S cerevisiae proteomes. J Proteome Res 5:1996–2000
11. Dunker AK, Brown CJ, Lawson JD, Iakoucheva LM, Obradovic Z (2002) Intrinsic disorder and protein function. Biochemistry 41:6573–6582
12. Tompa P, Szász C, Buday L (2005) Structural disorder throws new light on moonlighting. Trends Biochem Sci 30:484–489
13. Iakoucheva LM, Radivojac P, Brown CJ, O'Connor TR, Sikes JG, Obradovic Z, Dunker AK (2004) The importance of intrinsic disorder for protein phosphorylation. Nucleic Acids Res 32:1037–1049

14. Schweers O, Schonbrunn-Hanebeck E, Marx A, Mandelkow E (1994) Structural studies of tau protein and Alzheimer paired helical filaments show no evidence for beta-structure. J Biol Chem 269:24290–24297
15. Bell S, Klein C, Muller L, Hansen S, Buchner J (2002) p53 contains large unstructured regions in its native state. J Mol Biol 322:917–927
16. Weinreb PH, Zhen W, Poon AW, Conway KA, Lansbury PT Jr (1996) NACP, a protein implicated in Alzheimer's disease and learning, is natively unfolded. Biochemistry 35:13709–13715
17. Lopez Garcia F, Zahn R, Riek R, Wuthrich K (2000) NMR structure of the bovine prion protein. Proc Natl Acad Sci USA 97:8334–8339
18. Mark WY, Liao JC, Lu Y, Ayed A, Laister R, Szymczyna B, Chakrabartty A, Arrowsmith CH (2005) Characterization of segments from the central region of BRCA1: an intrinsically disordered scaffold for multiple protein-protein and protein-DNA interactions? J Mol Biol 345:275–287
19. Sickmeier M, Hamilton JA, LeGall T, Vacic V, Cortese MS, Tantos A, Szabo B, Tompa P, Chen J, Uversky VN et al. (2007) DisProt: the database of disordered proteins. Nucleic Acids Res 35:D786–D793
20. Ferron F, Longhi S, Canard B, Karlin D (2006) A practical overview of protein disorder prediction methods. Proteins 65:1–14
21. Dosztányi Z, Sándor M, Tompa P, Simon I (2007) Prediction of protein disorder at the domain level. Curr Protein Pept Sci 8:161–171
22. Dyson HJ, Wright PE (2004) Unfolded proteins and protein folding studied by NMR. Chem Rev 104:3607–3622
23. Cortese MS, Baird JP, Uversky VN, Dunker AK (2005) Uncovering the unfoldome: enriching cell extracts for unstructured proteins by acid treatment. J Proteome Res 4:1610–1618
24. Galea CA, Pagala VR, Obenauer JC, Park CG, Slaughter CA, Kriwacki RW (2006) Proteomic studies of the intrinsically unstructured mammalian proteome. J Proteome Res 5:2839–2848
25. Tsvetkov P, Asher G, Paz A, Reuven N, Sussman JL, Silman I, Shaul Y (2008) Operational definition of intrinsically unstructured protein sequences based on susceptibility to the 20S proteasome. Proteins 70:1357–1366
26. Csizmók V, Bokor M, Bánki P, Klement E, Medzihradszky KF, Friedrich P, Tompa K, Tompa P (2005) Primary contact sites in intrinsically unstructured proteins: the case of calpastatin and microtubule-associated protein 2. Biochemistry 44:3955–3964
27. Szollosi E, Bokor M, Bodor A, Perczel A, Klement E, Medzihradszky KF, Tompa K, Tompa P (2008) Intrinsic structural disorder of DF31, a Drosophila protein of chromatin decondensation and remodeling activities. J Proteome Res 7:2291–2299
28. von Ossowski I, Eaton JT, Czjzek M, Perkins SJ, Frandsen TP, Schulein M, Panine P, Henrissat B, Receveur-Brechot V (2005) Protein disorder: conformational distribution of the flexible linker in a chimeric double cellulase. Biophys J 88:2823–2832
29. Zhu F, Kapitan J, Tranter GE, Pudney PD, Isaacs NW, Hecht L, Barron LD (2007) Residual structure in disordered peptides and unfolded proteins from multivariate analysis and ab initio simulation of Raman optical activity data. Proteins 70:823–833
30. Watts JD, Cary PD, Sautiere P, Crane-Robinson C (1990) Thymosins: both nuclear and cytoplasmic proteins. Eur J Biochem 192:643–651
31. Dobson CM (1993) Flexible friends Current Biology 3:530–532
32. Kriwacki RW, Hengst L, Tennant L, Reed SI, Wright PE (1996) Structural studies of p21Waf1/Cip1/Sdi1 in the free and Cdk2-bound state: conformational disorder mediates binding diversity. Proc Natl Acad Sci USA 93:11504–11509
33. Daughdrill GW, Hanely LJ, Dahlquist FW (1998) The C-terminal half of the anti-sigma factor FlgM contains a dynamic equilibrium solution structure favoring helical conformations. Biochemistry 37:1076–1082

34. Sivakolundu SG, Bashford D, Kriwacki RW (2005) Disordered p27Kip1 exhibits intrinsic structure resembling the Cdk2/cyclin A-bound conformation. J Mol Biol 353: 1118–1128
35. Libich DS, Harauz G (2008) Backbone dynamics of the 18.5 kDa isoform of myelin basic protein reveals transient alpha-helices and a calmodulin-binding site. Biophys J 94: 4847–4866
36. Sugase K, Dyson HJ, Wright PE (2007) Mechanism of coupled folding and binding of an intrinsically disordered protein. Nature 447:1021–1025
37. McNulty BC, Young GB, Pielak GJ (2006) Macromolecular crowding in the Escherichia coli periplasm maintains alpha-synuclein disorder. J Mol Biol 355:893–897
38. Dunker AK, Lawson JD, Brown CJ, Romero P, Oh JS, Oldfield CJ, Campen AM, Ratliff CM, Hipps KW, Ausio J et al. (2001) Intrinsically disordered protein. J Mol Graphics Modelling 19:26–59
39. Prilusky J, Felder CE, Zeev-Ben-Mordehai T, Rydberg EH, Man O, Beckmann JS, Silman I, Sussman JL (2005) FoldIndex: a simple tool to predict whether a given protein sequence is intrinsically unfolded. Bioinformatics 21:3435–3438
40. Garner E, Cannon P, Romero P, Obradovic Z, Dunker AK (1998) Predicting disordered regions from amino acid sequence: Common Themes despite differing structural characterization. Genome Inform Ser Workshop Genome Inform 9:201–213
41. Peng K, Radivojac P, Vucetic S, Dunker AK, Obradovic Z (2006) Length-dependent prediction of protein intrinsic disorder. BMC Bioinformatics 7:208
42. Bordoli L, Kiefer F, Schwede1 T (2007) Assessment of disorder predictions in CASP7. Proteins 69 (Suppl 8):129–136
43. Dosztányi Z, Csizmók V, Tompa P, Simon I (2005) IUPred: web server for the prediction of intrinsically unstructured regions of proteins based on estimated energy content. Bioinformatics 21:3433–3434
44. Dosztányi Z, Csizmók V, Tompa P, Simon I (2005) The pairwise energy content estimated from amino acid composition discriminates between folded and instrinsically unstructured proteins. J Mol Biol 347:827–839
45. Galzitskaya OV, Garbuzynskiy SO, Lobanov MY (2006) Fold Unfold: web server for the prediction of disordered regions in protein chain. Bioinformatics 22:2948–2949
46. Schlessinger A, Punta M, Rost B (2007) Natively unstructured regions in proteins identified from contact predictions. Bioinformatics 23:2376–2384
47. Ellis RJ (2001) Macromolecular crowding: obvious but under appreciated. Trends Biochem Sci 26:597–604
48. Minton AP (2005) Models for excluded volume interaction between an unfolded protein and rigid macromolecular cosolutes: macromolecular crowding and protein stability revisited. Biophys J 88:971–985
49. Baskakov I, Bolen DW (1998) Forcing thermodynamically unfolded proteins to fold. J Biol Chem 273:4831–4834
50. Qu Y, Bolen DW (2002) Efficacy of macromolecular crowding in forcing proteins to fold. Biophys Chem 101–102:155–165
51. Flaugh SL, Lumb KJ (2001) Effects of macromolecular crowding on the intrinsically disordered proteins c-Fos and p27(Kip1). Biomacromolecules 2:538–540
52. Morar AS, Olteanu A, Young GB, Pielak GJ (2001) Solvent-induced collapse of alpha-synuclein and acid-denatured cytochrome c. Protein Sci 10:2195–2199
53. Dedmon MM, Patel CN, Young GB, Pielak GJ (2002) FlgM gains structure in living cells. Proc Natl Acad Sci USA 99:12681–12684
54. Bodart JF, Wieruszeski JM, Amniai L, Leroy A, Landrieu I, Rousseau-Lescuyer A, Vilain JP, Lippens G (2008) NMR observation of Tau in Xenopus oocytes. J Magn Reson 192: 252–257
55. Ellis RJ, Minton AP (2006) Protein aggregation in crowded environments. Biol Chem 387:485–497

56. Uversky VN, E MC, Bower KS, Li J, Fink AL (2002) Accelerated alpha-synuclein fibrillation in crowded milieu. FEBS Lett 515:99–103
57. Chiti F, Dobson CM (2006) Protein misfolding, functional amyloid, and human disease. Annu Rev Biochem 75:333–366
58. Fernandez A, Kardos J, Scott LR, Goto Y, Berry RS (2003) Structural defects and the diagnosis of amyloidogenic propensity. Proc Natl Acad Sci USA 100:6446–6451
59. Chiti F, Stefani M, Taddei N, Ramponi G, Dobson CM (2003) Rationalization of the effects of mutations on peptide and protein aggregation rates. Nature 424:805–808
60. Fernandez-Escamilla AM, Rousseau F, Schymkowitz J, Serrano L (2004) Prediction of sequence-dependent and mutational effects on the aggregation of peptides and proteins. Nat Biotechnol 22:1302–1306
61. Linding R, Schymkowitz J, Rousseau F, Diella F, Serrano L (2004) A comparative study of the relationship between protein structure and beta-aggregation in globular and intrinsically disordered proteins. J Mol Biol 342:345–353
62. Williams RM, Obradovic Z, Mathura V, Braun W, Garner EC, Young J, Takayama S, Brown CJ, Dunker AK (2001) The protein non-folding problem: amino acid determinants of intrinsic order and disorder. Pac Symp Biocomput 6:89–100
63. Monsellier E, Chiti F (2007) Prevention of amyloid-like aggregation as a driving force of protein evolution. EMBO Rep 8:737–742
64. Parrini C, Taddei N, Ramazzotti M, Degl'Innocenti D, Ramponi G, Dobson CM, Chiti F (2005) Glycine residues appear to be evolutionarily conserved for their ability to inhibit aggregation. Structure 13:1143–1151
65. Rousseau F, Serrano L, Schymkowitz JW (2006) How evolutionary pressure against protein aggregation shaped chaperone specificity. J Mol Biol 355:1037–1047
66. Kirkitadze MD, Condron MM, Teplow DB (2001) Identification and characterization of key kinetic intermediates in amyloid beta-protein fibrillogenesis. J Mol Biol 312:1103–1119
67. Dedmon MM, Lindorff-Larsen K, Christodoulou J, Vendruscolo M, Dobson CM (2005) Mapping long-range interactions in alpha-synuclein using spin-label NMR and ensemble molecular dynamics simulations. J Am Chem Soc 127:476–477
68. Vitalis A, Wang X, Pappu RV (2007) Quantitative characterization of intrinsic disorder in polyglutamine: insights from analysis based on polymer theories. Biophys J 93:1923–1937
69. Prusiner SB (1998) Prions. Proc Natl Acad Sci USA 95:13363–13383
70. Stahl N, Baldwin MA, Teplow DB, Hood L, Gibson BW, Burlingame AL, Prusiner SB (1993) Structural studies of the scrapie prion protein using mass spectrometry and amino acid sequencing. Biochemistry 32:1991–2002
71. Donne DG, Viles JH, Groth D, Mehlhorn I, James TL, Cohen FE, Prusiner SB, Wright PE, Dyson HJ (1997) Structure of the recombinant full-length hamster prion protein PrP(29-231): the N terminus is highly flexible Proc Natl Acad Sci USA 94:13452–13457
72. Chien P, Weissman JS, DePace AH (2004) Emerging principles of conformation-based prion inheritance. Annu Rev Biochem 73:617–656
73. Fowler DM, Koulov AV, Balch WE, Kelly JW (2007) Functional amyloid–from bacteria to humans. Trends Biochem Sci 32:217–224
74. Wickner RB, Taylor KL, Edskes HK, Maddelein ML (2000) Prions: Portable prion domains. Curr Biol 10:R335–R337
75. Lindquist S (1997) Mad cows meet psi-chotic yeast: the expansion of the prion hypothesis. Cell 89:495–498
76. Nelson R, Sawaya MR, Balbirnie M, Madsen AO, Riekel C, Grothe R, Eisenberg D (2005) Structure of the cross-beta spine of amyloid-like fibrils. Nature 435:773–778
77. Li L, Lindquist S (2000) Creating a protein-based element of inheritance. Science 287:661–664
78. Si K, Giustetto M, Etkin A, Hsu R, Janisiewicz AM, Miniaci MC, Kim JH, Zhu H, Kandel ER (2003) A neuronal isoform of CPEB regulates local protein synthesis and stabilizes synapse-specific long-term facilitation in aplysia. Cell 115:893–904

79. Si K, Lindquist S, Kandel ER (2003) A neuronal isoform of the aplysia CPEB has prion-like properties. Cell 115:879–891
80. Come JH, Fraser PE, Lansbury PT Jr (1993) A kinetic model for amyloid formation in the prion diseases: importance of seeding. Proc Natl Acad Sci USA 90:5959–5963
81. Jarrett JT, Lansbury PT Jr (1993) Seeding "one-dimensional crystallization" of amyloid: a pathogenic mechanism in Alzheimer's disease and scrapie? Cell 73:1055–1058
82. Uversky VN, Fink AL (2004) Conformational constraints for amyloid fibrillation: the importance of being unfolded. Biochim Biophys Acta 1698:131–153
83. Booth DR, Sunde M, Bellotti V, Robinson CV, Hutchinson WL, Fraser PE, Hawkins PN, Dobson CM, Radford SE, Blake CC at al. (1997) Instability, unfolding and aggregation of human lysozyme variants underlying amyloid fibrillogenesis. Nature 385:787–793
84. Lashuel HA, Wurth C, Woo L, Kelly JW (1999) The most pathogenic transthyretin variant, L55P, forms amyloid fibrils under acidic conditions and protofilaments under physiological conditions. Biochemistry 38:13560–13573
85. Wall J, Schell M, Murphy C, Hrncic R, Stevens FJ, Solomon A (1999) Thermodynamic instability of human lambda 6 light chains: correlation with fibrillogenicity. Biochemistry 38:14101–14108
86. Guijarro JI, Sunde M, Jones JA, Campbell ID, Dobson CM (1998) Amyloid fibril formation by an SH3 domain. Proc Natl Acad Sci USA 95:4224–4228
87. Litvinovich SV, Brew SA, Aota S, Akiyama SK, Haudenschild C, Ingham KC (1998) Formation of amyloid-like fibrils by self-association of a partially unfolded fibronectin type III module. J Mol Biol 280:245–258
88. Dobson CM (1999) Protein misfolding, evolution and disease. Trends Biochem Sci 24:329–332
89. Chiti F, Taddei N, Stefani M, Dobson CM, Ramponi G (2001) Reduction of the amyloidogenicity of a protein by specific binding of ligands to the native conformation. Protein Sci 10:879–886
90. Uversky VN, Li J, Fink AL (2001) Evidence for a partially folded intermediate in alpha-synuclein fibril formation. J Biol Chem 276:10737–10744
91. Kayed R, Bernhagen J, Greenfield N, Sweimeh K, Brunner H, Voelter W, Kapurniotu A (1999) Conformational transitions of islet amyloid polypeptide (IAPP) in amyloid formation in vitro. J Mol Biol 287:781–796
92. Sunde M, Blake C (1997) The structure of amyloid fibrils by electron microscopy and X-ray diffraction. Adv Protein Chem 50:123–159
93. Rousseau F, Schymkowitz J, Serrano L (2006) Protein aggregation and amyloidosis: confusion of the kinds? Curr Opin Struct Biol 16:118–126
94. Perutz MF, Staden R, Moens L, De Baere I (1993) Polar zippers. Curr Biol 3:249–253
95. Ferguson N, Berriman J, Petrovich M, Sharpe TD, Finch JT, Fersht AR (2003) Rapid amyloid fiber formation from the fast-folding WW domain FBP28. Proc Natl Acad Sci USA 100:9814–9819
96. Petkova AT, Ishii Y, Balbach JJ, Antzutkin ON, Leapman RD, Delaglio F, Tycko R (2002) A structural model for Alzheimer's beta -amyloid fibrils based on experimental constraints from solid state NMR. Proc Natl Acad Sci USA 99:16742–16747
97. Torok M, Milton S, Kayed R, Wu P, McIntire T, Glabe CG, Langen R (2002) Structural and dynamic features of Alzheimer's Abeta peptide in amyloid fibrils studied by site-directed spin labeling. J Biol Chem 277:40810–40815
98. Chen M, Margittai M, Chen J, Langen R (2007) Investigation of alpha-synuclein fibril structure by site-directed spin labeling. J Biol Chem 282:24970–24979
99. Cobb NJ, Sonnichsen FD, McHaourab H, Surewicz WK (2007) Molecular architecture of human prion protein amyloid: a parallel, in-register beta-structure. Proc Natl Acad Sci USA 104:18946–18951
100. Hoshino M, Katou H, Hagihara Y, Hasegawa K, Naiki H, Goto Y (2002) Mapping the core of the beta(2)-microglobulin amyloid fibril by H/D exchange. Nat Struct Biol 9:332–336

101. Kajava AV, Aebi U, Steven AC (2005) The parallel superpleated beta-structure as a model for amyloid fibrils of human amylin. J Mol Biol 348:247–252
102. Krishnan R, Lindquist SL (2005) Structural insights into a yeast prion illuminate nucleation and strain diversity. Nature 435:765–772
103. Govaerts C, Wille H, Prusiner SB, Cohen FE (2004) Evidence for assembly of prions with left-handed beta-helices into trimers. Proc Natl Acad Sci USA 101:8342–8347

Chapter 2
Intrinsic Disorder in Proteins Associated with Neurodegenerative Diseases

Vladimir N. Uversky

Abstract Neurodegenerative diseases constitute a set of pathological conditions originating from the slow, irreversible and systemic cell loss within the various regions of the brain and/or the spinal cord. Depending on the affected region, the outcomes of the neurodegeneration are very broad, starting from the problems with movements and ending with dementia. Neurodegenerative diseases are proteinopathies associated with misbehavior and disarrangement of a specific protein, affecting its processing, functioning, and/or folding. Many proteins associated with human neurodegenerative diseases are intrinsically disordered; i.e., they lack stable tertiary and/or secondary structure under physiological conditions *in vitro*. The major goal of this chapter is to uncover intriguing interconnections between intrinsic disorder and human neurodegenerative diseases.

2.1 Neurodegenerative Diseases as Proteinopathies

The large class of human neurodegenerative disorders includes many acquired neurological diseases with distinct phenotypic and pathologic expressions, all characterized by the pathological conditions in which cells of the brain and spinal cord are lost. The name for these diseases is derived from a Greek word νρο-, *néuro-*, "nerval" and a Latin verb *dēgenerāre*, "to decline" or "to worsen". As neurons are not readily regenerated, their deterioration leads over time to dysfunction and disabilities. Neurodegenerative diseases can be divided into two groups according to their phenotypic effects: (i) Conditions causing problems with movements; and (ii) Conditions affecting memory and leading to dementia. Neurodegeneration is a slow process, which begins long before the patient experiences any symptoms.

V.N. Uversky (✉)
Center for Computational Biology and Bioinformatics, Department of Biochemistry and Molecular Biology, Institute for Intrinsically Disordered Protein Research, Indiana University School of Medicine, Indianapolis, Indiana 46202, USA
e-mail: vuversky@iupui.edu

Table 2.1 IDPs and associated neurodegenerative diseases

Protein	Disease(s)	Disorder by prediction (%)	Disorder by experiment
Aβ	Alzheimer's disease Dutch hereditary cerebral hemorrhage with amyloidosis Congophilic angiopathy	16.9 (28.6)	NMR and far-UV CD analyses revealed that the monomeric peptide is highly unfolded
Tau	Tauopathies Alzheimer's disease Corticobasal degeneration Pick's disease Progressive supranuclear palsy	77.6 (99.1)	Tau protein was shown to be in a random coil-like conformation according to far-UV CD, FTIR, X-ray scattering and biochemical assays
Prion protein	Prion diseases Creutzfeld-Jacob disease Gerstmann-Sträussler-Schneiker syndrome Fatal familial insomnia Kuru Bovine spongiform encephalopathy Scrapie Chronic wasting disease	55.8 (61.0)	According to NMR and far-UV CD, the N-terminal region (from amino acid 23 to 126) is largely unstructured in the isolated molecule in solution
α-Synuclein	Synucleinopathies Parkinson's disease Lewy body variant of Alzheimer's disease Diffuse Lewy body disease Dementia with Lewy bodies Multiple system atrophy Neurodegeneration with brain iron accumulation type I	90.7 (37.1)	Highly unfolded structure of entire protein is confirmed by NMR, FTIR, SAXS, far-UV CD, gel filtration, dynamic light scattering, FRET, limited proteolysis, aberrant mobility in SDS-PAGE
β-Synuclein	Parkinson's disease Diffuse Lewy body disease	87.3 (52.2)	Highly unfolded conformation is confirmed by NMR, FTIR, SAXS, far-UV CD and gel filtration
γ-Synuclein	Parkinson's disease Diffuse Lewy body disease	100 (56.8)	Highly unfolded conformation is confirmed by NMR, FTIR, SAXS, far-UV CD and gel filtration
Huntingtin	Huntington's disease	35.5 (30.4)	The far-UV CD spectra of poly(Gln) peptides with repeat lengths of 5, 15, 28 and 44 residues were shown to be nearly identical and were consistent with a high degree of random coil structure
DRPLA protein (atrophin-1)	Hereditary dentatorubral-pallidoluysian atrophy	89.5 (84.2)	Aberrant electrophoretic mobility. Apparent molecular mass estimated by SDS-PAGE is ~1.6-fold higher than the predicted molecular mass

Table 2.1 (continued)

Protein	Disease(s)	Disorder by prediction (%)	Disorder by experiment
Androgen receptor	Kennedy's disease or X-linked spinal and bulbar muscular atrophy	53.9 (46.7)	Far-UV CD, gel-filtration, limited proteolysis, ANS binding and urea-induced unfolding studies revealed that the AF1 transactivation domain is in the molten globule state
Ataxin-1	Spinocerebellar ataxia 1 Neuronal intranuclear inclusion disease	76.8 (73.4)	
Ataxin-2	Spinocerebellar ataxia 2	93.8 (76.9)	Ataxin-2 contains two globular domains, Lsm and LsmAD, in an acidic region (amino acid 254–475). The rest of ataxin-2 outside of the Lsm and LsmAD domains is predicted to be intrinsically disordered
Ataxin-3	Spinocerebellar ataxia 3	52.1 (47.1)	Far-UV CD and NMR spectroscopies suggest that ataxin-3 is only partially folded. The far-UV CD signal of the full-length protein is dominated by the Josephin motif (N-terminal domain 1–198), with the C-terminal portion of the protein making a smaller contribution, consistent with its largely unstructured conformation.
P/Q-type calcium channel α1A subunit	Spinocerebellar ataxia 6	53.0 (49.3)	Aberrant electrophoretic mobility
Ataxin-7	Spinocerebellar ataxia 7	89.5 (70.2)	Aberrant electrophoretic mobility. Apparent molecular mass estimated by SDS-PAGE is 1.15-fold higher than that calculated from amino acid sequence
TATA-box-binding protein	Spinocerebellar ataxia 17	53.9 (52.5)	Aberrant electrophoretic mobility. A protein with the calculated molecular mass of 37.7 kDa was shown to possess an apparent molecular mass of ~49 kDa.

Table 2.1 (continued)

Protein	Disease(s)	Disorder by prediction (%)	Disorder by experiment
ABri	Familial British dementia	29.4 (23.5)	Far-UV CD and NMR spectroscopy revealed that ABri is in the random coil-like conformation at slightly acidic pH
ADan	Familial Danish dementia	29.4 (23.5)	Far-UV CD revealed that ADan showed mostly random coil structure
Glial fibrillary acidic protein	Alexander's disease	82.4 (68.5)	Extremely high susceptibility to proteolysis
Mitochondrial DNA polymerase γ	Alpers disease	37.1 (36.7)	Aberrant electrophoretic mobility
DNA excision repair protein ERCC-6	Cockayne syndrome	56.8 (47.8)	Aberrant electrophoretic mobility
Survival motor neuron protein	Spinal muscular atrophy	69.7 (60.2)	Aberrant electrophoretic mobility

Disorder was predicted by two predictors, PONDR® VSL2 and VLXT (given in parenthesis), respectively.

It can take months or even years before visible outcomes of this degeneration are felt and diagnosed. Symptoms are usually noticed when many cells die or fail to function and a part of the brain begins to cease functioning properly. For example, the symptoms of Parkinson's disease (PD) become apparent after more than ∼70% dopaminergic neurons die in *substantia nigra* (a small area of cells in the mid-brain affected by PD.

Until recently, a link between Alzheimer's disease (AD), prion diseases, PD, Huntington's disease (HD), and several other neurodegenerative disorders was elusive. However, recent fascinating advances in molecular biology, immunopathology and genetics indicated that these diseases might share a common pathophysiologic mechanism, where disarrangement of a specific protein processing, functioning, and/or folding takes place. Therefore, neurodegenerative disorders represent a set of proteinopathies, which can be classified and grouped based on the causative proteins. In fact, from this viewpoint neurodegenerative disorders represent a subset of a broader class of human diseases known as protein conformational or protein misfolding diseases. These disorders arise from the failure of a specific peptide or protein to adopt its native functional conformational state. The obvious consequences of misfolding are protein aggregation (and/or fibril formation), loss of function, and gain of toxic function. Some proteins have an intrinsic propensity to assume a pathologic conformation, which becomes evident with aging or at persistently high concentrations. Interactions (or impaired interactions) with some endogenous factors (e.g., chaperones, intracellular or extracellular matrixes, other proteins, small molecules) can change conformation of a pathogenic protein and increase its propensity to misfold. Misfolding can originate from point mutation(s) or result from an exposure to internal or external toxins, impaired posttranslational modifications (phosphorylation, advanced glycation, deamidation, racemization, etc.), an increased probability of degradation, impaired trafficking, lost binding partners or oxidative damage. All these factors can act independently or in association with one another. Table 2.1 lists some of the intrinsically disordered proteins (IDPs) involved in various neurodegenerative disorders. As the major focus of this chapter is the neurodegenerative mechanisms of IDPs, subsequent paragraphs are devoted to the brief introduction of these interesting members of the protein kingdom.

2.2 Introducing Intrinsically Disordered Proteins

2.2.1 Concept

Evidence is rapidly accumulating that many protein regions and even entire proteins lack stable tertiary and/or secondary structure in solution, existing instead as dynamic ensembles of interconverting structures. These naturally flexible proteins are known by different names, including intrinsically disordered [1], natively denatured [2], natively unfolded [3], intrinsically unstructured [4], and natively disordered proteins [5]. These proteins are called "intrinsically disordered" from now on. By "intrinsic disorder" it is meant that the protein exists as a structural ensemble, either at the secondary or at the tertiary level. In other words, in contrast to ordered proteins whose 3-D structure is relatively stable and Ramachandran angles vary slightly around their equilibrium positions with occasional cooperative conformational switches, IDPs or intrinsically disordered regions (IDRs) exist as dynamic ensembles in which the atom positions and backbone Ramachandran angles vary significantly over time with no specific equilibrium values and typically undergo non-cooperative conformational changes. To some extent conformational behavior and structural features of IDPs and IDRs resemble those of non-native states of "normal" globular proteins, which may exist in at least four different conformations: ordered, molten globule, pre-molten globule, and coil-like [6–9]. Using this analogy, IDPs and IDRs might contain collapsed-disorder (i.e., where intrinsic disorder is present in a molten globular form) and extended-disorder (i.e., regions where intrinsic disorder is present in a form of random coil or pre-molten globule) under physiological conditions *in vitro* [1, 5, 7].

2.2.2 Experimental Techniques for IDP Detection

The disorder in IDPs has been detected by several physicochemical methods elaborated to characterize protein self-organization. The list includes but is not limited to X-ray crystallography [10], NMR spectroscopy [5, 11–15], near-UV circular dichroism (CD) [16], far-UV CD [17–20], ORD [17, 20], FTIR [20], Raman spectroscopy and Raman optical activity [21], different fluorescence techniques [22, 23], numerous hydrodynamic techniques (including gel-filtration, viscometry, small angle X-ray scattering (SAXS), small angle neutron scattering (SANS), sedimentation, and dynamic and static light scattering) [22, 23], rate of proteolytic degradation [24–28], aberrant mobility in SDS-gel electrophoresis [29, 30], low conformational stability [22, 31–34], H/D exchange [23], immunochemical methods [35, 36], interaction with molecular chaperones [22], electron microscopy or atomic force microscopy [22, 23], the charge state analysis of electrospray ionization mass-spectrometry [37]. For more detailed reviews on methods used to detect intrinsic disorder see [5, 12, 23, 38].

2.2.3 Sequence Peculiarities of IDPs and Predictors of Intrinsic Disorder

IDPs and IDRs differ from structured globular proteins and domains with regard to many attributes, including amino acid composition, sequence complexity, hydrophobicity, charge, flexibility, and type and rate of amino acid substitutions over evolutionary time. For example, IDPs are significantly depleted in a number of so-called order-promoting residues, including bulky hydrophobic (Ile, Leu, and Val) and aromatic amino acid residues (Trp, Tyr, and Phe), which would normally form the hydrophobic core of a folded globular protein, and also possess low content of Cys and Asn residues. On the other hands, IDPs were shown to be substantially enriched in so called disorder-promoting amino acids: Ala, Arg, Gly, Gln, Ser, Pro, Glu, and Lys [1, 39–41]. Many of the mentioned differences were utilized to develop numerous disorder predictors, including PONDR® (Predictor of Naturally Disordered Regions) [39, 42], charge-hydropathy plots (CH-plots) [20], NORSp [43], Glob-Plot [44, 45], FoldIndex© [46], IUPred [47], DisoPred [48–50] to name a few. It is important to remember that comparing several predictors on an individual protein of interest or on a protein dataset can provide additional insight regarding the predicted disorder if any exists.

2.2.4 Abundance of IDPs and Their Functions

Application of various disorder predictors to different proteomes revealed that intrinsic disorder is highly abundant in nature and the overall amount of disorder in proteins increases from bacteria to archaea to eukaryota, with over a half of the eukaryotic proteins containing long predicted IDRs [50–52]. One explanation for this trend is a change in the cellular requirements for certain protein functions, particularly cellular signaling. In support of this hypothesis, an analysis of a eukaryotic signal protein database indicated that the majority of known signal transduction proteins were predicted to contain significant regions of disorder [53].

Although IDPs fail to form unique 3D-structures under physiological conditions, they might carry out important biological functions, the fact which was recently confirmed by several comprehensive studies [1, 4, 5, 7, 14, 20, 30, 38, 53–65]. Furthermore, sites of posttranslational modifications (acetylation, hydroxylation, ubiquitination, methylation, phosphorylation, etc.) and proteolytic attack are frequently associated with regions of intrinsic disorder [64]. The functional diversity provided by IDRs was suggested to complement functions of ordered protein regions [62–64].

IDPs have specific functions that can be grouped into four broad classes: (i) molecular recognition; (ii) molecular assembly; (iii) protein modification; and (iv) entropic chain activities [53]. Despite (or may be due to) their high flexibility, IDPs are involved in regulation, signaling and control pathways in which interactions with multiple partners and high-specificity/low-affinity interactions are often requisite [55, 65]. In a living organism, proteins participate in complex interactions, which represent the mechanistic foundation of the organism's physiology and function. Regulation, recognition and cell signaling involve the coordinated actions of many players. To achieve this coordination, each participant must have a valid identification ("ID") that is easily recognized by the others. For proteins, these "IDs" are often within IDRs [55, 65].

Another very important feature of the IDPs is their unique capability to fold under the variety of conditions [4, 6, 11, 14, 20, 30, 38, 53, 55, 57, 61, 65, 66]. In fact, the folding of these proteins can be brought about by interaction with other proteins, nucleic acids, membranes or small molecules. It also can be driven by changes in the protein environment. The resulting conformations could be either relatively non-compact (i.e., remain substantially disordered) or be tightly folded.

2.3 Abundance of IDPs in Neurodegenerative Diseases. Evidence from the Bioinformatics Analyses

Because of the fact that IDPs play a number of crucial roles in numerous biological processes, it was not too surprising to find that some of them are involved in human diseases. An incomplete list of human neurodegenerative diseases associated with IDPs includes AD (deposition of amyloid-β, tau-protein, α-synuclein fragment NAC [67, 68]; Niemann-Pick disease type C, subacute sclerosing panencephalitis, argyrophilic grain disease, myotonic dystrophy, and motor neuron disease with neurofibrillary tangles (NFTs) (accumulation of tau-protein in form of NFTs [69]); Down's syndrome (nonfilamentous amyloid-β deposits [70]); PD, dementia with Lewy body (LB), diffuse LB disease, LB variant of AD, multiple system atrophy

(MSA) and Hallervorden-Spatz disease (deposition of α-synuclein in a form of LB, or Lewy neurites (LNs) [71]); prion diseases (deposition of PrP^{SC} [72]); and a family of polyQ diseases, a group of neurodegenerative disorders caused by expansion of GAC trinucleotide repeats coding for PolyQ in the gene products [73].

Table 2.1 and Fig. 2.1 illustrates that some individual proteins involved in human neurodegenerative diseases are either completely disordered or contain long disordered regions. Figure 2.1 represents the results of the comparison of the compositions of proteins from Table 2.1 with the composition of ordered proteins from the Protein Data Bank (PDB). The corresponding data for the DisProt [74] are shown for comparison. Calculations were done using a normalization procedure elaborated for analysis of IDPs [1, 75]. In brief, compositional profiling is based on the evaluation of the $(C_{s1} - C_{s2})/C_{s2}$ values, where C_{s1} is a content of a given residue in a set of interest (proteins associated with neurodegenerative diseases or typical IDPs from DisProt), whereas C_{s2} is the corresponding value for the set of ordered proteins. In this presentation, negative values correspond to residues which are depleted in a given dataset in comparison with a set of ordered proteins, whereas the positive values correspond to the residues which are over-represented in the set.

Figure 2.1 shows that in general all proteins in Table 2.1 are highly different from typical ordered proteins and generally follow the trend for IDPs (with some exceptions). Proteins associated with neurodegenerative diseases are in general depleted

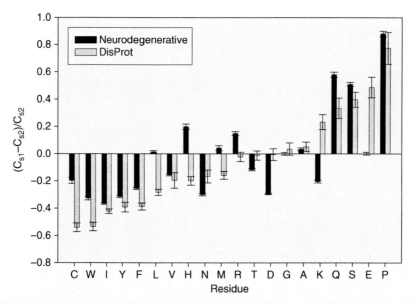

Fig. 2.1 Compositional profiling of proteins involved in neurodegenerative disease. Analyzed proteins are listed in Table 2.1. Enrichment or depletion in each amino acid type appears as a positive or negative bar, respectively. Amino acids are indicated by the single-letter code and ordered according to their disorder promoting strength. Error bars are also shown. Corresponding data for well-characterized IDPs from DisProt are also shown

in major order-promoting residues. This includes C, W, I, Y, F, V and N. They are highly enriched in the major disorder-promoting residues (Q, S, R, and P). There are also some deviations from the behavior of "typical" disordered proteins. This includes the high abundance of L and H and the depletion in T, D, and K. This suggests that proteins listed in Table 2.1 are in general characterized by a high level of intrinsic disorder.

This fact raises the question of how abundant are the IDPs in various neurodegenerative conditions. To answer this question, a set of 689 proteins related to neurodegenerative diseases was collected and analyzed using an approach elaborated to analyze the abundance of intrinsic disorder in cancer-related proteins [66]. In that study, 79% of cancer-associated and 66% of cell-signaling proteins were found to contain predicted regions of disorder of 30 residues or longer [66]. In contrast, only 13% of proteins from a set of proteins with well-defined ordered structures contained such long regions of predicted disorder. In agreement with these bioinformatics studies, the presence of intrinsic disorder has been directly observed in many cancer-associated proteins.

The overall results of the analogous analysis for proteins associated with neurodegenerative disease are shown in Fig. 2.2, which represents percentages of proteins with ≥30 consecutive residues predicted to be disordered in various datasets, including cancer-related proteins, signaling proteins, ordered proteins from PDB, eukaryotic proteins and proteins involved in various neurodegenerative diseases. This figure illustrates that intrinsic disorder is highly prevalent in neurodegenerative

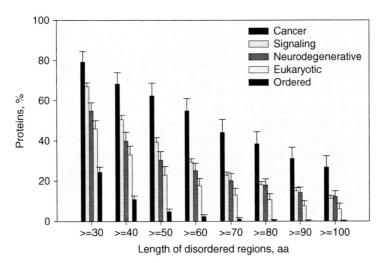

Fig. 2.2 Abundance of intrinsic disorder in proteins associated with neurodegenerative diseases. Percentages of disease associated proteins with ≥30 to ≥100 consecutive residues predicted to be disordered. The error bars represent 95% confidence intervals and were calculated using 1,000 bootstrap re-sampling. Corresponding data for signaling and ordered proteins are shown for the comparison. Analyzed sets of disease-related proteins included 1786 proteins associated with cancer and 689 proteins involved in the neurodegenerative disease

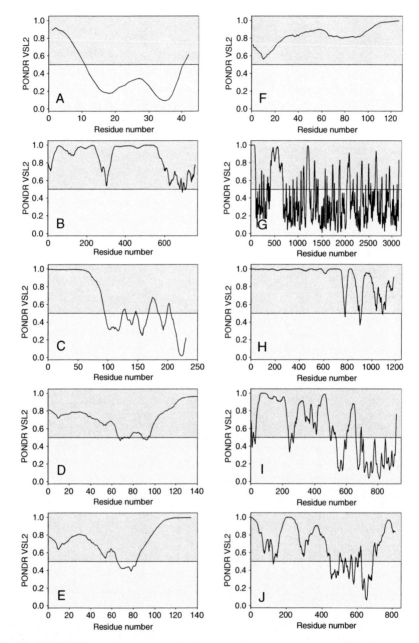

Fig. 2.3 (continued)

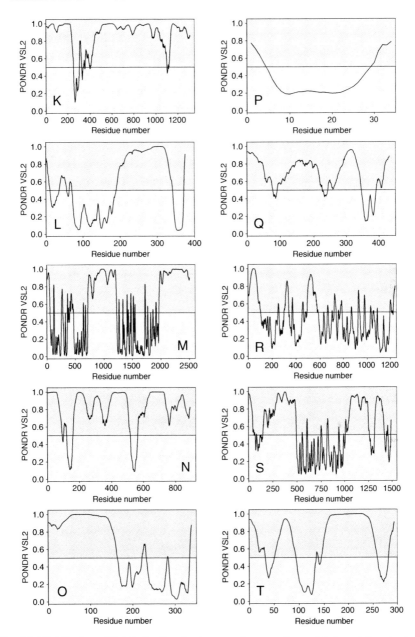

Fig. 2.3 Distribution of intrinsic disorder in neurodegeneration-related proteins. Analyzed proteins are listed in Table 2.1: A, Aβ; B, Tau protein; C, Prion protein; D, α-Synuclein; E, β-Synuclein; F, γ-Synuclein; G, Huntingtin; H, DRPLA protein (atrophin-1); I, Androgen receptor; J, Ataxin-1; K, Ataxin-2; L, Ataxin-3; M, P/Q-type calcium channel α1A subunit; N, Ataxin-7; O, TATA-box-binding protein; P, ABri; Q, Glial fibrillary acidic protein; R, Mitochondrial DNA polymerase γ; S, DNA excision repair protein ERCC-6; T, Survival motor neuron protein. Shaded areas in each plot correspond to the scores associated with intrinsic disorder

disease-related proteins, being comparable with that of signaling and cancer-related proteins and significantly exceeds the level of intrinsic disorder in eukaryotic proteins from SWISS-PROT and in non-homologous, structured proteins from the PDB. Thus, intrinsic disorder is very common in neurodegeneration-associated proteins. To further illustrate this concept, Table 2.1 represents some of the IDPs and their corresponding neuropathological conditions. Many of these proteins were structurally characterized and experimental evidence on the presence of intrinsic disorder in some of these proteins is also listed. Table 2.1 shows that there is a great agreement between experimental and computational data. Finally, the results of the disorder prediction by two predictors, PONDR® VSL2 and VLXT, are shown. Figure 2.3 represents plots of the PONDR® VSL2 predicted disorder distribution within the sequences of 20 neurodegenerative disease-related IDPs. It clearly shows that these proteins are very divers: their length range from 34 to 3144 amino acid residues, the amount of predicted disorder range from 16.7 to 100%, and the profiles of disorder distribution are very different. Therefore, computational analysis showed that the majority of the proteins involved into the pathogenesis of neurodegenerative disease are intrinsically disordered. Subsequent sections consider illustrative examples of some of the most important neurodegenerative IDPs and their corresponding diseases.

2.4 Intrinsic Disorder in Proteins Associated with Neurodegenerative Diseases

2.4.1 Amyloid β-Protein and Alzheimer's Disease

AD is the most prevalent age-dependent dementia, causing cognitive decline among people of age 65 and older. It currently affects 4.5 million Americans and is projected to afflict 13.2 million by the year 2050 in the US alone [76]. AD ranks third in total health care cost after heart disease and cancer. The national direct and indirect annual cost of AD approaches 100 billion dollars per year [77].

AD was described for the first time in 1907 by a German physician Alois Alzheimer [78]. AD is the most common aging-related neurological disorder, which constitutes about two thirds of cases of dementia overall [79, 80] and is characterized by slow, progressive memory loss and dementia due to a gradual neurodegeneration particularly in the cortex and hippocampus [81]. The clinical hallmarks are progressive impairment in memory, judgment, decision making, orientation to physical surroundings, and language [82]. From the initial symptoms, disease progression can last up to 25 years, although typically the duration ranges from 8 to 10 years.

Sporadic AD is a disease of the elderly; most patients are diagnosed after 65 years of age. About 10% of AD cases present under age 65 and have been referred to as having early onset AD. Three causative autosomal dominant mutations have been described – the amyloid β-protein precursor (APP) gene mutation on chromosome

21, the presenilin 1 gene mutation on chromosome 14 and the presenilin 2 gene mutation on chromosome 1. These autosomal dominant forms comprise only about 2% of all AD [83]. Having an extra copy of the APP gene, as in case of Down's patients (trisomy 21), also leads to early pathological and clinical changes of AD.

AD is characterized pathologically by the accumulation of two types of proteinaceous inclusions, extracellular amyloid deposits, senile plaques, in the cerebral cortex and vasculature and intracellular NFTs (paired helical filaments, PHFs) [84]. Amyloid is a descriptive term for proteinaceous deposits that stain with Congo red and thioflavin S and demonstrate birefringence in polarized light. Amyloid deposits in AD contain the amyloid β-protein (Aβ), which is a 40–42 residue peptide, produced by endoproteolytic cleavage of the APP. PHFs are assembled from a hyperphosphorylated form of the microtubular protein tau (see next section).

APP, the parent molecule of Aβ, plays a role in synaptic stabilization and plasticity, regulation of neuronal survival, neuritic outgrowth and cell adhesion [85, 86]. Nexin-2, a secreted form of APP, inhibits coagulation factor XIa [87, 88]. C-terminal fragment of APP originating after the γ-secretase cleavage mediates nuclear signaling and modulate gene expression [89–92]. The Aβ fragment of the APP protein is a byproduct of APP processing. The normally prevailing α-secretase-mediated APP processing splits the large APP molecule in the middle of the Aβ sequence and does not produce pathogenic Aβ species. However, alternative cleavage by the β- and γ-secretases results in generation of the pathogenic Aβ fragment. Depending on the exact site of action of γ-secretase, several Aβ peptides with 39–43 amino acids are produced [93]. The longer moieties are more amyloidogenic [94]. Although β- and γ-secretase are active throughout the lifespan, plaques rarely form in young individuals, but after the age of 60 nearly all elderly develop some Aβ deposits [95, 96].

Many lines of evidence support the crucial role of Aβ in AD. Aggregated forms of the Aβ peptide with amyloid-like cross-β structure are neurotoxic to cortical cell cultures [97–100]. Some of the Aβ–derived diffusible ligands (small Aβ aggregates) kill mature neurons at nanomolar concentrations and cause neurological dysfunction in the hippocampus [101]. The two major Aβ peptides are the 40-residue $A\beta_{1-40}$ and the 42-residue $A\beta_{1-42}$, which differ in the absence or presence of two extra C-terminal residues (Ile41-Ala42). The N-terminal (residues 1–28) residues comprise a hydrophilic domain with a high proportion of charged residues (46%), whereas the C-terminal domain (residues 29–40 or 29–42) is completely hydrophobic and is presumably associated with the cell membrane of APP. Although the $A\beta_{1-40}$ and $A\beta_{1-42}$ peptides are ubiquitous in biological fluids of humans (at an approximate ratio of 9:1), it is thought that the longer $A\beta_{1-42}$ is more pathogenic, due to its higher quantities in the amyloid plaques of sporadic AD cases, its even higher quantities in patients afflicted with early onset AD [102, 103], and because of the greater *in vitro* tendency of the $A\beta_{1-42}$ to aggregate and precipitate as amyloid [104, 105]. Fibrillation of Aβ is associated with the development of the cascade of neuropathogenetic events, ending with the appearance of cognitive and behavioral features typical of AD.

Aβ appears to be unfolded at the beginning of the fibrillation under physiological conditions. NMR studies have shown that monomers of Aβ(1–40), or Aβ(1–42) possess no α-helical or β-sheet structure [106]; i.e., they exist predominately as random coil-like highly extended chains. Partial refolding to the pre-molten globule-like conformation has been detected at the earliest stages of Aβ fibrillation [106].

Besides AD, Aβ aggregation was implemented in several other neurodegenerative diseases (see Table 2.1). For example, the E22Q mutation of Aβ is associated with the rare disorder, hereditary cerebral hemorrhage with amyloidosis-Dutch type (HCHWA-D). HCHWA-D is characterized by severe cerebral amyloid angiopathy (CAA), which is characterized by extensive amyloid deposition in the small leptomeningeal arteries and cortical arterioles, leading to hemorrhagic strokes of mid-life onset, dementia and an early death of those afflicted in their fifth or sixth decade. Therefore, this disorder is an autosomal dominant form of vascular amyloidosis restricted to the leptomeninges and cerebral cortex. CAA severity tends to increase with age [107]. In HCHWA-D, parenchymal Aβ deposition is enhanced, with non-fibrillar membrane-bound Aβ$_{42}$ deposits evolving into relatively fibrillar diffuse plaques variously associated with reactive astrocytes, activated microglia, and degenerating neurites [107]. Although silver stain-positive, "senile plaque-like" structures found in the HCHWA-D brain were immunopositive for Aβ, yet these lesions lacked the dense amyloid cores present in typical AD plaques [108]. No NFTs are present in this disorder. The total Aβ production is not affected by E22Q mutation. However, the proteolytic degradation of Aβ and its transport across the blood–brain barrier as well as the Aβ$_{42}$:Aβ$_{40}$ ratio are altered. Aβ E22Q aggregates faster and fibrils formed by tthis mutant form are more stable than amyloid-like fibrils produced by the wild-type Aβ [107].

2.4.2 Tau Protein in Alzheimer's Disease and Other Tauopathies

The tau gene is located on chromosome 17. It encodes for a protein with four 31–32 amino acid tandem repeats close to its C-terminus. Tau protein is a vital structural element of the microtubular transport system in the nervous system. Its aggregation is implemented in AD and several other diseases collectively known as tauopathies (see Table 2.1). Tau protein represents a family of isoforms migrating as close bands of 55–62 kDa in SDS gel electrophoresis. Heterogeneity is due in part to alternative mRNA splicing leading to the appearance of one, two, three or four repeats in the C-terminal region [109, 110]. In normal cortex the three and four-repeat forms are equally expressed. In tauopathies the ratio is changed. AD is the only dementia with both three- and four-repeat tau [111].

In vitro, tau binds to microtubules, promotes microtubule assembly, and affects the dynamic instability of individual microtubules [112–116]. *In situ*, tau is highly enriched in the axons [117]. In living cells and brain tissue, tau protein has been estimated as comprising 0.025–0.25% of total protein [118, 119]. On the basis of its *in vitro* activity and its distribution, it is believed that tau regulates the organization of neuronal microtubules. Interest in tau dramatically increased with the discovery of its aggregation in neuronal cells in the progress of AD and various other neurodegenerative disorders, especially frontotemporal dementia [120, 121]. In these cases specific tau-containing NFTs (paired helical filaments, PHFs) are formed [121]. Hyperphosphorylation was shown to be a common characteristic of pathological tau [122]. Hyperphosphorylated tau isolated from patients with AD was shown to be unable to bind to microtubules and promote microtubule assembly. However, both of these activities were restored after enzymatic dephosphorylation of tau protein [123–126]. Although tau inclusions can be stained with hematoxylin-eosin and amyloid stains, they are much easier visualized after silver impregnation. The most sensitive and specific method is tau immunohistochemistry. There are three types of tau deposits in AD – NFTs, neuropil threads, and dystrophic neurites.

NFTs are composed of 22 nm PHF and each PHF is composed of 8–14 tau monomers [127]. They commonly affect the pyramidal cortical neurons and assume a flame-like shape. Extracellular NFTs are rare and are referred to as ghost tangles. They are presumed to be the remnants of dead neurons and are most commonly seen in the hippocampus. When surrounded by dystrophic neuritis, they are called tangle associated neuritic clusters [128]. Although NFTs correlate better with dementia severity than amyloid plaques [95], they can be absent in the neocortex in 10% of patients with AD and in as many as the 50% of mild AD cases [129].

Neuropil threads are most commonly seen in AD and only rarely identified in other tauopathies such as corticobasal dgeneration [130]. They are short tortuous neuronal dendirites filled with abnormal tau [128]. Dystrophic neurites are tau-containing dendritic structures that are seen in the periphery of the senile plaques.

Post-translational phosphorylation of tau is an additional source of microheterogeneity [131]. During brain development, tau is phosphorylated at many residues with GSK-3β, cdk 5, and MAPK [132]. *In vitro*, tau can be phosphorylated on multiple sites by several kinases, too (for a review, see [133]). Most of the *in vitro* phosphorylation sites are located within the microtubule interacting region (repeat domain) and sequences flanking the repeat domain. Many of these sites are also phosphorylated in PHF-tau [134, 135]. In fact, 10 major phosphorylation sites have been identified in tau isolated from PHFs from patients with AD [134]. Hyperphosphorylation was shown to be accompanied by the transformation from the unfolded state of tau into a partially folded conformation [136, 137], accelerating the self-assembly of this protein into paired helical filaments *in vitro* [124]. To analyze the potential role of tau hyperphosphorylation in tauopathies, mutated tau proteins have been produced, in which all 10 serine/threonine residues known to be highly phosphorylated in PHF-tau were substituted for negatively charged residues, thus producing a model for a defined and permanent hyperphosphorylation-like state of tau protein [138]. It has been demonstrated that, like hyperphosphorylation, glutamate substitutions induce compact structure elements and SDS-resistant conformational domains in tau protein, as well as lead to the dramatic acceleration of its fibrillation [138].

Prior the aggregation, tau protein was shown to be in a mostly random coil-like state. This conclusion followed from the conformational analysis of this protein by CD, Fourier transform infrared spectroscopy, small angle X-ray scattering and biochemical assays [139]. Analysis of the primary structure reveals a very low content of hydrophobic amino acids and a high content of charged residues, which was sufficient to explain the lack of folding [139]. Analysis of the hydrodynamic radii confirms a mostly disordered structure of various tau isoforms and tau domains. However, the protein was further unfolded in the presence of high concentrations of strong denaturant GdmCl, indicating the presence of some residual structure. This conclusion was supported by a FRET-based approach where the distances between different domains of tau were determined. The combined data show that tau is mostly disordered and flexible but tends to assume a hairpin-like overall fold which may be important in the transition to a pathological aggregate [139].

Intriguingly, purified recombinant tau isoforms do not detectably aggregate over days of incubation under physiological conditions. However, aggregation and fibrillization can be dramatically accelerated by the addition of anionic surfactants [140]. Based on the detailed analysis of tau fibrillation in the presence of anionic inducers using a set of spectroscopic techniques (CD and reactivity with thioflavin S and 8-anilino-1-naphthalenesulfonic acid (ANS) fluorescent probes) it has been established that the inducer stabilized a monomeric partially folded species with the structural characteristics of a pre-molten globule state [141]. The stabilization of this intermediate was sufficient to trigger the fibrilliation of full-length tau protein [141].

2.4.3 Prion Protein and Prion Diseases

Prion diseases are a group of incurable, fatal neurodegenerative maladies that afflict mammals. These diseases, collectively referred to as the transmissible spongiform encephalopathies (TSEs), are caused by the pathological deposition of the prion protein (PrP) in its aggregated form. TSEs include Creutzfeldt-Jakob disease (CJD), Gerstmann-Sträussler-Scheinker (GSS) disease, fatal familial insomnia (FFI) and kuru in humans, scrapie in sheep, bovine spongiform encephalopathy (BSE) in cattle, and chronic wasting disease (CWD) in mule deer and elk [142]. The most important aspect is the transmission of PrP aggregates from one individual or species to another, causing prion diseases. Prion diseases are unique among all illnesses in that they can manifest as sporadic, genetic or infectious maladies. Similar to many other neurodegenerative diseases, the sporadic form of prion disease accounts for ∼80–90% of cases whereas the genetic forms account for 10 to 20% [143, 144]. Infection by exogenous prions seems to be responsible for <1% of all human cases of prion disease [145].

The characteristic pathological features of TSEs are spongiform degeneration of the brain and accumulation of the abnormal, protease-resistant PrP isoform in the central nervous system, which sometimes forms amyloid-like plaques. The prion concept was introduced in 1982 in order to explain a vast body of scientific data, much of which argued the pathogen causing scrapie is devoid of nucleic acid but contains a protein that is essential for infectivity [146]. Prions are unprecedented infectious pathogens that give rise to invariably fatal neurodegenerative diseases via an entirely novel mechanism of disease.

Native prion protein (PrP^C) is attached to the extracellular plasma membrane surface by a glycosylphosphatidylinositol lipid anchor and undergoes endocytosis. The N-terminal region of about 100 amino acids in PrP^C (from amino acid 23 to 126) is largely unstructured in the isolated molecule in solution [147]. The C-terminal domain is folded into a largely α-helical conformation (three α-helices and a short antiparallel β-sheet) and stabilized by a single disulphide bond linking helices 2 and 3 [148]. The central event in the pathogenesis of prion diseases is believed to be a major conformational change of the C-terminal region of the PrP from an α-helical (PrP^C) to a β-sheet-rich isoform (PrP^{Sc}), and PrP^{Sc} propagates itself by causing the conversion of PrP^C to PrP^{Sc}. Although unstructured in the isolated molecule, the N-terminal region contains tight binding sites for Cu^{2+} ions and acquires structure following copper binding [149, 150].

Two pathological GSS-like mutations, Y145Stop and Q160Stop, result in C-terminal truncated isoforms. The truncation occurs just after the central region from amino acid 90 to 145, which was shown to be converted into β-sheet as a result of the PrP^C to PrP^{Sc} conversion [151, 152]. Structural properties and aggregation propensities of these mutants *in vitro* were analyzed by a variety of biophysical techniques [153]. It has been shown that although both proteins are substantially disordered, a continuous stretch of positive secondary chemical NMR shifts was found for residues 144–154 in Q160Stop protein, indicative of helical structure. This clearly demonstrated that although the vast majority of a polypeptide chain is substantially disordered, a significantly populated helix 1 is present in human Q160Stop protein [153]. Q160Stop protein was shown to fibrillate faster than shorter Y145Stop variant. Intriguingly, helix 1 was not converted to the β-sheet during the protein aggregation. Based on the results of this analysis it has been concluded that the highly charged helix 1 is involved in the aggregation of Q160Stop protein likely via the formation of intermolecular salt bridges [153].

Investigations of the steps required for prion propagation and neurodegeneration in transgenic mice expressing chimeric mouse–hamster–mouse or mouse–human–mouse PrP transgenics indicated that the last 50 residues in the disordered N-terminal region play a particularly important role in the interaction of PrP^C with PrP^{Sc} leading to the conversion of the former to the latter [154, 155]. Those residues are largely unordered or weakly helical in the full-length PrP^C [156, 157], but are predicted to be β-structure in PrP^{Sc} [146]. These observations emphasize a crucial role of the disordered N-terminal region in the modulation of PrP aggregation. Several kinetics studies have revealed the existence of partially folded intermediates for the PrP [146, 158, 159], and it is reasonable to assume that fibrillation requires partial unfolding of the C-terminal domain prior to self-association.

2.4.4 α-Synuclein and Synucleinopathies

Synucleinopathies (see Table 2.2) is a group of neurodegenerative disorders characterized by fibrillary aggregates of α-synuclein protein in the cytoplasm of selective populations of neurons and glia [160–163]. Clinically, synucleinopathies are

Table 2.2 Human neurodegenerative disorders with α-synuclein deposits

Diseases with neuronal inclusions
Normal aging
Parkinson's disease
  Idiopathic
  Neurotoxicant-induced (incidental)
  Familial
    With α-synuclein point mutations
    With α-synuclein gene triplication
    With mutations in other proteins
  Pure autonomic failure
  Lewy body dysphagia
Parkinsonism plus syndromes
  Sporadic
    Progressive supranuclear palsy
    Olivoponto cerebrellar atrophy (Shy-Dragger syndrome)
    Cortical-basal ganglionic degeneration
    Sporadic pallidal degeneration
    Bilateral striatopallido dentate calcinosis
    Parkinsonism with neuroacanthocytosis
  Familial
    Familial diffuse Lewy body disease
    Familial dementia with swollen achromatic neurons and cortico-basal inclusion bodies
    Frontotemporal dementia with parkinsonism linked to chromosome 17
    Associated with psychiatric disturbances
    Associated with respiratory disturbances
    Associated with dystonia
    Associated with myoclonus and seizures
    Familial progressive supranuclear palsy
Alzheimer's disease
  Sporadic
  Familial with APP mutation
  Familial with PS-1 mutation
  Familial with other mutations
  Familial British dementia
Lewy body diseases
  Dementia with Lewy bodies
    Pure form – transitional/limbic
    Pure form – neocortical
  Diffuse Lewy body disease
    Common form
    Pure form
  Lewy body variant of Alzheimer's disease
  Incidental Lewy body disease
  Lewy body dementia
  Senile dementia of Lewy body type
  Dementia associated with cortical Lewy bodies
Down's syndrome
Amyotrophic lateral sclerosis-parkinsonsim/dementia complex of Guam

Table 2.2 (continued)

Neuroaxonal dystrophies
Neurodegeneration with brain iron accumulation, type I (Hallervorden-Spatz syndrome or adult neuroaxonal dystrophy)
Motor neuron disease
Amyotrophic lateral sclerosis
Familial
Sporadic
Tauopathies
Frontotemporal degeneration/dementia
Pick's disease
Post-encephalitic parkinsonism
Dementia pugilistica
Argyrophilic grain disease
Corticobasal degeneration
Prion diseases
Transmissible spongiform encephalopathies
Sporadic
Creutzfeldt-Jakob disease
Familial
Familial Creutzfeldt-Jakob disease
Gertsmann-Straussler-Scheinker syndrome
Infectious
Iatrogenic Creutzfeldt-Jakob disease
Variant Creutzfeldt-Jakob disease
Kuru
Fatal familial insomnia
Ataxia telangiectatica
Meige's syndrome
Diseases with neuronal and glial inclusions
Multiple system atrophy
Shy-Drager syndrome
Striatonigral degeneration (MSA-P)
Olivopontocerebellar atrophy (MSA-C)

characterized by a chronic and progressive decline in motor, cognitive, behavioral, and autonomic functions, depending on the distribution of the lesions. Because of clinical overlap, differential diagnosis is sometimes very difficult [164]. Depending on the type of pathology, α-synuclein inclusions are present in neurons (both dopaminergic and non-dopaminargic), where they can be deposited in perikarya or in axonal processes of neurons, and in glia. At least five morphologically different α-synuclein containing inclusions have been determined: LBs, LNs (dystrophic neurites), glial cytoplasmic inclusions (GCIs), neuronal cytoplasmic inclusions and axonal spheroids. Some of the disorders associated with the α-synuclein depositions are discussed below to illustrate a wide range of pathological manifestations in synucleinopathies.

2.4.4.1 α-Synuclein and Parkinson's Disease

PD is the most common ageing-related movement disorder and second most common neurodegenerative disorder after AD. It is estimated that ~1.5 million Americans are affected by PD. Since only a small percentage of patients are diagnosed before age 50, PD is generally considered as an aging-related disease, and approximately one of every 100 persons over the age of 55 in the US suffers from this disorder [165]. PD is a slowly progressive disease that affects neurons of the *substantia nigra*, a small area of cells in the mid-brain. Gradual degeneration of the dopaminergic neurons causes a reduction in the dopamine content. This, in turn, can produce one or more of the classic signs of PD: resting tremor on one (or both) side(s) of the body; generalized slowness of movement (bradykinesia); stiffness of limbs (rigidity); and gait or balance problems (postural dysfunction). The *substantia nigra* consists of ~400,000 nerve cells, which begin to pigment after birth and are fully pigmented at age 18. The symptoms of PD become apparent after more than ~70% dopaminergic neurons die. The "normal" rate of nigral cell loss is ~2,400 per a year. Thus, if an unaffected person lives to be 100 years old he (she) will probably develop PD. In PD, the neuron loss is accelerated. Although, it is unknown why nerve cells loss accelerates, it appears to be due to a combination of genetic susceptibility and environmental factors. Some surviving nigral dopaminergic neurons contain cytosolic filamentous inclusions known as LBs when found in the neuronal cell body, or LNs when found in axons [166, 167].

Several observations implicate α-synuclein in the pathogenesis of PD. Autosomal dominant early-onset PD was shown to be induced in a small number of kindreds as a result of three different missense mutations in the α-synuclein gene, corresponding to A30P, E46K, and A53T substitutions in α-synuclein [168–170] or as a result of the hyper-expression of the wild type α-synuclein protein due to gene triplication [171–173]. Antibodies to α-synuclein detect this protein in LBs and LNs. A substantial portion of fibrillar material in these specific inclusions was shown to be comprised of α-synuclein, and insoluble α-synuclein filaments were recovered from purified LBs [174, 175]. The production of wild type α-synuclein in transgenic mice [176] or of WT, A30P, and A53T in transgenic flies [177], leads to motor deficits and neuronal inclusions reminiscent of PD. Under the particular conditions, cells transfected with α-synuclein might develop LB-like inclusions. Other important observations correlating α-synuclein and PD pathogenesis were reviewed in more detail elsewhere [71, 162, 178–181].

2.4.4.2 α-Synuclein in Dementia with Lewy Bodies and Other Lewy Body Disorders

Dementia with Lewy Bodies

Dementia with Lewy bodies (DLB), being the second most frequent neurodegenerative dementing disorder after AD, is a common form of late-onset dementia that exists in a pure form or overlaps with the neuropathological features of AD. This disease is characterized clinically by neuropsychiatric changes often with marked fluctuations in cognition and attention, hallucinations, and parkinsonism [182]. Similar to PD, neurophathological hallmarks of DLB are numerous LBs and LNs in the substantia nigra, which are strongly immunoreactive for α-synuclein [174]. However, unlike PD, DLB is characterized by large numbers of LBs and LNs in cortical brain areas [183]. It has been noted that filaments from LBs in DLB are decorated by α-synuclein antibodies [175, 184, 185], and that their morphology closely resembles that of filaments extracted from the *substantia nigra* of PD brains [175, 185]. DLB and PD with dementia, being different in the temporal course of the disease, share most of the same clinical and neuropathologic features and are often considered as belonging to a spectrum of the same disease [186–188]. It is well recognized now that the incidence of dementia in PD is higher than expected from aging alone [182], as dementia affects about 40% of PD patients [189], and the incidence of dementia in PD patients is up to six times greater than observed in normal aged matched control subjects [190].

Amyotrophic Lateral Sclerosis-Parkinsonsim/Dementia Complex of Guam

Guam disease is another example of PD and dementia junction. Guam disease is a neurodengenerative disorder with unusually high incidence among the Chamorro people of Guam [191–193]. The neurotoxic plant *Cycas circinalis*, a traditional source of food and medicine used by the Chamorro people, plays a role in the development of Guam amyotrophic lateral sclerosis-parkinsonism-dementia [193]. Intriguingly, recent studies revealed that in general three neurodegenerative disorders, amyotrophic lateral sclerosis, dementia, and PD, co-occur within families more often than expected by chance, suggesting that there may be a shared genetic susceptibility to these disorders [194].

Other Lewy Body Diseases

Several peripheral and central areas of the nervous system can be affected by the LB deposition. Besides already discussed *substantia nigra*, this includes *hypothalamic nuclei, nucleus basalis of Meynert, dorsal raphe, locus ceruleus, dorsal vagus nucleus*, and *intermediolateral nucleus* [195]. A 'neuritic' form of LB was also described in the *dorsal vagus nucleus, sympathetic ganglia*, and in *intramural autonomic ganglia* of the gastrointestinal tract, as well cases were demonstrated with extensive cortical and basal ganglia involvement [183, 196]. This broad spectrum of the nervous system regions potentially affected by LB formation produces great variability in the disease manifestation and LB pathology is also a characteristic feature of several rarer diseases, such as pure autonomic failure, LB dysphagia, incidental LB disease [179, 180]. Pure autonomic failure (also known as Bradbury-Eggleston syndrome) [197] and LB dysphagia [198] are the results of the predominant involvement of the peripheral nervous system with minimal central nervous system involvement. In incidental LB disease, ~5%–10% of asymptomatic individuals have insignificant numbers of LBs bodies, usually located in *substantia nigra* [199].

2.4.4.3 α-Synuclein and Alzheimer's Disease

Detailed analysis of the α-synuclein immunoreactivity in the brains from the patients with sporadic AD revealed the presence of α-synuclein-positive inclusions resembling LBs and LNs in ~50% cases studied [200]. α-Synuclein-positive LB-like intra-cytoplasmic inclusions were found in the amygdale, the temporal cortex, the parahippocampal gyrus, and in the parietal cortex, whereas LN-like inclusions were abundant in the amygdala, the CA2/3 region of hippocampus formation, parahippocampal gyrus, the temporal cortex, substantia nigra, locus ceruleus, the frontal cortex, and in the parietal cortex [200].

2.4.4.4 α-Synuclein and Down's Syndrome

Down's syndrome is a genetic disorder characterized by an extra chromosome 21 (trisomy 21, i.e., instead of having the normal 2 copies of chromosome 21, the Down's syndrome patient has 3 copies of this chromosome). The person with Down's syndrome has mild mental retardation, short stature, a flattened facial profile, a risk of multiple malformations (including heart malformations; duodenal atresia, where part of the small intestines is not developed and leukemia), and susceptibility to early-onset AD. Incidence of this disorder among the newborn is estimated at 3 in 1,000, whereas in the general population it is approximately 1 in 1,000. The difference reflects the early mortality. The analysis of Down's syndrome with Alzheimer pathology revealed presence of numerous LBs and LNs in the neurons of the limbic areas, predominantly of the amygdala. Similar lesions were less common in other regions of these brains [201, 202]. Importantly, in the vast majority of cases examined no LBs and LNs were detected in the *substantia nigra* and *locus ceruleus*, and there was no significant neuronal loss in the *substantia nigra*.

2.4.4.5 α-Synuclein and Multiple System Atrophy

MSA is an adult-onset progressive neurodegenerative disorder of unknown etiology which is characterized clinically by any combination of parkinsonian, autonomic, cerebellar or pyramidal symptoms and signs, and pathologically by cell loss, gliosis and GCIs in several brain and spinal cord structures. Most patients affected by MSA deteriorate rapidly and survival beyond ten years after disease onset is unusual. It is believed that the motor impairment in MSA results from L-dopa-unresponsive parkinsonism, cerebellar ataxia and pyramidal signs, with 80% of MSA cases showing predominant parkinsonism (MSA-P) due to underlying striatonigral degeneration, and the remaining 20% developing predominant cerebellar ataxia (MSA-C) associated with olivopontocerebellar atrophy [203]. Autonomic dysfunction including urogenital failure and orthostatic hypotension is common in both motor presentations, MSA-P and MSA-C, reflecting degenerative lesions of central autonomic pathways [204]. Distinguishing MSA-P from PD is problematic at early stages owing to PD-like features in MSA-P, including a transient L-dopa response in some patients [205]. MSA is less common than PD as epidemiological studies suggested a prevalence of 1.9–4.9 people per 100,000 and an incidence of 3 patients per 100,000 people per year [206–208]. Histologically, MSA is characterized by the variable neuron loss in the *striatum, substantia nigra pars compacta, cerebellum*, pons, inferior olives and intermediolateral column of the spinal cord [209]. The histological hallmark of MSA is the presence of argyrophilic fibrillary inclusions in the oligodendrocytes, referred to as GCIs, which are also known as Papp-Lantos bodies [210]. Fibrillary inclusions are also found in the neuronal somata, axons, and nucleus. Neuronal cytoplasmic inclusions are frequently found in the pontine and inferior olivary nuclei [211]. It has been established that α-synuclein is a major component of glial and neuronal inclusions in MSA [185, 211]. Although both LBs and GCIs contain α-synuclein, they are differently localized, with α-synuclein inclusions being neuronal in PD and DLB, and oligodendroglial in MSA. This suggests the existence of a unique pathogenic mechanism that ultimately lead to neuron loss via disturbance of axonal function [210]. In MSA, besides formation of GCIs α-synuclein also aggregates in the cytoplasm, axons and nuclei of neurons, and the nuclei of oligodendroglia. The relationship between GCIs and these additional α-synuclein deposition sites is not understood [210].

2.4.4.6 α-Synuclein and Neurodegeneration with Brain Iron Accumulation Type 1 (NBIA1)

Neurodegeneration with brain iron accumulation type 1 (NBIA1) (formerly know as Hallervorden-Spatz disease (HSD) or adult neuroaxonal dystrophy) represents a rare progressive neurodegenerative disorder that occurs in both sporadic as well as in familial forms. Clinically, NBIA1 is characterized by rigidity, dystonia, dyskinesia, and choreoathetosis [212–215], together with dysarthria, dysphagia, ataxia, and dementia [215–217]. Symptoms usually present in late adolescence or early adult life and this disease is persistently progressive [212, 216, 217]. The histopathologic hallmarks of NBIA1 include neuronal loss, neuraxonal spheroids, and iron deposition in the *globus pallidus* and *substantia nigra pars compacta*, as well as by the presence of the LB-like and GCI-like inclusions and dystrophic neuritis [216]. NBIA1 is characterized by an association of extrapyramidal movement disorders with neuroaxonal dystrophy (NAD) and iron accumulation in the basal ganglia. It represents a pantothenate kinase-associated neurodegeneration caused by the PNAK2 gene linked to chromosome 20p12.3–13 [218]. It has been shown that the LB-like inclusions throughout the cortex and brainstem, axonal swellings, and rare GCI-like inclusions of the midbrain clearly possess α-synuclein immunoreactivity [219–221]. Importantly, axonal spheroids were also shown to contain α-synuclein [221, 222].

2.4.4.7 Structural Properties of α-Synuclein

α-Synuclein, a protein that links various synucleinopathies, is one of the most studied IDPs. It possesses little or no ordered structure under the "physiological" conditions *in vitro* (i.e., conditions of neutral pH and low to moderate ionic strength)

[223]. For example, at neutral pH α-synuclein is characterized by far-UV CD and FTIR spectra typical of a substantially unfolded polypeptide chain with a low content of ordered secondary structure. The hydrodynamic properties of this protein are in a good agreement with the results of the far-UV CD and FTIR studies and show that α-synuclein, being essentially expanded, does not have a tightly packed globular structure, but is slightly more compact than expected for a random coil [223]. Based on the results of pulsed-field gradient NMR (which allows an estimation of the hydrodynamic radii), it has been concluded that α-synuclein is slightly collapsed [224]. In agreement with this conclusion, a high resolution NMR analysis of the protein revealed that α-synuclein is largely unfolded in a solution, but exhibits a region between residues 6 and 37 with a preference for helical conformation [225]. Interestingly, Raman optical activity spectra indicate that α-synuclein contains some helical poly-(L-proline) II-like conformation [226].

α-Synuclein, with its high propensity to aggregate, represents an ideal model for the amyloidogenic IDP and the molecular mechanisms underlying the amyloidogenesis of this protein were intensively studied. It has been shown that α-synuclein partially folds at acidic pH and high temperature; i.e., under conditions that enhanced dramatically the propensity of the protein to form amyloid-like fibrils [223]. Conformational behavior of α-synuclein under the variety of environments revealed that structure of this protein is extremely sensitive to the environment. It adopts a variety of structurally unrelated conformations including the substantially unfolded state, an amyloidogenic partially folded conformation, and different α-helical or β-structural species folded to a different degree, both monomeric and oligomeric [223]. Furthermore, it might form several morphologically different types of aggregates, including oligomers (spheres or doughnuts), amorphous aggregates, and amyloid-like fibrils [223]. Based on this astonishing conformational behavior the concept of a protein-chameleon was proposed, according to which the structure of α-synuclein to a dramatic degree depends on the environment and the choice between different conformations is determined by the peculiarities of protein surroundings [223].

2.4.4.8 α-Synuclein Maintains Disordered Structure Inside of a Living Cell

The cell's interior is crowded with small and large molecules [227, 228]. Recently, the effects of macromolecular crowding on α-synuclein was assessed by combining NMR data acquired in living *Escherichia coli* with *in vitro* NMR data [229]. The technique of in-cell NMR spectroscopy has been developed and refined to investigate proteins in living *Escherichia coli* [230–232]. Using this approach, it has been shown that crowded environment in the *E. coli* periplasm not only keeps α-synuclein disordered, but prevents a conformational change that is detected at 35°C in dilute solution [229]. Two disease-associated variants (A30P and A53T) behave in the same way in both dilute solution and in the *E. coli* periplasm. The authors reported the same stabilization *in vitro* upon crowding α-synuclein with 300 g/l of bovine serum albumin. Comparison of these *in vivo* and *in vitro* data suggests that crowding alone is sufficient to stabilize the intrinsically disordered, monomeric protein

[229]. This is a very important observation, which suggests that some IDPs, including α-synuclein, can maintain their disordered structure even in the highly crowded environment of a living cell.

2.4.5 β- and γ-Synucleins in Parkinson's Disease and Dementia with Lewy Bodies

Synucleins are members of a family of closely related presynaptic proteins that arise from three distinct genes, described currently only in vertebrates [233]. This family includes: α-synuclein, which also known as the non-amyloid component precursor protein, NACP, or synelfin [68, 234, 235]; β-synuclein, also referred to as phosphoneuro-protein 14 or PNP14 [235–237] and γ-synuclein, also known as breast cancer-specific gene 1 or BCSG1 and persyn [238–241].

Human β-synuclein is a 134-aa neuronal protein showing 78% identity to α-synuclein. The α- and β-synucleins share a conserved C-terminus with three identically placed tyrosine residues. However, β-synuclein is missing 11 residues within the specific non-amyloid component (NAC) region [242, 243]. The activity of β-synuclein may be regulated by phosphorylation [236]. This protein, like α-synuclein, is expressed predominantly in the brain, however, in contrast to α-synuclein, β-synuclein is distributed more uniformly throughout the brain [244, 245]. Besides the central nervous system β-synuclein was also found in Sertoli cells of the testis [246, 247], whereas α-synuclein was found in platelets [248].

The third member of the human synuclein family is the 127-aa γ-synuclein, which shares 60% similarity with α-synuclein at the amino acid sequence level [242, 243]. This protein is specifically lacks the tyrosine rich C-terminal signature of α- and β-synucleins [243]. γ-Synuclein is abundant in spinal cord and sensory ganglia [240]. Interestingly, this protein is more widely distributed within the neuronal cytoplasm than α-and β-synucleins, being present throughout the cell body and axons [240]. It was also found in metastatic breast cancer tissue [239] and epidermis [249].

It has recently been established that in addition to the traditional α-synuclein-containing LBs and LNs, the development of PD and DLB is accompanied by appearance of novel α-, β- and γ-synuclein-positive lesions at the axon terminals of hippocampus [250]. These pathological vesicular-like lesions located at the presynaptic axon terminals in the hippocampal dentate, hilar, and CA2/3 regions have been co-stained by antibodies to α- and β-synucleins, whereas antibodies to γ-synuclein detect previously unrecognized axonal spheroid-like inclusions in the hippocampal dentate molecular layer [250]. This broadens the concept of neurodegenerative "synucleinopathies" by implicating β- and γ-synucleins, in addition to α-synuclein, in the onset/progression of these two diseases.

Structural properties of the members of synuclein family have been compared using several physico-chemical methods [251]. All three proteins showed far-UV CD spectra typical of an unfolded polypeptide chain. Interestingly, α- and γ-synucleins possessed almost indistinguishable spectra, whereas the far UV-CD spectrum of β-synuclein showed a slightly increased degree of disorder. The increased unfoldedness of β-synuclein was further confirmed by hydrodynamic studies performed by size-exclusion chromatography and SAXS [251]. This emphasized the importance of the NAC region to maintain the residual partially collapsed structure in α- and γ-synucleins.

Conformational analysis revealed that α-, β-, and γ-synucleins are typical natively unfolded proteins that are able to adopt comparable partially folded conformations at acidic pH or at high temperature [251]. Although both α- and γ-synucleins were shown to form fibrils, β-synuclein did not fibrillate, being incubated under the same conditions [251]. However, even non-amyloidogenic β-synuclein can be forced to fibrillate in the presence of some metals (Zn^{2+}, Pb^{2+}, and Cu^{2+}) [252].

Intriguingly, the addition of either β- or γ-synuclein in a 1:1 molar ratio to α-synuclein solution substantially increased the duration of the lag-time and dramatically reduced the elongation rate of α-synuclein fibrillation [251]. Fibrillation was completely inhibited at a 4:1 molar excess of β- or γ-synuclein over α-synuclein [251]. β-Synuclein inhibited α-synuclein aggregation in an animal model, too [253]. This suggests that β- and γ-synucleins may act as regulators of α-synuclein fibrillation *in vivo*, potentially acting as chaperones. Therefore, one possible factor in the etiology of PD would be a decrease in the levels of β- or γ-synucleins [251].

2.4.6 Polyglutamine Repeat Diseases and Huntingtin, Ataxin-1, Ataxin-3, Androgen Receptor and Atrophin-1

2.4.6.1 Polyglutamine Repeat Diseases

Currently there are at least nine known hereditary diseases in which the expansion of a CAG repeat in the gene leads to neurodegeneration [254, 255]. Table 2.1 shows that these polyglutamine repeat diseases includes HD, Kennedy disease (also known as spinal and bulbar muscular atrophy, SBMA), spinocerebellar ataxia type 1 (SCA1), dentatorubral-pallidoluysian atrophy (DRPLA), spinocerebellar ataxia type 2 (SCA2), Machado-Joseph disease (MJD/SCA3), SCA6, SCA7 and SCA17. These diseases are accompanied by the progressive death of neurons, with insoluble, granular, and fibrous deposits being found in the cell nuclei of the affected neurons. The neurotoxicity in these diseases is due to the expansion of the $(CAG)_N$-encoded polyglutamine (polyGln) repeat, which lead to the formation of amyloid fibrils and neuronal death. In HD, the CAG repeat that encodes the polyQ region is part of exon 1 in the 3,140-residue huntingtin protein [256]. The polyQ repeat varies between 16 and 37 residues in healthy individuals, and individuals who are afflicted by disease have repeats of >38 residues.

The age of onset and the severity of the progression of SCA1, an autosomal-dominant neurodegenerative disorder characterized by ataxia and progressive motor deterioration, are directly correlated with the length of the polyQ segment in ataxin-1, a nuclear protein of ~800 residues [257–259]. When the number of glutamine residues in the polyQ tract exceeds a threshold (39–44 glutamine residues), ataxin-1 aggregates with granular or fibrillar morphologies accumulate intranuclearly and eventually lead to cell death [260, 261].

Kennedy disease (also known as spinal and bulbar muscular atrophy, SBMA) is linked to the expansion of a Gln-rich segment in the androgen receptor [262]; healthy individuals have a 15- to 31-residue polyQ segment, and individuals who are afflicted with the disease have 40–62 Gln residues. Intriguingly, in the human androgen receptor there are three polyglutamine repeats ranging in size from five to 22 residues, stretches of seven prolines and five alanines, and a polyglycine repeat of 24 residues. Polymorphisms in the length of the largest polyglutamine and the polyglycine repeats of the androgen receptor have been associated with a number of clinical disorders, including prostate cancer, benign prostatic hyperplasia, male infertility and rheumatoid arthritis [263].

The onset of the DRPLA, another progressive neurodegenerative disorder characterized by a distinctive pathology in the cerebellar and pallidal outflow pathways, is inversely correlated with the polyQ repeat size in the corresponding DRPLA protein (also known as atrophin-1), a product of the gene on chromosome 12p [264]. The repeat size varied from 7–23 in normal individuals and was expanded to 49–75 in DRPLA patients.

2.4.6.2 Huntingtin and Structure of Polyglutamine Stretches

Huntingtin, a protein with an estimated molecular mass of 350 kDa, contains a polyglutamine tract near its N terminus that when expanded beyond 37 glutamines causes HD [256]. The N terminus of wild-type huntingtin interacts with proteins involved in nuclear functions, including HYPA/FBP-11, which functions in pre-mRNA processing (splicesome function) [265], nuclear receptor co-repressor protein (NCoR) [266], which plays a role in the repression of gene activity, and p53 [267], a tumor suppressor involved in regulation of the cell cycle. Full-length huntingtin contains candidate binding sites for other proteins with nuclear functions. Huntingtin contains a PXDLS motif, a candidate-binding site for the transcriptional corepressor C-terminal binding protein (CtBP) [268], suggesting that huntingtin may play a role in transcriptional repression.

The localization and potential function of normal and mutant huntingtin in the nucleus was suggested to be important for understanding HD pathogenesis. For example, N-terminal mutant huntingtin was shown to be toxic when targeted to the nucleus of cultured striatal neurons [269]. Mutant huntingtin has been implicated in abnormal transcriptional repression in HD. In cellular systems, short N-terminal mutant huntingtin fragments disrupt transcriptional regulation, which occurs through a mechanism involving sequestration of transcription factors including p53 [267], TATA-box-binding protein (TBP) [270], and CREB-binding protein [271] into huntingtin-positive aggregates. These results suggest that the N terminus of mutant huntingtin may disrupt neuronal function in HD by interfering with nuclear organization and transcriptional regulation. Full-length huntingtin was shown to co-immunoprecipitate with the transcriptional corepressor C-terminal binding protein, and polyglutamine expansion in huntingtin reduced this interaction

[272]. Interestingly, although full-length wild-type and mutant huntingtin both repressed transcription when targeted to DNA, truncated N-terminal mutant huntingtin was shown to repress transcription, whereas the corresponding wild-type fragment did not [272].

Proteolytic cleavage of mutant huntingtin is suggested to play a key role in the pathogenesis of HD. Huntingtin was shown to be cleaved by caspases and calpains within a region between 460–600 amino acids from the N-terminus. Furthermore, two smaller N-terminal fragments produced by unknown protease have been described as cp-A and cp-B [273]. In fact, based on the analysis of human HD patients, animal models and cell-based models of HD it has been suggested that truncated polyglutamine-containing fragments are more toxic than full-length huntingtin [274].

The mechanistic hypothesis linking CAG repeat expansion to toxicity involves the tendency of longer polyGln sequences, regardless of protein context, to form insoluble aggregates [73, 275–282]. To help evaluate various possible mechanisms, the biophysical properties of a series of simple polyglutamine peptides have been analyzed. The far-UV CD spectra of poly(Gln) peptides with repeat lengths of 5, 15, 28 and 44 residues were shown to be nearly identical and were consistent with a high degree of random coil structure, suggesting that the length-dependence of disease is not related to a conformational change in the monomeric states of expanded poly(Gln) sequences [280]. In contrast, there was a dramatic acceleration in the spontaneous formation of ordered, amyloid-like aggregates for poly(Gln) peptides with repeat lengths of greater than 37 residues. Several studies established the role of partially folded intermediates of polyglutamine-repeat proteins as key species in fibrillation [281, 283, 284].

Huntingtin was shown to interact with more than 200 proteins [285]. One of these huntingtin interactors, huntingtin yeast-two hybrid protein K (HYPK) was recently identified as a typical IDP using a set of biophysical and biochemical techniques [285]. Among the experimental data supporting this conclusion there were aberrant electrophoretic mobility [the molecular weight of HYPK determined by gel electrophoresis was found to be about 1.3-folds (~22 kDa) higher than that obtained from mass spectrometric analysis (16.9 kDa)]; increased hydrodynamic dimensions [in size exclusion chromatography experiment, HYPK was eluted as a protein with the hydrodynamic radius which was ~1.5-folds (23 Å) higher than that expected for globular proteins of equivalent mass (17.3 Å)]; random coil characteristics of far-UV CD spectra; and highly sensitive to limited proteolysis by trypsin and papain [285]. Subsequent analysis of HYPK revealed that this huntingtin interacting protein was able to reduce aggregates and apoptosis induced by N-terminal huntingtin with 40 glutamines in Neuro2a cells and exhibited chaperone-like activity [286].

2.4.6.3 Dentatorubral-Pallidoluysian Atrophy Protein (Atrophin-1)

Investigations of the DRPLA gene (encoding for atrophin-1) indicate that it is widely expressed in brain and other tissues as a 4.5-kb transcript with an open reading frame encoding 1184 amino acids [287–289]. The rat atrophin-1 coding sequence is 88% identical to the coding sequence of human atrophin-1 at the level of DNA and 94% identical at the protein level, but encodes a shorter glutamine repeat that is followed by a series of alternating glutamine and proline residues [290, 291]. The predicted molecular mass of the atrophin-1 gene product is 124 kDa, yet atrophin-1 appeared to migrate at about 200 kDa [292]. The anomalous electrophoretic mobility is considered as one of the characteristic features of IDPs [23, 29, 30]. In fact, the apparent molecular masses of IDPs determined by this technique are often 1.2–1.8 times higher than real one calculated from sequence data or measured by mass spectrometry [23, 29, 30]. Our analysis revealed that the abnormality degree of the electrophoretic mobility of an IDP is directly proportional to the amount of intrinsic disorder present in its sequence (Uversky, personal communication). It has been suggested that IDPs bind less SDS than "normal" proteins. This explains their abnormal mobility in SDS polyacrylamide gel electrophoresis experiments, resulting in the observed increase in the apparent molecular masses.

2.4.6.4 Androgen Receptor

CD analysis of a region of the androgen receptor N-terminal domain lacking the largest polyglutamine stretch, but containing the remaining repeats, showed that it lacked stable tertiary structure in aqueous solutions [263]. Detailed conformational studies using a combination of experimental and computational techniques revealed that the AF1 transactivation domain is in the molten globule-like conformation [293]. In fact, this region of the receptor was predicted to contain long disordered regions, when analyzed by amino acid composition, PONDR®, RONN, and Glob-Plot. However, this domain was predicted to have compact globular structure when analyzed by a charge-hydropathy plot (CH-plot, [20]). This discrepancy between the CH-plot and PONDR®-based predictions for the androgen receptor AF1 suggests that this domain possesses properties consistent with a dynamic conformation and to fall into a "collapsed disorder class" of proteins, typical of the molten globule folding intermediate [20, 52]. This conclusion was confirmed by the analysis of a hydrophobic fluorescence probe, ANS binding and by size-exclusion chromatography [293]. The results of this analysis suggest that native androgen receptor AF1 exists in a collapsed disordered conformation, distinct from extended disordered (random coil) and a stable globular fold [293].

2.4.6.5 Ataxin-2

SCA2 is an autosomal-dominantly inherited, neurodegenerative disorder, caused by the expansion of an unstable CAG/polyglutamine repeat located at the N-terminus of ataxin-2 protein. The age of onset of SCA2 is in the third to fourth decade. The characteristic phenotypic features of SCA2 are the degeneration of specific vulnerable neuron populations and the presence of intracellular aggregations of the mutant protein in affected neurons. Ataxin-2 has 1312 residues (including 22 glutamines of the polyQ stretch) and a molecular mass of ~140 kDa. Ataxin-2 is a highly basic protein except for one acidic region (amino acid 254–475) containing

46 acidic amino acids [294]. This region consists of two predicted globular domains, Lsm (Like Sm, amino acid 254–345) and LsmAD (Lsm-associated domain, amino acid 353–475). The LsmAD domain contains a clathrin-mediated *trans*-Golgi signal (YDS, amino acid 414–416) and an endoplasmic reticulum (ER) exit signal (ERD, amino acid 426–428). This domain is composed mainly of α-helices according to the results from secondary structure prediction servers. The rest of ataxin-2 outside of the Lsm and LsmAD domains is only weakly conserved in eukaryotic ataxin-2 homologues and is predicted to be intrinsically disordered [294].

2.4.6.6 Ataxin-3

Human ataxin-3, the protein related to Spinocerebellar ataxia type 3 or Machado–Joseph disease (SCA3/MJD), is a ubiquitously expressed 41 kDa protein whose polyQ tract contains 12–40 glutamines in normal individuals and 55–84 glutamines in the pathogenic form [255]. Ataxin-3 is present in the genomes of several species, from nematodes to human, including plants [295]. Alignment of the ataxin-3 family shows a conserved N-terminal block that corresponds to the sequence motif named Josephin (residues 1–198 in the human protein) [295]. The C-terminus is non-conserved throughout different species and contains long stretches of low complexity regions which include the polyQ tract, preceded by a highly charged region [295].

Human ataxin-3 was analyzed by a range of biophysical and biochemical techniques, including limited proteolysis, CD and NMR spectroscopies [296]. The deconvolution of the far-UV CD spectra indicated that ataxin-3 contained 32% α-helix, 17% β-sheet, 20% β-turn, and 31% random coil. Based on this results, it has been concluded that the high percentage of random coil conformation estimated by this analysis suggests the presence of unstructured portions of the molecule alongside one or more folded regions [296]. This conclusion was further supported by the 2D ^{15}N NMR spectra (HSQC), which were shown to contain two main resonance types: well dispersed resonances typical of a folded conformation and sharp highly overlapped peaks typical of a random coil conformation. Furthermore, limited proteolysis revealed that the intact protein was almost completely digested after 1 min of incubation with a series of proteases and a protease-resistant N-terminal domain was generated [296]. These data indicated that ataxin-3 is composed of a structured N-terminal domain, followed by a flexible tail.

2.4.6.7 P/Q-Type Calcium Channel α1A Subunit (CACNA1A)

The underlying mutation in SCA6, a dominantly inherited neurodegenerative disease characterized by progressive ataxia and dyasrthria caused by cerebellar atrophy, is an expansion of the trinucleotide CAG repeat in exon 47 of the *CACNA1A* gene which encodes the α1A subunit of the P/Q type voltage-dependent calcium channel [297]. Unlike many other polyglutamine diseases the expanded SCA6 alleles unusually have small expansions (21–30 repeats compared to generally >40 repeats in other polyglutamine diseases) [297]. The product of the *CACNA1A* gene, P/Q-type

Calcium Channel α1A Subunit (CACNA1A), is a protein with 2505 residues and a calculated molecular mass of 282.4 kDa. It has been found that the CACNA1A is processed in such a way that a C-terminal polyglutamine-containing fragment which is less soluble and more toxic than the truncated polyglutamine stretch itself is produced [298]. This protein was predicted to have several long IDRs (see Fig. 2.2).

2.4.6.8 Ataxin-7

Spinocerebellar ataxia type 7 (SCA7) is characterized by cone-rod dystrophy retinal degeneration and is caused by a polyglutamine expansion within ataxin-7. It has been recently reported that report that ataxin-7 is a component of the mammalian STAGA (SPT3-TAF9-ADA-GCN5 acetyltransferase) transcription coactivator complex [299]. In this complex, ataxin-7 interacts directly with the GCN5 histone acetyltransferase component of STAGA, and mediates a direct interaction of STAGA with the CRX (cone-rod homeobox) transactivator of photoreceptor genes. Furthermore, poly(Q)-expanded ataxin-7 was ncorporated into STAGA and inhibited the nucleosomal histone acetylation function of STAGA GCN5. Based on these results it has been suggested that the normal function of ataxin-7 may intersect with its pathogenic mechanism [299]. Ataxin-7 has 892 amino acid residues and a molecular mass of 95.4 kDa. However, at the SDS-PAGE this protein migrates at about 110 kDa [299]. In other words, the apparent molecular mass of ataxin-7 determined by gel electrophoresis was found to be about 1.15-folds higher than that expected from amino acid sequence. This suggests that ataxin-7 possesses significant amount of intrinsic disorder.

2.4.6.9 TATA-Box-Binding Protein

SCA17 is characterized by the heterogeneous clinical phenotype, including ataxia, dementia, psychiatric symptoms, and, in some cases, epilepsy. Neurodegeneration in SCA17 is frequently widespread (atrophy of the striatum, thalamus, cerebral cortex, inferior olive, and nucleus accumbens have been reported), being most prominent in the cerebellum [300]. Ubiquinated intranuclear inclusions were found in postmortem brain tissue from SCA17 patients as a result of immunohistochemical examination [300]. SCA17 originates from the polyglutamin expansion of the TBP, which normally contains the polyQ tract of 25–42 glutamine residues, but is expanded >42 glutamines in SCA17 [300]. TBP is required for transcriptional initiation by the three major RNA polymerases (RNAP I, II, and III) in eukaryotic nuclei. Being a component of distinct multi-subunit transcriptional complexes, TBP is involved in the expression of most eukaryotic genes [301]. TBP is a 339 amino acid residues-long protein, which can be divided on two functional domains. The C-terminal domain is highly conserved among eukaryotes and mediates virtually all of the transcriptionally relevant interactions involving TBP [302], whereas the N-terminal domain is evolutionarily divergent and shows sequence conservation only in vertebrates. It has been demonstrated that polyQ expansion caused abnormal interaction of TBP with the general transcription factor TFIIB and induced neurodegeneration in transgenic SCA17 mice [303]. Furthermore, polyQ expansion was shown to reduce the *in vitro* binding of TBP to DNA. The mutant TBP fragments lacking an intact C-terminal DNA-binding domain were shown to be present in transgenic SCA17 mouse brains. PolyQ-expanded TBP with a deletion spanning part of the DNA-binding domain did not bind DNA *in vitro* but formed nuclear aggregates and inhibited TATA-dependent transcription activity in cultured cells [304]. SDS-PAGE analysis of the murine TBP revealed that this protein is characterized by the apparent molecular mass of ∼37 kDa, which exceeds the predicted molecular mass of 34.7 kDa [304]. The difference between observed and calculated molecular masses was even higher for a truncated TBP fragment that lacks an intact C-terminal domain [304]. Similarly, human TBP, a protein with the calculated molecular mass of 37.7 kDa, was shown to possess an apparent molecular mass of ∼49 kDa [305].

2.4.7 *ABri Peptide and Familial British Dementia*

The ABri is a 34 residue peptide that is the major component of amyloid deposits in familial British dementia (FBD), which is an autosomal dominant disorder with onset at around the fifth decade of life and full penetrance by age 60 characterized by the presence of amyloid deposits in cerebral blood vessels and brain parenchyma that coexist with NFTs in limbic areas [306]. FBD patients develop progressive dementia, spasticity, and cerebellar ataxia. The protein subunit (termed ABri) is an example of an amyloid molecule created *de novo* by the abolishment of the stop codon in its precursor, a protein comprised of 266 amino acid residues (BRI-266) that is codified by a single gene, BRI, located on the long arm of chromosome 13 [307, 308]. The FBD has a single nucleotide change (TGA→AGA, codon 267) that results in an arginine residue substitution for the stop codon in the wild-type precursor molecule and a longer open reading frame of 277 amino acid residues in a disease-related protein (BRI-277 instead of BRI-266). The ABri amyloid peptide is formed by the 34 C-terminal amino acid residues of the mutant precursor protein BRI-277, presumably generated from furin-like processing [309]. Thus, the point mutation at the stop codon of BRI results in the generation of the 34 residue ABri peptide (instead of the shorter 23 residue wild type peptide), which is deposited as amyloid fibrils causing neuronal dysfunction and dementia [310]. It has been emphasized that athough FBD and AD share almost identical neurofibrillar pathology and neuronal loss that co-localize with amyloid deposits, the primary sequences of the amyloid proteins (ABri and Aβ) differ. Therefore, ABri and Aβ amyloid deposition in the brain can trigger similar neuropathological changes (neuronal loss and dementia) and thus may be a key event in the initiation of neurodegeneration [310].

Using far-UV CD and NMR spectroscopy it has been recently established that ABri is in the random coil-like conformation at slightly acidic pH [310]. The solution pH was shown to play an important role in promoting the amyloid-like β-sheet structure and the characteristic fibril morphology of ABri and this protein forms amyloid fibrils at pH 4.9 with no distinct fibril morphology being observed at neutral and slightly basic pH (pH 7.1–8.3), except for smaller spherical aggregates that gradually disappeared and assembled into larger amorphous aggregates [310]. It has been also pointed out that at pH 4.9 the ABri undergoes relatively slow β-aggregation, where it is possible for fibril formation to occur, similar to the behavior of the amyloid Aβ peptide [310].

2.4.8 ADan in Familial Danish Dementia

Familial Danish dementia (FDD) is a neurodegenerative disorder linked to a genetic defect in the *BRI2* gene. Similar to FBD, FDD results form the genetic alterations in this gene and the deposited amyloid protein, ADan, is the C-terminal proteolytic fragment of a genetically altered BRI2 precursor molecule [311]. The amyloid peptides ABri and ADan originate as a result of two different genetic defects at, or immediately before, the BRI2 stop codon with a common final outcome in both diseases: regardless of the nucleotide changes, the ordinarily occurring stop codon is either non-existent (in FBD) or out of frame (in FDD) causing the genesis of an extended precursor featuring a C-terminal piece that does not exist in normal conditions (reviewed in [312]). ABri and ADan are released by a furin-like proteolytic processing. Both these peptides are 34-residues-long, which share 100% homology on the first 22 residues, a completely different 12 amino acid C-terminus and have no sequence identity to any other known amyloid protein. Despite the structural differences among the corresponding amyloid subunits FDD and FBD show striking clinical and neuropathological similarities with AD, including the presence of NFTs, parenchymal amyloid and pre-amyloid deposits and CAA co-localizing with inflammatory markers, reactive microglia and activation products of the complement system (reviewed in [312]). Structural analysis revealed that similar to Aβ and ABri, ADan is a typical natively unfolded protein, which is characterized by a random coil structure in a wide pH range and is prone to form fibrils in a pH-dependent manner [313].

2.4.9 Glial Fibrillary Acidic Protein and Alexander Disease

Alexander disease is a specific astrocytic disease caused by a dominant heterozygous mutation in glial fibrillary acidic protein (GFAP) [314, 315]. A major pathological hallmark of Alexander disease is a presence of specific inclusion bodies called Rosenthal fibers (RFs) in astrocytes that are formed by the mutant GFAP [316]. Besides mutant GFAP, these inclusions contains small heat shock proteins, including αB-crystallin and HSP27[317]. Clinically, the phenotype of Alexander disease depends on the age of onset. The infantile form severely affects the entire central nervous system, with rapid progression and is characterized by megalencephaly, epilepsy, motor impairment, cognitive decline, and extensive loss of white matter with frontal predominance. However, the adult form progresses slowly and is characterized by predominant rhombencephalic degeneration without epilepsy, cognitive impairment, and little, if any, leukodystrophy. The juvenile form is intermediate in severity [318, 319]. It has been shown that GFAP is characterized by an extremely high susceptibility to proteolysis [320]. Electrophoretic analysis of GFAP produced an apparent molecular mass of 54 kDa, which exceeds the calculated molecular mass of 49.9 kDa [321].

2.4.10 Mitochondrial DNA Polymerase γ and Alpers Disease

Alpers disease, also known as progressive neuronal degeneration of childhood, is characterized by developmental regression, intractable epilepsy, progressive neurologic deterioration, liver disease, and death usually before 10 years of age [322–324]. Neuropathologic changes include patchy neuronal loss and gliosis, particularly in the striate cortex [325], whereas the liver shows steatosis, cellular necrosis, focal inflammation, and fibrosis [326]. Alpers disease is attributed to mutations in the catalytic subunit of the mitochondrial DNA (mtDNA) polymerase gene polymerase γ (*POLG1*) [327]. POLG is the only known DNA polymerase in the mitochondrion, which is responsible for ~1% of the total cellular DNA polymerase activity. The human POLG holoenzyme comprises a 140 kDa catalytic subunit (POLGα) and a 55 kDa accessory subunit (POLGβ). POLGα is a member of a DNA polymerase family with separate polymerase and 3′–5′ exonuclease domains thus exhibiting both DNA polymerase and 3′–5′ exonuclease activities. POLGβ increases DNA-binding affinity, stimulates the catalytic activities and enhances the processivity of the holoenzyme [328]. The region of POLGα (444–820 fragment) that lies between the exonuclease and polymerase is known as spacer. Its size and sequence in POLGα are substantially different from those of other members of the DNA polymerase family. In POLGα, this large interdomain region is likely to participate in DNA-template binding and guidance, as well as in subunit interactions. Importantly, spacer mutations were found frequently in the infantile Alpers syndrome, affecting most severely the brain and the liver [327, 329]. These reports emphasize the exceptional variability of POLGα-associated neurological phenotypes and the specific role for spacer mutations in the most severe neurological manifestations [330]. POLGα was shown to possess an apparent molecular mass of 145–147 kDa [331], whereas its theoretical molecular mass is 139.5 kDa.

2.4.11 DNA Excision Repair Protein ERCC-6 and Cockayne Syndrome

Cockayne syndrome (CS) (also called Weber-Cockayne syndrome, or Neill-Dingwall Syndrome) is a rare, autosomal recessive disorder. Affected individuals suffer from postnatal growth failure resulting in cachectic dwarfism, photosensitivity, skeletal abnormalities, mental retardation and progressive neurological degeneration, retinopathy, cataracts and sensorineural hearing loss [332–334]. Two complementation groups of CS (CS-A and CS-B) have been identified, the corresponding genes, *CSA* and *CSB*, have been cloned [335, 336] and their products biochemically characterized. The majority of CS cases are caused by defects in the CS complementation group B protein. CSA is a 44 kDa protein and belongs to the 'WD repeat' family of proteins [336], which exhibit structural and regulatory roles but no enzymatic activity. The *CSB* gene product is a 168 kDa protein [335], also known as DNA excision repair protein ERCC-6, belongs to the SWI/SNF family of proteins, which all contain seven sequence motifs conserved between two superfamilies of DNA and RNA helicases and which have roles in transcription regulation, chromosome stability, and DNA repair. The involvement of CSB in transcription, transcription-coupled repair of DNA, and base-excision repair might be simultaneous. However, it is suggested that some interregulation, depending on cellular status, takes place. This regulation is done via posttranslational modifications of CSB and changes in function and localization of its interaction partners. These many roles of CSB explain the multisystem manifestations of the CS phenotype [334].

In vitro studies demonstrate that CSB exists in a quaternary complex composed of RNA pol II, CSB, DNA and the RNA transcript. The CSB protein contains an acidic amino acid stretch (~60% of the residues in a 39-amino-acid stretch are acidic), a glycine-rich region and two putative nuclear localization signal (NLS) sequences [337]. The cellular and molecular phenotypes of CS include increased sensitivity to oxidative and UV-induced DNA lesions. The CSB protein plays a crucial role in transcription-coupled repair. The corresponding CS-B cells are defective in the repair of the transcribed strand of active genes, both after exposure to UV and in the presence of oxidative DNA lesions [337]. According to SDS-PAGE analysis, the CSB protein has an apparent molecular mass of ~200 kDa [337], whereas its theoretical molecular mass calculated from amino acid sequence is 168.4 kDa.

2.4.12 Survival of Motor Neurons Protein and Spinal Muscular Atrophy

Spinal muscular atrophy (SMA) is an autosomal recessive disease with a carrier frequency of about 1 in 50. SMA is the most common genetic cause of childhood mortality and leads to muscle weakness and atrophy due to the degeneration of motor neurons from the spinal cord [338]. The disease is mapped to the survival of motor neurons (SMN) 1 (*smn1*) gene, which carries mutations in over 98% of all SMA patients [339]. SMN mutants (SMNDelta7 and SMN-Y272C) found in patients with SMA not only lack antiapoptotic activity but also are potently proapoptotic, causing increased neuronal apoptosis and animal mortality. The SMN protein is a part of a larger protein complex that is present both in the nucleus and the cytoplasm. In the nucleus, SMN protein localizes to spots that are rich in small nuclear ribonucleoprotein particles (snRNPs). In the cytoplasm, the SMN protein plays an important role in the assembly of these snRNPs [340]. The SMN protein interacts with core components of the snRNPs, Sm proteins. SMA-causing mutations in a

C-terminal region and in the central Tudor domain of the SMN protein have been shown to affect the Sm interaction. Mutations in the C-terminal region may interfere with the Sm interaction indirectly, since this region is also required for SMN protein oligomerization [341]. The SMN protein Tudor domain has been shown to directly bind to the arginine-glycine (RG) rich tails of the Sm proteins *in vitro* [342, 343]. Furthermore, the type I SMA causing point mutation E134K in the SMN protein Tudor domain abolishes Sm binding *in vitro* and interferes with snRNP assembly *in vivo* [343].

In vivo the RG-rich tails of the SmB, SmD1 and SmD3 proteins are posttranslationally modified and contain symmetrically dimethylated arginine residues [344, 345]. This modification strongly enhances the affinity of the SMN/Sm interaction and has been implicated in the regulation of uridine-rich snRNP assembly [345, 346]. Many other proteins, including coilin, RNA helicase A, fibrillarin and heterogeneous nuclear ribonucleoproteins, interact with the SMN complex and contain RG-rich domains that can potentially be methylated [347, 348], suggesting that the SMN protein Tudor domain could have an additional function in the regulation of these interactions. The crystal structure of the SMN protein Tudor domain comprising residues 82–147 was solved to high (1.8 Å) resolution [349, 350]. The crystal structure consists of a five-stranded β-sheet that forms a β-barrel. Comparison of the crystal structure and an NMR structure revealed that the backbone conformation of both structures is very similar. However, differences were observed for the cluster of conserved aromatic side-chains in the symmetrically dimethylated arginine residues (sDMA) binding pocket, suggesting that the SMN protein Tudor domain adopts two different conformations in the sDMA binding pocket [349]. Full-length SMN protein (calculated molecular mass 31.8 Kda) was shown to possess an apparent molecular mass of 39 kDa [351].

2.5 Concluding Remarks: Another Illustration of the D^2 Concept

Intrinsic disorder is highly abundant among proteins associated with human neurodegenerative diseases. This provides a strong factual support to a D^2 (disorder in disorders) concept. The validity of this concept in neurodegeneration is illustrated at several levels, starting from the results of the bioinformatics analysis of an extended set of proteins associated with various neurodegenerative conditions and ending with the extensive data for a number of well-characterized neurodegeneration-related proteins. High degree of association between intrinsic disorder and neurodegenerative diseases is due to the unique structural and functional peculiarities of IDPs and IDRs. IDPs/IDRs are among major cellular regulators, recognizers and signal transducers. Their functionality and misbehavior are modulated via a number of posttranslational modifications (i.e., tau protein). Many IDPs/IDRs can fold (completely or partially) upon interaction with corresponding binding partners. They possess multiple binding specificity and they are able to participate in one-to-many and many-to-one interactions.

Acknowledgments This work was supported in part by the grants R01 LM007688-01A1 and GM071714-01A2 from the National Institutes of Health and the Programs of the Russian Academy of Sciences for the "Molecular and cellular biology" and "Fundamental science for medicine". The support of the IUPUI Signature Centers Initiative is gratefully acknowledged.

Abbreviations

8-anilino-1-naphthalenesulfonic acid (ANS)
Alzheimer's disease (AD)
Amyloid β-protein (Aβ)
Amyloid β-protein precursor (APP)
Cerebral amyloid angiopathy (CAA)
Cockayne syndrome (CS)
Circular dichroism (CD)
Dentatorubral-pallidoluysian atrophy (DRPLA)
Familial British dementia (FBD)
Familial Danish dementia (FDD)
Glial cytoplasmic inclusion (GCI)
Glial fibrillary acidic protein (GFAP)
Hemorrhage with amyloidosis-Dutch type (HCHWA-D).
Huntington's disease (HD)
Huntingtin yeast-two hybrid protein K (HYPK)
Intrinsically disordered protein (IDP)
Intrinsically disordered region (IDR)
Lewy body (LB)
Lewy neurites (LNs)
Mitochondrial DNA polymerase γ (POLG)
Multiple system atrophy (MSA)
Neurodegeneration with brain iron accumulation type 1 (NBIA1)
Neurofibrillary tangle (NFT)
Paired helical filament (PHF)
Parkinson's disease (PD)
Protein Data Bank (PDB)
Prion protein (Prp)
Prion protein, native (Prp^c)
Prion protein, scrapie (Prp^{sc})
Small nuclear ribonucleoprotein particles (snRNPs)
Spinal Muscular Atrophy (SMA)
Spinocerebellar ataxia (SCA)
SPT3-TAF9-ADA-GCN5 acetyltransferase (STAGA)
Survival of Motor Neurons (SMN)
TATA-box-binding protein (TBP)
Transmissible spongiform encephalopathy (TSE)

References

1. Dunker AK, Lawson JD, Brown CJ, Williams RM, Romero P, Oh JS, Oldfield CJ, Campen AM, Ratliff CM, Hipps KW et al (2001) Intrinsically disordered protein. J Mol Graph Model 19(1):26–59.
2. Schweers O, Schonbrunn-Hanebeck E, Marx A, Mandelkow E (1994) Structural studies of tau protein and Alzheimer paired helical filaments show no evidence for beta-structure. J Biol Chem 269(39):24290–24297.
3. Weinreb PH, Zhen W, Poon AW, Conway KA, Lansbury PT, Jr. (1996) NACP, a protein implicated in Alzheimer's disease and learning, is natively unfolded. Biochemistry 35(43):13709–13715.
4. Wright PE, Dyson HJ (1999) Intrinsically unstructured proteins: re-assessing the protein structure-function paradigm. J Mol Biol 293(2):321–331.
5. Daughdrill GW, Pielak GJ, Uversky VN, Cortese MS, Dunker AK (2005) Natively disordered proteins. In: Buchner J, Kiefhaber T (eds.) Handbook of Protein Folding. Wiley-VCH, Verlag GmbH & Co. KGaA, Weinheim, Germany. 271–353.
6. Fink AL (2005) Natively unfolded proteins. Curr Opin Struct Biol 15(1):35–41.
7. Uversky VN (2003) Protein folding revisited. A polypeptide chain at the folding-misfolding-nonfolding cross-roads: which way to go? Cell Mol Life Sci 60(9):1852–1871.
8. Uversky VN, Ptitsyn OB (1994) "Partly folded" state, a new equilibrium state of protein molecules: four-state guanidinium chloride-induced unfolding of beta-lactamase at low temperature. Biochemistry 33(10):2782–2791.
9. Uversky VN, Ptitsyn OB (1996) Further evidence on the equilibrium "pre-molten globule state": four-state guanidinium chloride-induced unfolding of carbonic anhydrase B at low temperature. J Mol Biol 255(1):215–228.
10. Ringe D, Petsko GA (1986) Study of protein dynamics by X-ray diffraction. Meth Enzymol 131:389–433.
11. Dyson HJ, Wright PE (2002) Insights into the structure and dynamics of unfolded proteins from nuclear magnetic resonance. Adv Protein Chem 62:311–340.
12. Bracken C, Iakoucheva LM, Romero PR, Dunker AK (2004) Combining prediction, computation and experiment for the characterization of protein disorder. Curr Opin Struct Biol 14(5):570–576.
13. Dyson HJ, Wright PE (2004) Unfolded proteins and protein folding studied by NMR. Chem Rev 104(8):3607–3622.
14. Dyson HJ, Wright PE (2005) Intrinsically unstructured proteins and their functions. Nat Rev Mol Cell Biol 6(3):197–208.
15. Dyson HJ, Wright PE (2005) Elucidation of the protein folding landscape by NMR. Meth Enzymol 394:299–321.
16. Fasman GD (1996) Circular dichroism and the conformational analysis of biomolecules. Plenum Press, New York.
17. Adler AJ, Greenfield NJ, Fasman GD (1973) Circular dichroism and optical rotatory dispersion of proteins and polypeptides. Meth Enzymol 27:675–735.
18. Provencher SW, Glockner J (1981) Estimation of globular protein secondary structure from circular dichroism. Biochemistry 20(1):33–37.
19. Woody RW (1995) Circular dichroism. Meth Enzymol 246:34–71.
20. Uversky VN, Gillespie JR, Fink AL (2000) Why are "natively unfolded" proteins unstructured under physiologic conditions? Proteins 41(3):415–427.
21. Smyth E, Syme CD, Blanch EW, Hecht L, Vasak M, Barron LD (2001) Solution structure of native proteins with irregular folds from Raman optical activity. Biopolymers 58(2):138–151.
22. Uversky VN (1999) A multiparametric approach to studies of self-organization of globular proteins. Biochemistry (Mosc) 64(3):250–266.
23. Receveur-Brechot V, Bourhis JM, Uversky VN, Canard B, Longhi S (2006) Assessing protein disorder and induced folding. Proteins 62(1):24–45.

24. Markus G (1965) Protein substrate conformation and proteolysis. Proc Natl Acad Sci U S A 54:253–258.
25. Mikhalyi E (1978) Application of proteolytic enzymes to protein structure studies. CRC Press, Boca Raton
26. Hubbard SJ, Eisenmenger F, Thornton JM (1994) Modeling studies of the change in conformation required for cleavage of limited proteolytic sites. Protein Sci 3: 757–768.
27. Fontana A, de Laureto PP, de Filippis V, Scaramella E, Zambonin M (1997) Probing the partly folded states of proteins by limited proteolysis. Fold Des 2:R17–R26.
28. Fontana A, de Laureto PP, Spolaore B, Frare E, Picotti P, Zambonin M (2004) Probing protein structure by limited proteolysis. Acta Biochim Pol 51(2):299–321.
29. Iakoucheva LM, Kimzey AL, Masselon CD, Smith RD, Dunker AK, Ackerman EJ (2001) Aberrant mobility phenomena of the DNA repair protein XPA. Protein Sci 10: 1353–1362.
30. Tompa P (2002) Intrinsically unstructured proteins. Trends Biochem Sci 27(10):527–533.
31. Privalov PL (1979) Stability of proteins: small globular proteins. Adv Protein Chem 33: 167–241.
32. Ptitsyn O (1995) Molten globule and protein folding. Adv Protein Chem 47:83–229.
33. Ptitsyn OB, Uversky VN (1994) The molten globule is a third thermodynamical state of protein molecules. FEBS Lett 341:15–18.
34. Uversky VN, Ptitsyn OB (1996) All-or-none solvent-induced transitions between native, molten globule and unfolded states in globular proteins. Fold Des 1(2):117–122.
35. Westhof E, Altschuh D, Moras D, Bloomer AC, Mondragon A, Klug A, Van Regenmortel MH (1984) Correlation between segmental mobility and the location of antigenic determinants in proteins. Nature 311(5982):123–126.
36. Berzofsky JA (1985) Intrinsic and extrinsic factors in protein antigenic structure. Science 229(4717):932–940.
37. Kaltashov IA, Mohimen A (2005) Estimates of protein surface areas in solution by electrospray ionization mass spectrometry. Anal Chem 77(16):5370–5379.
38. Uversky VN (2002) Natively unfolded proteins: a point where biology waits for physics. Protein Sci 11(4):739–756.
39. Romero P, Obradovic Z, Li X, Garner EC, Brown CJ, Dunker AK (2001) Sequence complexity of disordered protein. Proteins 42(1):38–48.
40. Williams RM, Obradovic Z, Mathura V, Braun W, Garner EC, Young J, Takayama S, Brown CJ, Dunker AK (2001) The protein non-folding problem: amino acid determinants of intrinsic order and disorder. Pac Symp Biocomput:89–100.
41. Radivojac P, Iakoucheva LM, Oldfield CJ, Obradovic Z, Uversky VN, Dunker AK (2007) Intrinsic disorder and functional proteomics. Biophys J 92(5):1439–1456.
42. Li X, Romero P, Rani M, Dunker AK, Obradovic Z (1999) Predicting protein disorder for N-, C-, and internal regions. Genome Inform Ser Workshop Genome Inform 10:30–40.
43. Liu J, Rost B (2003) NORSp: Predictions of long regions without regular secondary structure. Nucleic Acids Res 31(13):3833–3835.
44. Linding R, Jensen LJ, Diella F, Bork P, Gibson TJ, Russell RB (2003) Protein disorder prediction: implications for structural proteomics. Structure 11(11):1453–1459.
45. Linding R, Russell RB, Neduva V, Gibson TJ (2003) GlobPlot: exploring protein sequences for globularity and disorder. Nucleic Acids Res 31(13):3701–3708.
46. Prilusky J, Felder CE, Zeev-Ben-Mordehai T, Rydberg EH, Man O, Beckmann JS, Silman I, Sussman JL (2005) FoldIndex: a simple tool to predict whether a given protein sequence is intrinsically unfolded. Bioinformatics 21(16):3435–3438.
47. Dosztanyi Z, Csizmok V, Tompa P, Simon I (2005) IUPred: web server for the prediction of intrinsically unstructured regions of proteins based on estimated energy content. Bioinformatics 21(16):3433–3434.
48. Jones DT, Ward JJ (2003) Prediction of disordered regions in proteins from position specific score matrices. Proteins 53 (Suppl 6):573–578.

49. Ward JJ, McGuffin LJ, Bryson K, Buxton BF, Jones DT (2004) The DISOPRED server for the prediction of protein disorder. Bioinformatics 20(13):2138–2139.
50. Ward JJ, Sodhi JS, McGuffin LJ, Buxton BF, Jones DT (2004) Prediction and functional analysis of native disorder in proteins from the three kingdoms of life. J Mol Biol 337(3):635–645.
51. Dunker AK, Obradovic Z, Romero P, Garner EC, Brown CJ (2000) Intrinsic protein disorder in complete genomes. Genome Inform Ser Workshop Genome Inform 11:161–171.
52. Oldfield CJ, Cheng Y, Cortese MS, Brown CJ, Uversky VN, Dunker AK (2005) Comparing and combining predictors of mostly disordered proteins. Biochemistry 44(6):1989–2000.
53. Dunker AK, Brown CJ, Lawson JD, Iakoucheva LM, Obradovic Z (2002) Intrinsic disorder and protein function. Biochemistry 41(21):6573–6582.
54. Dunker AK, Brown CJ, Obradovic Z (2002) Identification and functions of usefully disordered proteins. Adv Protein Chem 62:25–49.
55. Dunker AK, Cortese MS, Romero P, Iakoucheva LM, Uversky VN (2005) Flexible nets. The roles of intrinsic disorder in protein interaction networks. FEBS J 272(20):5129–5148.
56. Dunker AK, Garner E, Guilliot S, Romero P, Albrecht K, Hart J, Obradovic Z, Kissinger C, Villafranca JE (1998) Protein disorder and the evolution of molecular recognition: theory, predictions and observations. Pac Symp Biocomput:473–484.
57. Dunker AK, Obradovic Z (2001) The protein trinity–linking function and disorder. Nat Biotechnol 19(9):805–806.
58. Tompa P (2005) The interplay between structure and function in intrinsically unstructured proteins. FEBS Lett 579(15):3346–3354.
59. Tompa P, Csermely P (2004) The role of structural disorder in the function of RNA and protein chaperones. FASEB J 18(11):1169–1175.
60. Tompa P, Szasz C, Buday L (2005) Structural disorder throws new light on moonlighting. Trends Biochem Sci 30(9):484–489.
61. Uversky VN (2002) What does it mean to be natively unfolded? Eur J Biochem 269(1):2–12.
62. Xie H, Vucetic S, Iakoucheva LM, Oldfield CJ, Dunker AK, Uversky VN, Obradovic Z (2007) Functional anthology of intrinsic disorder. 1. Biological processes and functions of proteins with long disordered regions. J Proteome Res 6(5):1882–1898.
63. Vucetic S, Xie H, Iakoucheva LM, Oldfield CJ, Dunker AK, Obradovic Z, Uversky VN (2007) Functional anthology of intrinsic disorder. 2. Cellular components, domains, technical terms, developmental processes, and coding sequence diversities correlated with long disordered regions. J Proteome Res 6(5):1899–1916.
64. Xie H, Vucetic S, Iakoucheva LM, Oldfield CJ, Dunker AK, Obradovic Z, Uversky VN (2007) Functional anthology of intrinsic disorder. 3. Ligands, post-translational modifications, and diseases associated with intrinsically disordered proteins. J Proteome Res 6(5):1917–1932.
65. Uversky VN, Oldfield CJ, Dunker AK (2005) Showing your ID: intrinsic disorder as an ID for recognition, regulation and cell signaling. J Mol Recognit 18(5):343–384.
66. Iakoucheva LM, Brown CJ, Lawson JD, Obradovic Z, Dunker AK (2002) Intrinsic disorder in cell-signaling and cancer-associated proteins. J Mol Biol 323:573–584.
67. Glenner GG, Wong CW (1984) Alzheimer's disease: initial report of the purification and characterization of a novel cerebrovascular amyloid protein. Biochem Biophys Res Commun 120(3):885–890.
68. Ueda K, Fukushima H, Masliah E, Xia Y, Iwai A, Yoshimoto M, Otero DA, Kondo J, Ihara Y, Saitoh T (1993) Molecular cloning of cDNA encoding an unrecognized component of amyloid in Alzheimer disease. Proc Natl Acad Sci U S A 90(23):11282–11286.
69. Lee VM, Balin BJ, Otvos L, Jr., Trojanowski JQ (1991) A68: a major subunit of paired helical filaments and derivatized forms of normal Tau. Science 251(4994):675–678.
70. Wisniewski KE, Dalton AJ, McLachlan C, Wen GY, Wisniewski HM (1985) Alzheimer's disease in Down's syndrome: clinicopathologic studies. Neurology 35(7):957–961.

71. Dev KK, Hofele K, Barbieri S, Buchman VL, van der Putten H (2003) Part II: alpha-synuclein and its molecular pathophysiological role in neurodegenerative disease. Neuropharmacology 45(1):14–44.
72. Prusiner SB (2001) Shattuck lecture–neurodegenerative diseases and prions. N Engl J Med 344(20):1516–1526.
73. Zoghbi HY, Orr HT (1999) Polyglutamine diseases: protein cleavage and aggregation. Curr Opin Neurobiol 9(5):566–570.
74. Sickmeier M, Hamilton JA, LeGall T, Vacic V, Cortese MS, Tantos A, Szabo B, Tompa P, Chen J, Uversky VN et al (2007) DisProt: the database of disordered proteins. Nucleic Acids Res 35(Database issue):D786–793.
75. Vacic V, Uversky VN, Dunker AK, Lonardi S (2007) Composition Profiler: a tool for discovery and visualization of amino acid composition differences. BMC Bioinformatics 8:211.
76. Hebert LE, Scherr PA, Bienias JL, Bennett DA, Evans DA (2003) Alzheimer disease in the US population: prevalence estimates using the 2000 census. Arch Neurol 60(8): 1119–1122.
77. Schumock GT (1998) Economic considerations in the treatment and management of Alzheimer's disease. Am J Health Syst Pharm 55(Suppl 2):S17–S21.
78. Alzheimer A (1907) Über eine eigenartige Eskrankung der Nirnrinde. Allg Z Psychiatr Psych-Gerichtl 64:146–148.
79. Helmer C, Joly P, Letenneur L, Commenges D, Dartigues JF (2001) Mortality with dementia: results from a French prospective community-based cohort. Am J Epidemiol 154(7):642–648.
80. Aronson MK, Ooi WL, Geva DL, Masur D, Blau A, Frishman W (1991) Dementia. Age-dependent incidence, prevalence, and mortality in the old. Arch Intern Med 151(5):989–992.
81. McKhann G, Drachman D, Folstein M, Katzman R, Price D, Stadlan EM (1984) Clinical diagnosis of Alzheimer's disease: report of the NINCDS-ADRDA Work Group under the auspices of Department of Health and Human Services Task Force on Alzheimer's disease. Neurology 34(7):939–944.
82. Nussbaum RL, Ellis CE (2003) Alzheimer's disease and Parkinson's disease. N Engl J Med 348(14):1356–1364.
83. Campion D, Dumanchin C, Hannequin D, Dubois B, Belliard S, Puel M, Thomas-Anterion C, Michon A, Martin C, Charbonnier F et al (1999) Early-onset autosomal dominant Alzheimer disease: prevalence, genetic heterogeneity, and mutation spectrum. Am J Hum Genet 65(3):664–670.
84. Clark CM, Ewbank D, Lee VM-Y, Trojanowski JQ (1998) Molecular pathology of Alzheimer's disease: neuronal cytoskeletal abnormalities. In: Growdon JH, Rossor MN (eds.) The dementias Vol 19 of Blue books of practical neurology. Butterworth–Heinemann, Boston. 285–304.
85. Van Gassen G, Annaert W, Van Broeckhoven C (2000) Binding partners of Alzheimer's disease proteins: are they physiologically relevant? Neurobiol Dis 7(3):135–151.
86. Mattson MP (2004) Pathways towards and away from Alzheimer's disease. Nature 430(7000):631–639.
87. Van Nostrand WE, Schmaier AH, Wagner SL (1992) Potential role of protease nexin-2/amyloid beta-protein precursor as a cerebral anticoagulant. Ann N Y Acad Sci 674: 243–252.
88. Schmaier AH, Dahl LD, Rozemuller AJ, Roos RA, Wagner SL, Chung R, Van Nostrand WE (1993) Protease nexin-2/amyloid beta protein precursor. A tight-binding inhibitor of coagulation factor IXa. J Clin Invest 92(5):2540–2545.
89. Gao Y, Pimplikar SW (2001) The gamma – secretase-cleaved C-terminal fragment of amyloid precursor protein mediates signaling to the nucleus. Proc Natl Acad Sci U S A 98(26):14979–14984.
90. Cao X, Sudhof TC (2004) Dissection of amyloid-beta precursor protein-dependent transcriptional transactivation. J Biol Chem 279(23):24601–24611.

91. von Rotz RC, Kohli BM, Bosset J, Meier M, Suzuki T, Nitsch RM, Konietzko U (2004) The APP intracellular domain forms nuclear multiprotein complexes and regulates the transcription of its own precursor. J Cell Sci 117(Pt 19):4435–4448.
92. Leissring MA, Murphy MP, Mead TR, Akbari Y, Sugarman MC, Jannatipour M, Anliker B, Muller U, Saftig P, De Strooper B et al (2002) A physiologic signaling role for the gamma – secretase-derived intracellular fragment of APP. Proc Natl Acad Sci U S A 99(7):4697–4702.
93. Mesulam MM (2000) Aging, Alzheimer's disease and dementia. In: Mesulam MM (ed.) Principles of Behavioral and Cognitive Neurology. 2 edn. Oxford University Press, Oxford. 439–510.
94. Kayed R, Head E, Thompson JL, McIntire TM, Milton SC, Cotman CW, Glabe CG (2003) Common structure of soluble amyloid oligomers implies common mechanism of pathogenesis. Science 300(5618):486–489.
95. Berg L, McKeel DW, Jr., Miller JP, Storandt M, Rubin EH, Morris JC, Baty J, Coats M, Norton J, Goate AM et al (1998) Clinicopathologic studies in cognitively healthy aging and Alzheimer's disease: relation of histologic markers to dementia severity, age, sex, and apolipoprotein E genotype. Arch Neurol 55(3):326–335.
96. Price JL, Morris JC (1999) Tangles and plaques in nondemented aging and "preclinical" Alzheimer's disease. Ann Neurol 45(3):358–368.
97. Simmons LK, May PC, Tomaselli KJ, Rydel RE, Fuson KS, Brigham EF, Wright S, Lieberburg I, Becker GW, Brems DN et al (1994) Secondary structure of amyloid beta peptide correlates with neurotoxic activity in vitro. Mol Pharmacol 45(3):373–379.
98. Walsh DM, Klyubin I, Fadeeva JV, Rowan MJ, Selkoe DJ (2002) Amyloid-beta oligomers: their production, toxicity and therapeutic inhibition. Biochem Soc Trans 30(4):552–557.
99. Walsh DM, Klyubin I, Fadeeva JV, Cullen WK, Anwyl R, Wolfe MS, Rowan MJ, Selkoe DJ (2002) Naturally secreted oligomers of amyloid beta protein potently inhibit hippocampal long-term potentiation in vivo. Nature 416(6880):535–539.
100. Hartley DM, Walsh DM, Ye CP, Diehl T, Vasquez S, Vassilev PM, Teplow DB, Selkoe DJ (1999) Protofibrillar intermediates of amyloid beta-protein induce acute electrophysiological changes and progressive neurotoxicity in cortical neurons. J Neurosci 19(20):8876–8884.
101. Klein WL, Krafft GA, Finch CE (2001) Targeting small Abeta oligomers: the solution to an Alzheimer's disease conundrum? Trends Neurosci 24(4):219–224.
102. Gravina SA, Ho L, Eckman CB, Long KE, Otvos L, Jr., Younkin LH, Suzuki N, Younkin SG (1995) Amyloid beta protein (A beta) in Alzheimer's disease brain. Biochemical and immunocytochemical analysis with antibodies specific for forms ending at A beta 40 or A beta 42(43). J Biol Chem 270(13):7013–7016.
103. Hardy J (1997) Amyloid, the presenilins and Alzheimer's disease. Trends Neurosci 20(4):154–159.
104. Barrow CJ, Zagorski MG (1991) Solution structures of beta peptide and its constituent fragments: relation to amyloid deposition. Science 253(5016):179–182.
105. Harper JD, Lansbury PT, Jr. (1997) Models of amyloid seeding in Alzheimer's disease and scrapie: mechanistic truths and physiological consequences of the time-dependent solubility of amyloid proteins. Annu Rev Biochem 66:385–407.
106. Kirkitadze MD, Condron MM, Teplow DB (2001) Identification and characterization of key kinetic intermediates in amyloid beta-protein fibrillogenesis. J Mol Biol 312(5):1103–1119.
107. Maat-Schieman M, Roos R, van Duinen S (2005) Hereditary cerebral hemorrhage with amyloidosis-Dutch type. Neuropathology 25(4):288–297.
108. van Duinen SG, Castano EM, Prelli F, Bots GT, Luyendijk W, Frangione B (1987) Hereditary cerebral hemorrhage with amyloidosis in patients of Dutch origin is related to Alzheimer disease. Proc Natl Acad Sci U S A 84(16):5991–5994.
109. Himmler A (1989) Structure of the bovine tau gene: alternatively spliced transcripts generate a protein family. Mol Cell Biol 9(4):1389–1396.
110. Himmler A, Drechsel D, Kirschner MW, Martin DW, Jr. (1989) Tau consists of a set of proteins with repeated C-terminal microtubule-binding domains and variable N-terminal domains. Mol Cell Biol 9(4):1381–1388.

111. Goedert M (2003) Introduction to the Tauopathies. In: Dickson DW (ed.) Neurodegeneration: the molecular pathology of dementia and movement disorders. ISN Neuropath Press, Basel. 82–85.
112. Cleveland DW, Hwo SY, Kirschner MW (1977) Purification of tau, a microtubule-associated protein that induces assembly of microtubules from purified tubulin. J Mol Biol 116(2): 207–225.
113. Cleveland DW, Hwo SY, Kirschner MW (1977) Physical and chemical properties of purified tau factor and the role of tau in microtubule assembly. J Mol Biol 116(2): 227–247.
114. Drechsel DN, Hyman AA, Cobb MH, Kirschner MW (1992) Modulation of the dynamic instability of tubulin assembly by the microtubule-associated protein tau. Mol Biol Cell 3(10):1141–1154.
115. Brandt R, Lee G (1993) The balance between tau protein's microtubule growth and nucleation activities: implications for the formation of axonal microtubules. J Neurochem 61(3):997–1005.
116. Brandt R, Lee G (1993) Functional organization of microtubule-associated protein tau. Identification of regions which affect microtubule growth, nucleation, and bundle formation in vitro. J Biol Chem 268(5):3414–3419.
117. Binder LI, Frankfurter A, Rebhun LI (1985) The distribution of tau in the mammalian central nervous system. J Cell Biol 101(4):1371–1378.
118. Drubin DG, Feinstein SC, Shooter EM, Kirschner MW (1985) Nerve growth factor-induced neurite outgrowth in PC12 cells involves the coordinate induction of microtubule assembly and assembly-promoting factors. J Cell Biol 101(5 Pt 1):1799–1807.
119. Khatoon S, Grundke-Iqbal I, Iqbal K (1992) Brain levels of microtubule-associated protein tau are elevated in Alzheimer's disease: a radioimmuno-slot-blot assay for nanograms of the protein. J Neurochem 59(2):750–753.
120. Crowther RA, Goedert M (2000) Abnormal tau-containing filaments in neurodegenerative diseases. J Struct Biol 130(2–3):271–279.
121. Delacourte A, Buee L (1997) Normal and pathological Tau proteins as factors for microtubule assembly. Int Rev Cytol 171:167–224.
122. Vulliet R, Halloran SM, Braun RK, Smith AJ, Lee G (1992) Proline-directed phosphorylation of human Tau protein. J Biol Chem 267(31):22570–22574.
123. Lu Q, Wood JG (1993) Functional studies of Alzheimer's disease tau protein. J Neurosci 13(2):508–515.
124. Alonso AC, Grundke-Iqbal I, Iqbal K (1996) Alzheimer's disease hyperphosphorylated tau sequesters normal tau into tangles of filaments and disassembles microtubules. Nat Med 2(7):783–787.
125. Alonso AC, Zaidi T, Grundke-Iqbal I, Iqbal K (1994) Role of abnormally phosphorylated tau in the breakdown of microtubules in Alzheimer disease. Proc Natl Acad Sci U S A 91(12):5562–5566.
126. Iqbal K, Zaidi T, Bancher C, Grundke-Iqbal I (1994) Alzheimer paired helical filaments. Restoration of the biological activity by dephosphorylation. FEBS Lett 349(1):104–108.
127. Friedhoff P, von Bergen M, Mandelkow EM, Davies P, Mandelkow E (1998) A nucleated assembly mechanism of Alzheimer paired helical filaments. Proc Natl Acad Sci U S A 95(26):15712–15717.
128. Duyckaerts C, Dickson DW (2003) Neuropathology of Alzheimer's disease. In: Dickson DW (ed.) Neurodegeneration: the molecular pathology of dementia and movement disorders. ISN Neuropath Press, Basel. 47–65.
129. Tiraboschi P, Hansen LA, Thal LJ, Corey-Bloom J (2004) The importance of neuritic plaques and tangles to the development and evolution of AD. Neurology 62(11):1984–1989.
130. Feany MB, Dickson DW (1996) Neurodegenerative disorders with extensive tau pathology: a comparative study and review. Ann Neurol 40(2):139–148.
131. Kenessey A, Yen SH (1993) The extent of phosphorylation of fetal tau is comparable to that of PHF-tau from Alzheimer paired helical filaments. Brain Res 629(1):40–46.

132. Watanabe A, Hasegawa M, Suzuki M, Takio K, Morishima-Kawashima M, Titani K, Arai T, Kosik KS, Ihara Y (1993) In vivo phosphorylation sites in fetal and adult rat tau. J Biol Chem 268(34):25712–25717.
133. Billingsley ML, Kincaid RL (1997) Regulated phosphorylation and dephosphorylation of tau protein: effects on microtubule interaction, intracellular trafficking and neurodegeneration. Biochem J 323 (Pt 3):577–591.
134. Morishima-Kawashima M, Hasegawa M, Takio K, Suzuki M, Yoshida H, Watanabe A, Titani K, Ihara Y (1995) Hyperphosphorylation of tau in PHF. Neurobiol Aging 16(3):365–371; discussion 371–380.
135. Morishima-Kawashima M, Hasegawa M, Takio K, Suzuki M, Yoshida H, Titani K, Ihara Y (1995) Proline-directed and non-proline-directed phosphorylation of PHF-tau. J Biol Chem 270(2):823–829.
136. Uversky VN, Winter S, Galzitskaya OV, Kittler L, Lober G (1998) Hyperphosphorylation induces structural modification of tau-protein. FEBS Lett 439(1–2):21–25.
137. Hagestedt T, Lichtenberg B, Wille H, Mandelkow EM, Mandelkow E (1989) Tau protein becomes long and stiff upon phosphorylation: correlation between paracrystalline structure and degree of phosphorylation. J Cell Biol 109(4 Pt 1):1643–1651.
138. Eidenmuller J, Fath T, Hellwig A, Reed J, Sontag E, Brandt R (2000) Structural and functional implications of tau hyperphosphorylation: information from phosphorylation-mimicking mutated tau proteins. Biochemistry 39(43):13166–13175.
139. von Bergen M, Barghorn S, Jeganathan S, Mandelkow EM, Mandelkow E (2006) Spectroscopic approaches to the conformation of tau protein in solution and in paired helical filaments. Neurodegener Dis 3(4–5):197–206.
140. Chirita CN, Necula M, Kuret J (2003) Anionic micelles and vesicles induce tau fibrillization in vitro. J Biol Chem 278(28):25644–25650.
141. Chirita CN, Congdon EE, Yin H, Kuret J (2005) Triggers of full-length tau aggregation: a role for partially folded intermediates. Biochemistry 44(15):5862–5872.
142. Aronoff-Spencer E, Burns CS, Avdievich NI, Gerfen GJ, Peisach J, Antholine WE, Ball HL, Cohen FE, Prusiner SB, Millhauser GL (2000) Identification of the Cu2+ binding sites in the N-terminal domain of the prion protein by EPR and CD spectroscopy. Biochemistry 39(45):13760–13771.
143. Masters CL, Harris JO, Gajdusek DC, Gibbs CJ, Jr., Bernouilli C, Asher DM (1978) Creutzfeldt-Jakob disease: patterns of worldwide occurrence and the significance of familial and sporadic clustering. Ann Neurol 5:177–188.
144. Will RG, Alpers MP, Dormont D, Schonberger LB (2004) Infectious and sporadic prion diseases. In: Prusiner SB (ed.) Prion Biology and Diseases. 2 edn. Cold Spring Harbor Laboratory Press, Cold Spring Harbor. 629–671.
145. Asante EA, Linehan JM, Desbruslais M, Joiner S, Gowland I, Wood AL, Welch J, Hill AF, Lloyd SE, Wadsworth JD et al (2002) BSE prions propagate as either variant CJD-like or sporadic CJD-like prion strains in transgenic mice expressing human prion protein. EMBO J 21(23):6358–6366.
146. Prusiner SB, Scott MR, DeArmond SJ, Cohen FE (1998) Prion protein biology. Cell 93(3):337–348.
147. Riek R, Hornemann S, Wider G, Glockshuber R, Wuthrich K (1997) NMR characterization of the full-length recombinant murine prion protein, mPrP(23-231). FEBS Lett 413(2):282–288.
148. Burns CS, Aronoff-Spencer E, Dunham CM, Lario P, Avdievich NI, Antholine WE, Olmstead MM, Vrielink A, Gerfen GJ, Peisach J et al (2002) Molecular features of the copper binding sites in the octarepeat domain of the prion protein. Biochemistry 41(12):3991–4001.
149. Wildegger G, Liemann S, Glockshuber R (1999) Extremely rapid folding of the C-terminal domain of the prion protein without kinetic intermediates. Nat Struct Biol 6(6):550–553.

150. Hosszu LL, Baxter NJ, Jackson GS, Power A, Clarke AR, Waltho JP, Craven CJ, Collinge J (1999) Structural mobility of the human prion protein probed by backbone hydrogen exchange. Nat Struct Biol 6(8):740–743.
151. Peretz D, Williamson RA, Matsunaga Y, Serban H, Pinilla C, Bastidas RB, Rozenshteyn R, James TL, Houghten RA, Cohen FE et al (1997) A conformational transition at the N terminus of the prion protein features in formation of the scrapie isoform. J Mol Biol 273(3):614–622.
152. Vanik DL, Surewicz KA, Surewicz WK (2004) Molecular basis of barriers for interspecies transmissibility of mammalian prions. Mol Cell 14(1):139–145.
153. Watzlawik J, Skora L, Frense D, Griesinger C, Zweckstetter M, Schulz-Schaeffer WJ, Kramer ML (2006) Prion protein helix1 promotes aggregation but is not converted into beta-sheet. J Biol Chem 281(40):30242–30250.
154. Scott M, Groth D, Foster D, Torchia M, Yang SL, DeArmond SJ, Prusiner SB (1993) Propagation of prions with artificial properties in transgenic mice expressing chimeric PrP genes. Cell 73(5):979–988.
155. Telling GC, Scott M, Mastrianni J, Gabizon R, Torchia M, Cohen FE, DeArmond SJ, Prusiner SB (1995) Prion propagation in mice expressing human and chimeric PrP transgenes implicates the interaction of cellular PrP with another protein. Cell 83(1):79–90.
156. Donne DG, Viles JH, Groth D, Mehlhorn I, James TL, Cohen FE, Prusiner SB, Wright PE, Dyson HJ (1997) Structure of the recombinant full-length hamster prion protein PrP(29-231): the N terminus is highly flexible. Proc Natl Acad Sci U S A 94(25): 13452–13457.
157. James TL, Liu H, Ulyanov NB, Farr-Jones S, Zhang H, Donne DG, Kaneko K, Groth D, Mehlhorn I, Prusiner SB et al (1997) Solution structure of a 142-residue recombinant prion protein corresponding to the infectious fragment of the scrapie isoform. Proc Natl Acad Sci U S A 94(19):10086–10091.
158. Apetri AC, Surewicz WK (2002) Kinetic intermediate in the folding of human prion protein. J Biol Chem 277(47):44589–44592.
159. Martins SM, Chapeaurouge A, Ferreira ST (2003) Folding intermediates of the prion protein stabilized by hydrostatic pressure and low temperature. J Biol Chem 278(50):50449–50455.
160. Goedert M (1999) Filamentous nerve cell inclusions in neurodegenerative diseases: tauopathies and alpha-synucleinopathies. Philos Trans R Soc Lond B Biol Sci 354(1386):1101–1118.
161. Spillantini MG, Goedert M (2000) The alpha-synucleinopathies: Parkinson's disease, dementia with Lewy bodies, and multiple system atrophy. Ann N Y Acad Sci 920:16–27.
162. Trojanowski JQ, Lee VM (2003) Parkinson's disease and related alpha-synucleinopathies are brain amyloidoses. Ann N Y Acad Sci 991:107–110.
163. Galvin JE, Lee VM, Trojanowski JQ (2001) Synucleinopathies: clinical and pathological implications. Arch Neurol 58(2):186–190.
164. Marti MJ, Tolosa E, Campdelacreu J (2003) Clinical overview of the synucleinopathies. Mov Disord 18 Suppl 6:S21–S27.
165. Olanow CW, Tatton WG (1999) Etiology and pathogenesis of Parkinson's disease. Annu Rev Neurosci 22:123–144.
166. Forno LS (1996) Neuropathology of Parkinson's disease. J Neuropathol Exp Neurol 55(3):259–272.
167. Lewy FH (1912) Paralysis Agitans. Pathologische Anatomie. In: Lewandowski M (ed.) Handbuch der Neurologie. Springer, Berlin. 920–933.
168. Zarranz JJ, Alegre J, Gomez-Esteban JC, Lezcano E, Ros R, Ampuero I, Vidal L, Hoenicka J, Rodriguez O, Atares B et al (2004) The new mutation, E46K, of alpha-synuclein causes Parkinson and Lewy body dementia. Ann Neurol 55(2):164–173.
169. Polymeropoulos MH, Lavedan C, Leroy E, Ide SE, Dehejia A, Dutra A, Pike B, Root H, Rubenstein J, Boyer R et al (1997) Mutation in the alpha-synuclein gene identified in families with Parkinson's disease. Science 276(5321):2045–2047.

170. Kruger R, Kuhn W, Muller T, Woitalla D, Graeber M, Kosel S, Przuntek H, Epplen JT, Schols L, Riess O (1998) Ala30Pro mutation in the gene encoding alpha-synuclein in Parkinson's disease. Nat Genet 18(2):106–108.
171. Singleton A, Gwinn-Hardy K, Sharabi Y, Li ST, Holmes C, Dendi R, Hardy J, Crawley A, Goldstein DS (2004) Association between cardiac denervation and parkinsonism caused by alpha-synuclein gene triplication. Brain 127(Pt 4):768–772.
172. Singleton AB, Farrer M, Johnson J, Singleton A, Hague S, Kachergus J, Hulihan M, Peuralinna T, Dutra A, Nussbaum R et al (2003) alpha-Synuclein locus triplication causes Parkinson's disease. Science 302(5646):841.
173. Farrer M, Kachergus J, Forno L, Lincoln S, Wang DS, Hulihan M, Maraganore D, Gwinn-Hardy K, Wszolek Z, Dickson D et al (2004) Comparison of kindreds with parkinsonism and alpha-synuclein genomic multiplications. Ann Neurol 55(2):174–179.
174. Spillantini MG, Schmidt ML, Lee VM, Trojanowski JQ, Jakes R, Goedert M (1997) Alpha-synuclein in Lewy bodies. Nature 388(6645):839–840.
175. Spillantini MG, Crowther RA, Jakes R, Hasegawa M, Goedert M (1998) alpha-Synuclein in filamentous inclusions of Lewy bodies from Parkinson's disease and dementia with lewy bodies. Proc Natl Acad Sci U S A 95(11):6469–6473.
176. Masliah E, Rockenstein E, Veinbergs I, Mallory M, Hashimoto M, Takeda A, Sagara Y, Sisk A, Mucke L (2000) Dopaminergic loss and inclusion body formation in alpha-synuclein mice: implications for neurodegenerative disorders. Science 287(5456):1265–1269.
177. Feany MB, Bender WW (2000) A Drosophila model of Parkinson's disease. Nature 404(6776):394–398.
178. Dickson DW (2001) Alpha-synuclein and the Lewy body disorders. Curr Opin Neurol 14(4):423–432.
179. Goedert M (2001) Parkinson's disease and other alpha-synucleinopathies. Clin Chem Lab Med 39(4):308–312.
180. Goedert M (2001) Alpha-synuclein and neurodegenerative diseases. Nat Rev Neurosci 2(7):492–501.
181. Uversky VN, Fink AL (2002) Biophysical properties of human alpha-synuclein and its role in Parkinson's disease. In: Pandalai SG (ed.) Recent Research Developments in Proteins. Transworld Research Network, Kerala, India. 153–186.
182. Galpern WR, Lang AE (2006) Interface between tauopathies and synucleinopathies: a tale of two proteins. Ann Neurol 59(3):449–458.
183. Kosaka K (1978) Lewy bodies in cerebral cortex, report of three cases. Acta Neuropathol (Berl) 42(2):127–134.
184. Baba M, Nakajo S, Tu PH, Tomita T, Nakaya K, Lee VM, Trojanowski JQ, Iwatsubo T (1998) Aggregation of alpha-synuclein in Lewy bodies of sporadic Parkinson's disease and dementia with Lewy bodies. Am J Pathol 152(4):879–884.
185. Spillantini MG, Crowther RA, Jakes R, Cairns NJ, Lantos PL, Goedert M (1998) Filamentous alpha-synuclein inclusions link multiple system atrophy with Parkinson's disease and dementia with Lewy bodies. Neurosci Lett 251(3):205–208.
186. McKeith IG, Galasko D, Kosaka K, Perry EK, Dickson DW, Hansen LA, Salmon DP, Lowe J, Mirra SS, Byrne EJ et al (1996) Consensus guidelines for the clinical and pathologic diagnosis of dementia with Lewy bodies (DLB): report of the consortium on DLB international workshop. Neurology 47(5):1113–1124.
187. McKeith I, Mintzer J, Aarsland D, Burn D, Chiu H, Cohen-Mansfield J, Dickson D, Dubois B, Duda JE, Feldman H et al (2004) Dementia with Lewy bodies. Lancet Neurol 3(1):19–28.
188. Emre M (2004) Dementia in Parkinson's disease: cause and treatment. Curr Opin Neurol 17(4):399–404.
189. Mayeux R, Denaro J, Hemenegildo N, Marder K, Tang MX, Cote LJ, Stern Y (1992) A population-based investigation of Parkinson's disease with and without dementia. Relationship to age and gender. Arch Neurol 49(5):492–497.

190. Aarsland D, Andersen K, Larsen JP, Lolk A, Nielsen H, Kragh-Sorensen P (2001) Risk of dementia in Parkinson's disease: a community-based, prospective study. Neurology 56(6):730–736.
191. Plato CC, Cruz MT, Kurland LT (1969) Amyotrophic lateral sclerosis-Parkinsonism dementia complex of Guam: further genetic investigations. Am J Hum Genet 21(2):133–141.
192. Schmitt HP, Emser W, Heimes C (1984) Familial occurrence of amyotrophic lateral sclerosis, parkinsonism, and dementia. Ann Neurol 16(6):642–648.
193. Spencer PS, Nunn PB, Hugon J, Ludolph AC, Ross SM, Roy DN, Robertson RC (1987) Guam amyotrophic lateral sclerosis-parkinsonism-dementia linked to a plant excitant neurotoxin. Science 237(4814):517–522.
194. Majoor-Krakauer D, Ottman R, Johnson WG, Rowland LP (1994) Familial aggregation of amyotrophic lateral sclerosis, dementia, and Parkinson's disease: evidence of shared genetic susceptibility. Neurology 44(10):1872–1877.
195. den Hartogjager WA, Bethlem J (1960) The distribution of Lewy bodies in the central and autonomic nervous systems in idiopathic paralysis agitans. J Neurol Neurosurg Psychiatry 23:283–290.
196. Kosaka K, Mehraein P (1979) Dementia-Parkinsonism syndrome with numerous Lewy bodies and senile plaques in cerebral cortex. Arch Psychiatr Nervenkr 226(4): 241–250.
197. Hague K, Lento P, Morgello S, Caro S, Kaufmann H (1997) The distribution of Lewy bodies in pure autonomic failure: autopsy findings and review of the literature. Acta Neuropathol (Berl) 94(2):192–196.
198. Jackson M, Lennox G, Balsitis M, Lowe J (1995) Lewy body dysphagia. J Neurol Neurosurg Psychiatry 58(6):756–758.
199. Hansen LA, Galasko D (1992) Lewy body disease. Curr Opin Neurol Neurosurg 5(6): 889–894.
200. Arai Y, Yamazaki M, Mori O, Muramatsu H, Asano G, Katayama Y (2001) Alpha-synuclein-positive structures in cases with sporadic Alzheimer's disease: morphology and its relationship to tau aggregation. Brain Res 888(2):287–296.
201. Lippa CF, Schmidt ML, Lee VM, Trojanowski JQ (1999) Antibodies to alpha-synuclein detect Lewy bodies in many Down's syndrome brains with Alzheimer's disease. Ann Neurol 45(3):353–357.
202. Marui W, Iseki E, Ueda K, Kosaka K (2000) Occurrence of human alpha-synuclein immunoreactive neurons with neurofibrillary tangle formation in the limbic areas of patients with Alzheimer's disease. J Neurol Sci 174(2):81–84.
203. Stefanova N, Tison F, Reindl M, Poewe W, Wenning GK (2005) Animal models of multiple system atrophy. Trends Neurosci 28(9):501–506.
204. Wenning GK, Colosimo C, Geser F, Poewe W (2004) Multiple system atrophy. Lancet Neurol 3(2):93–103.
205. Wenning GK, Ben-Shlomo Y, Hughes A, Daniel SE, Lees A, Quinn NP (2000) What clinical features are most useful to distinguish definite multiple system atrophy from Parkinson's disease? J Neurol Neurosurg Psychiatry 68(4):434–440.
206. Bower JH, Maraganore DM, McDonnell SK, Rocca WA (1997) Incidence of progressive supranuclear palsy and multiple system atrophy in Olmsted County, Minnesota, 1976 to 1990. Neurology 49(5):1284–1288.
207. Chrysostome V, Tison F, Yekhlef F, Sourgen C, Baldi I, Dartigues JF (2004) Epidemiology of multiple system atrophy: a prevalence and pilot risk factor study in Aquitaine, France. Neuroepidemiology 23(4):201–208.
208. Schrag A, Ben-Shlomo Y, Quinn NP (1999) Prevalence of progressive supranuclear palsy and multiple system atrophy: a cross-sectional study. Lancet 354(9192):1771–1775.
209. Daniel S (1999) The neuropathology and neurochemistry of multiple system atrophy. In: Mathias CJ, Bannister R (eds.) Autonomic Failure. Oxford University Press, Oxford. 321–328.

210. Papp MI, Kahn JE, Lantos PL (1989) Glial cytoplasmic inclusions in the CNS of patients with multiple system atrophy (striatonigral degeneration, olivopontocerebellar atrophy and Shy-Drager syndrome). J Neurol Sci 94(1–3):79–100.
211. Wakabayashi K, Takahashi H (2006) Cellular pathology in multiple system atrophy. Neuropathology 26(4):338–345.
212. Taylor TD, Litt M, Kramer P, Pandolfo M, Angelini L, Nardocci N, Davis S, Pineda M, Hattori H, Flett PJ et al (1996) Homozygosity mapping of Hallervorden-Spatz syndrome to chromosome 20p12.3-p13. Nat Genet 14(4):479–481.
213. Malandrini A, Cesaretti S, Mulinari M, Palmeri S, Fabrizi GM, Villanova M, Parrotta E, Montagnani A, Montagnani M, Anichini M et al (1996) Acanthocytosis, retinitis pigmentosa, pallidal degeneration. Report of two cases without serum lipid abnormalities. J Neurol Sci 140(1–2):129–131.
214. Sugiyama H, Hainfellner JA, Schmid-Siegel B, Budka H (1993) Neuroaxonal dystrophy combined with diffuse Lewy body disease in a young adult. Clin Neuropathol 12(3): 147–152.
215. Swaiman KF (1991) Hallervorden-Spatz syndrome and brain iron metabolism. Arch Neurol 48(12):1285–1293.
216. Dooling EC, Schoene WC, Richardson EP, Jr. (1974) Hallervorden-Spatz syndrome. Arch Neurol 30(1):70–83.
217. Jankovic J, Kirkpatrick JB, Blomquist KA, Langlais PJ, Bird ED (1985) Late-onset Hallervorden-Spatz disease presenting as familial parkinsonism. Neurology 35(2): 227–234.
218. Zhou B, Westaway SK, Levinson B, Johnson MA, Gitschier J, Hayflick SJ (2001) A novel pantothenate kinase gene (PANK2) is defective in Hallervorden-Spatz syndrome. Nat Genet 28(4):345–349.
219. Tu PH, Galvin JE, Baba M, Giasson B, Tomita T, Leight S, Nakajo S, Iwatsubo T, Trojanowski JQ, Lee VM (1998) Glial cytoplasmic inclusions in white matter oligodendrocytes of multiple system atrophy brains contain insoluble alpha-synuclein. Ann Neurol 44(3):415–422.
220. Wakabayashi K, Yoshimoto M, Fukushima T, Koide R, Horikawa Y, Morita T, Takahashi H (1999) Widespread occurrence of alpha-synuclein/NACP-immunoreactive neuronal inclusions in juvenile and adult-onset Hallervorden-Spatz disease with Lewy bodies. Neuropathol Appl Neurobiol 25(5):363–368.
221. Galvin JE, Giasson B, Hurtig HI, Lee VM, Trojanowski JQ (2000) Neurodegeneration with brain iron accumulation, type 1 is characterized by alpha-, beta-, and gamma-synuclein neuropathology. Am J Pathol 157(2):361–368.
222. Neumann M, Adler S, Schluter O, Kremmer E, Benecke R, Kretzschmar HA (2000) Alpha-synuclein accumulation in a case of neurodegeneration with brain iron accumulation type 1 (NBIA-1, formerly Hallervorden-Spatz syndrome) with widespread cortical and brainstem-type Lewy bodies. Acta Neuropathol (Berl) 100(5):568–574.
223. Uversky VN (2003) A protein-chameleon: conformational plasticity of alpha-synuclein, a disordered protein involved in neurodegenerative disorders. J Biomol Struct Dyn 21(2): 211–234.
224. Morar AS, Olteanu A, Young GB, Pielak GJ (2001) Solvent-induced collapse of alpha-synuclein and acid-denatured cytochrome c. Protein Sci 10(11):2195–2199.
225. Eliezer D, Kutluay E, Bussell R, Jr., Browne G (2001) Conformational properties of alpha-synuclein in its free and lipid-associated states. J Mol Biol 307(4):1061–1073.
226. Syme CD, Blanch EW, Holt C, Jakes R, Goedert M, Hecht L, Barron LD (2002) A Raman optical activity study of rheomorphism in caseins, synucleins and tau. New insight into the structure and behaviour of natively unfolded proteins. Eur J Biochem 269(1):148–156.
227. Fulton AB (1982) How crowded is the cytoplasm? Cell 30(2):345–347.
228. Yancey PH, Clark ME, Hand SC, Bowlus RD, Somero GN (1982) Living with water stress: evolution of osmolyte systems. Science 217(4566):1214–1222.

229. McNulty BC, Young GB, Pielak GJ (2006) Macromolecular crowding in the Escherichia coli periplasm maintains alpha-synuclein disorder. J Mol Biol 355(5):893–897.
230. Serber Z, Dotsch V (2001) In-cell NMR spectroscopy. Biochemistry 40(48):14317–14323.
231. Serber Z, Ledwidge R, Miller SM, Dotsch V (2001) Evaluation of parameters critical to observing proteins inside living Escherichia coli by in-cell NMR spectroscopy. J Am Chem Soc 123(37):8895–8901.
232. Dedmon MM, Patel CN, Young GB, Pielak GJ (2002) FlgM gains structure in living cells. Proc Natl Acad Sci U S A 99(20):12681–12684.
233. Clayton DF, George JM (1999) Synucleins in synaptic plasticity and neurodegenerative disorders. J Neurosci Res 58(1):120–129.
234. Maroteaux L, Campanelli JT, Scheller RH (1988) Synuclein: a neuron-specific protein localized to the nucleus and presynaptic nerve terminal. J Neurosci 8(8):2804–2815.
235. Jakes R, Spillantini MG, Goedert M (1994) Identification of two distinct synucleins from human brain. FEBS Lett 345(1):27–32.
236. Nakajo S, Tsukada K, Omata K, Nakamura Y, Nakaya K (1993) A new brain-specific 14-kDa protein is a phosphoprotein. Its complete amino acid sequence and evidence for phosphorylation. Eur J Biochem 217(3):1057–1063.
237. Tobe T, Nakajo S, Tanaka A, Mitoya A, Omata K, Nakaya K, Tomita M, Nakamura Y (1992) Cloning and characterization of the cDNA encoding a novel brain-specific 14-kDa protein. J Neurochem 59(5):1624–1629.
238. Ji H, Liu YE, Jia T, Wang M, Liu J, Xiao G, Joseph BK, Rosen C, Shi YE (1997) Identification of a breast cancer-specific gene, BCSG1, by direct differential cDNA sequencing. Cancer Res 57(4):759–764.
239. Ninkina NN, Alimova-Kost MV, Paterson JW, Delaney L, Cohen BB, Imreh S, Gnuchev NV, Davies AM, Buchman VL (1998) Organization, expression and polymorphism of the human persyn gene. Hum Mol Genet 7(9):1417–1424.
240. Buchman VL, Hunter HJ, Pinon LG, Thompson J, Privalova EM, Ninkina NN, Davies AM (1998) Persyn, a member of the synuclein family, has a distinct pattern of expression in the developing nervous system. J Neurosci 18(22):9335–9341.
241. Lavedan C, Leroy E, Dehejia A, Buchholtz S, Dutra A, Nussbaum RL, Polymeropoulos MH (1998) Identification, localization and characterization of the human gamma-synuclein gene. Hum Genet 103(1):106–112.
242. Lucking CB, Brice A (2000) Alpha-synuclein and Parkinson's disease. Cell Mol Life Sci 57(13–14):1894–1908.
243. Clayton DF, George JM (1998) The synucleins: a family of proteins involved in synaptic function, plasticity, neurodegeneration and disease. Trends Neurosci 21(6):249–254.
244. Shibayama-Imazu T, Okahashi I, Omata K, Nakajo S, Ochiai H, Nakai Y, Hama T, Nakamura Y, Nakaya K (1993) Cell and tissue distribution and developmental change of neuron specific 14 kDa protein (phosphoneuroprotein 14). Brain Res 622(1–2):17–25.
245. Nakajo S, Shioda S, Nakai Y, Nakaya K (1994) Localization of phosphoneuroprotein 14 (PNP 14) and its mRNA expression in rat brain determined by immunocytochemistry and in situ hybridization. Brain Res Mol Brain Res 27(1):81–86.
246. Nakajo S, Tsukada K, Kameyama H, Furuyama Y, Nakaya K (1996) Distribution of phosphoneuroprotein 14 (PNP 14) in vertebrates: its levels as determined by enzyme immunoassay. Brain Res 741(1–2):180–184.
247. Shibayama-Imazu T, Ogane K, Hasegawa Y, Nakajo S, Shioda S, Ochiai H, Nakai Y, Nakaya K (1998) Distribution of PNP 14 (beta-synuclein) in neuroendocrine tissues: localization in Sertoli cells. Mol Reprod Dev 50(2):163–169.
248. Hashimoto M, Yoshimoto M, Sisk A, Hsu LJ, Sundsmo M, Kittel A, Saitoh T, Miller A, Masliah E (1997) NACP, a synaptic protein involved in Alzheimer's disease, is differentially regulated during megakaryocyte differentiation. Biochem Biophys Res Commun 237(3):611–616.

249. Ninkina NN, Privalova EM, Pinon LG, Davies AM, Buchman VL (1999) Developmentally regulated expression of persyn, a member of the synuclein family, in skin. Exp Cell Res 246(2):308–311.
250. Galvin JE, Uryu K, Lee VM, Trojanowski JQ (1999) Axon pathology in Parkinson's disease and Lewy body dementia hippocampus contains alpha-, beta-, and gamma-synuclein. Proc Natl Acad Sci U S A 96(23):13450–13455.
251. Uversky VN, Li J, Souillac P, Millett IS, Doniach S, Jakes R, Goedert M, Fink AL (2002) Biophysical properties of the synucleins and their propensities to fibrillate: inhibition of alpha-synuclein assembly by beta- and gamma-synucleins. J Biol Chem 277(14):11970–11978.
252. Yamin G, Munishkina LA, Karymov MA, Lyubchenko YL, Uversky VN, Fink AL (2005) Forcing the non-amyloidogenic beta-synuclein to fibrillate. Biochemistry 44 (25):9096–9107.
253. Hashimoto M, Rockenstein E, Mante M, Mallory M, Masliah E (2001) beta-Synuclein inhibits alpha-synuclein aggregation: a possible role as an anti-parkinsonian factor. Neuron 32(2):213–223.
254. Cummings CJ, Zoghbi HY (2000) Trinucleotide repeats: mechanisms and pathophysiology. Annu Rev Genomics Hum Genet 1:281–328.
255. Cummings CJ, Zoghbi HY (2000) Fourteen and counting: unraveling trinucleotide repeat diseases. Hum Mol Genet 9(6):909–916.
256. Perutz MF (1996) Glutamine repeats and inherited neurodegenerative diseases: molecular aspects. Curr Opin Struct Biol 6(6):848–858.
257. Ross CA (2002) Polyglutamine pathogenesis: emergence of unifying mechanisms for Huntington's disease and related disorders. Neuron 35(5):819–822.
258. Bates G (2003) Huntingtin aggregation and toxicity in Huntington's disease. Lancet 361(9369):1642–1644.
259. Soto C (2003) Unfolding the role of protein misfolding in neurodegenerative diseases. Nat Rev Neurosci 4(1):49–60.
260. Yue S, Serra HG, Zoghbi HY, Orr HT (2001) The spinocerebellar ataxia type 1 protein, ataxin-1, has RNA-binding activity that is inversely affected by the length of its polyglutamine tract. Hum Mol Genet 10(1):25–30.
261. Klement IA, Skinner PJ, Kaytor MD, Yi H, Hersch SM, Clark HB, Zoghbi HY, Orr HT (1998) Ataxin-1 nuclear localization and aggregation: role in polyglutamine-induced disease in SCA1 transgenic mice. Cell 95(1):41–53.
262. Gusella J, MacDonald M (2002) No post-genetics era in human disease research. Nat Rev Genet 3(1):72–79.
263. McEwan IJ (2001) Structural and functional alterations in the androgen receptor in spinal bulbar muscular atrophy. Biochem Soc Trans 29(Pt 2):222–227.
264. Nagafuchi S, Yanagisawa H, Ohsaki E, Shirayama T, Tadokoro K, Inoue T, Yamada M (1994) Structure and expression of the gene responsible for the triplet repeat disorder, dentatorubral and pallidoluysian atrophy (DRPLA). Nat Genet 8(2):177–182.
265. Faber PW, Barnes GT, Srinidhi J, Chen J, Gusella JF, MacDonald ME (1998) Huntingtin interacts with a family of WW domain proteins. Hum Mol Genet 7(9):1463–1474.
266. Boutell JM, Thomas P, Neal JW, Weston VJ, Duce J, Harper PS, Jones AL (1999) Aberrant interactions of transcriptional repressor proteins with the Huntington's disease gene product, huntingtin. Hum Mol Genet 8(9):1647–1655.
267. Steffan JS, Kazantsev A, Spasic-Boskovic O, Greenwald M, Zhu YZ, Gohler H, Wanker EE, Bates GP, Housman DE, Thompson LM (2000) The Huntington's disease protein interacts with p53 and CREB-binding protein and represses transcription. Proc Natl Acad Sci U S A 97(12):6763–6768.
268. Schaeper U, Boyd JM, Verma S, Uhlmann E, Subramanian T, Chinnadurai G (1995) Molecular cloning and characterization of a cellular phosphoprotein that interacts with a conserved C-terminal domain of adenovirus E1A involved in negative modulation of oncogenic transformation. Proc Natl Acad Sci U S A 92(23):10467–10471.

269. Saudou F, Finkbeiner S, Devys D, Greenberg ME (1998) Huntingtin acts in the nucleus to induce apoptosis but death does not correlate with the formation of intranuclear inclusions. Cell 95(1):55–66.
270. Huang CC, Faber PW, Persichetti F, Mittal V, Vonsattel JP, MacDonald ME, Gusella JF (1998) Amyloid formation by mutant huntingtin: threshold, progressivity and recruitment of normal polyglutamine proteins. Somat Cell Mol Genet 24(4):217–233.
271. Nucifora FC, Jr., Sasaki M, Peters MF, Huang H, Cooper JK, Yamada M, Takahashi H, Tsuji S, Troncoso J, Dawson VL et al (2001) Interference by huntingtin and atrophin-1 with cbp-mediated transcription leading to cellular toxicity. Science 291(5512): 2423–2428.
272. Kegel KB, Meloni AR, Yi Y, Kim YJ, Doyle E, Cuiffo BG, Sapp E, Wang Y, Qin ZH, Chen JD et al (2002) Huntingtin is present in the nucleus, interacts with the transcriptional corepressor C-terminal binding protein, and represses transcription. J Biol Chem 277(9):7466–7476.
273. Ratovitski T, Nakamura M, D'Ambola J, Chighladze E, Liang Y, Wang W, Graham R, Hayden MR, Borchelt DR, Hirschhorn RR et al (2007) N-terminal proteolysis of full-length mutant huntingtin in an inducible PC12 cell model of Huntington's disease. Cell Cycle 6(23):2970–2981.
274. Sun B, Fan W, Balciunas A, Cooper JK, Bitan G, Steavenson S, Denis PE, Young Y, Adler B, Daugherty L et al (2002) Polyglutamine repeat length-dependent proteolysis of huntingtin. Neurobiol Dis 11(1):111–122.
275. Ross CA, Wood JD, Schilling G, Peters MF, Nucifora FC, Jr., Cooper JK, Sharp AH, Margolis RL, Borchelt DR (1999) Polyglutamine pathogenesis. Philos Trans R Soc Lond B Biol Sci 354(1386):1005–1011.
276. Preisinger E, Jordan BM, Kazantsev A, Housman D (1999) Evidence for a recruitment and sequestration mechanism in Huntington's disease. Philos Trans R Soc Lond B Biol Sci 354(1386):1029–1034.
277. Wanker EE (2000) Protein aggregation and pathogenesis of Huntington's disease: mechanisms and correlations. Biol Chem 381(9–10):937–942.
278. McCampbell A, Taylor JP, Taye AA, Robitschek J, Li M, Walcott J, Merry D, Chai Y, Paulson H, Sobue G et al (2000) CREB-binding protein sequestration by expanded polyglutamine. Hum Mol Genet 9(14):2197–2202.
279. McCampbell A, Fischbeck KH (2001) Polyglutamine and CBP: fatal attraction? Nat Med 7(5):528–530.
280. Chen S, Berthelier V, Yang W, Wetzel R (2001) Polyglutamine aggregation behavior in vitro supports a recruitment mechanism of cytotoxicity. J Mol Biol 311(1):173–182.
281. Chen S, Berthelier V, Hamilton JB, O'Nuallain B, Wetzel R (2002) Amyloid-like features of polyglutamine aggregates and their assembly kinetics. Biochemistry 41(23): 7391–7399.
282. Perutz MF, Pope BJ, Owen D, Wanker EE, Scherzinger E (2002) Aggregation of proteins with expanded glutamine and alanine repeats of the glutamine-rich and asparagine-rich domains of Sup35 and of the amyloid beta-peptide of amyloid plaques. Proc Natl Acad Sci U S A 99(8):5596–5600.
283. Chow MK, Paulson HL, Bottomley SP (2004) Destabilization of a non-pathological variant of ataxin-3 results in fibrillogenesis via a partially folded intermediate: a model for misfolding in polyglutamine disease. J Mol Biol 335(1):333–341.
284. Poirier MA, Li H, Macosko J, Cai S, Amzel M, Ross CA (2002) Huntingtin spheroids and protofibrils as precursors in polyglutamine fibrilization. J Biol Chem 277(43):41032–41037.
285. Raychaudhuri S, Majumder P, Sarkar S, Giri K, Mukhopadhyay D, Bhattacharyya NP (2007) Huntingtin interacting protein HYPK is intrinsically unstructured. Proteins.
286. Raychaudhuri S, Sinha M, Mukhopadhyay D, Bhattacharyya NP (2008) HYPK, a Huntingtin interacting protein, reduces aggregates and apoptosis induced by N-terminal Huntingtin with 40 glutamines in Neuro2a cells and exhibits chaperone-like activity. Hum Mol Genet 17(2):240–255.

287. Koide R, Ikeuchi T, Onodera O, Tanaka H, Igarashi S, Endo K, Takahashi H, Kondo R, Ishikawa A, Hayashi T et al (1994) Unstable expansion of CAG repeat in hereditary dentatorubral-pallidoluysian atrophy (DRPLA). Nat Genet 6(1):9–13.
288. Margolis RL, Li SH, Young WS, Wagster MV, Stine OC, Kidwai AS, Ashworth RG, Ross CA (1996) DRPLA gene (atrophin-1) sequence and mRNA expression in human brain. Brain Res Mol Brain Res 36(2):219–226.
289. Onodera O, Oyake M, Takano H, Ikeuchi T, Igarashi S, Tsuji S (1995) Molecular cloning of a full-length cDNA for dentatorubral-pallidoluysian atrophy and regional expressions of the expanded alleles in the CNS. Am J Hum Genet 57(5):1050–1060.
290. Loev SJ, Margolis RL, Young WS, Li SH, Schilling G, Ashworth RG, Ross CA (1995) Cloning and expression of the rat atrophin-I (DRPLA disease gene) homologue. Neurobiol Dis 2(3):129–138.
291. Schmitt I, Epplen JT, Riess O (1995) Predominant neuronal expression of the gene responsible for dentatorubral-pallidoluysian atrophy (DRPLA) in rat. Hum Mol Genet 4(9):1619–1624.
292. Yazawa I, Nukina N, Hashida H, Goto J, Yamada M, Kanazawa I (1995) Abnormal gene product identified in hereditary dentatorubral-pallidoluysian atrophy (DRPLA) brain. Nat Genet 10(1):99–103.
293. Lavery DN, McEwan IJ (2008) Structural characterization of the native NH2-terminal transactivation domain of the human androgen receptor: a collapsed disordered conformation underlies structural plasticity and protein-induced folding. Biochemistry 47(11): 3360–3369.
294. Albrecht M, Golatta M, Wullner U, Lengauer T (2004) Structural and functional analysis of ataxin-2 and ataxin-3. Eur J Biochem 271(15):3155–3170.
295. Albrecht M, Hoffmann D, Evert BO, Schmitt I, Wullner U, Lengauer T (2003) Structural modeling of ataxin-3 reveals distant homology to adaptins. Proteins 50(2):355–370.
296. Masino L, Musi V, Menon RP, Fusi P, Kelly G, Frenkiel TA, Trottier Y, Pastore A (2003) Domain architecture of the polyglutamine protein ataxin-3: a globular domain followed by a flexible tail. FEBS Lett 549(1–3):21–25.
297. Zhuchenko O, Bailey J, Bonnen P, Ashizawa T, Stockton DW, Amos C, Dobyns WB, Subramony SH, Zoghbi HY, Lee CC (1997) Autosomal dominant cerebellar ataxia (SCA6) associated with small polyglutamine expansions in the alpha 1A-voltage-dependent calcium channel. Nat Genet 15(1):62–69.
298. Kubodera T, Yokota T, Ohwada K, Ishikawa K, Miura H, Matsuoka T, Mizusawa H (2003) Proteolytic cleavage and cellular toxicity of the human alpha1A calcium channel in spinocerebellar ataxia type 6. Neurosci Lett 341(1):74–78.
299. Palhan VB, Chen S, Peng GH, Tjernberg A, Gamper AM, Fan Y, Chait BT, La Spada AR, Roeder RG (2005) Polyglutamine-expanded ataxin-7 inhibits STAGA histone acetyltransferase activity to produce retinal degeneration. Proc Natl Acad Sci U S A 102(24): 8472–8477.
300. Nakamura K, Jeong SY, Uchihara T, Anno M, Nagashima K, Nagashima T, Ikeda S, Tsuji S, Kanazawa I (2001) SCA17, a novel autosomal dominant cerebellar ataxia caused by an expanded polyglutamine in TATA-binding protein. Hum Mol Genet 10(14):1441–1448.
301. Hochheimer A, Tjian R (2003) Diversified transcription initiation complexes expand promoter selectivity and tissue-specific gene expression. Genes Dev 17(11):1309–1320.
302. Burley SK (1996) The TATA box binding protein. Curr Opin Struct Biol 6(1):69–75.
303. Friedman MJ, Shah AG, Fang ZH, Ward EG, Warren ST, Li S, Li XJ (2007) Polyglutamine domain modulates the TBP-TFIIB interaction: implications for its normal function and neurodegeneration. Nat Neurosci 10(12):1519–1528.
304. Friedman MJ, Wang CE, Li XJ, Li S (2008) Polyglutamine expansion reduces the association of TATA-binding protein with DNA and induces DNA binding-independent neurotoxicity. J Biol Chem 283(13):8283–8290.
305. Lescure A, Lutz Y, Eberhard D, Jacq X, Krol A, Grummt I, Davidson I, Chambon P, Tora L (1994) The N-terminal domain of the human TATA-binding protein plays a role

in transcription from TATA-containing RNA polymerase II and III promoters. EMBO J 13(5):1166–1175.
306. Plant GT, Revesz T, Barnard RO, Harding AE, Gautier-Smith PC (1990) Familial cerebral amyloid angiopathy with nonneuritic amyloid plaque formation. Brain 113 (Pt 3): 721–747.
307. Vidal R, Frangione B, Rostagno A, Mead S, Revesz T, Plant G, Ghiso J (1999) A stop-codon mutation in the BRI gene associated with familial British dementia. Nature 399(6738): 776–781.
308. Ghiso JA, Holton J, Miravalle L, Calero M, Lashley T, Vidal R, Houlden H, Wood N, Neubert TA, Rostagno A et al (2001) Systemic amyloid deposits in familial British dementia. J Biol Chem 276(47):43909–43914.
309. Kim SH, Wang R, Gordon DJ, Bass J, Steiner DF, Lynn DG, Thinakaran G, Meredith SC, Sisodia SS (1999) Furin mediates enhanced production of fibrillogenic ABri peptides in familial British dementia. Nat Neurosci 2(11):984–988.
310. Srinivasan R, Jones EM, Liu K, Ghiso J, Marchant RE, Zagorski MG (2003) pH-dependent amyloid and protofibril formation by the ABri peptide of familial British dementia. J Mol Biol 333(5):1003–1023.
311. Vidal R, Revesz T, Rostagno A, Kim E, Holton JL, Bek T, Bojsen-Moller M, Braendgaard H, Plant G, Ghiso J et al (2000) A decamer duplication in the 3′ region of the BRI gene originates an amyloid peptide that is associated with dementia in a Danish kindred. Proc Natl Acad Sci U S A 97(9):4920–4925.
312. Rostagno A, Tomidokoro Y, Lashley T, Ng D, Plant G, Holton J, Frangione B, Revesz T, Ghiso J (2005) Chromosome 13 dementias. Cell Mol Life Sci 62(16):1814–1825.
313. Surolia I, Reddy GB, Sinha S (2006) Hierarchy and the mechanism of fibril formation in ADan peptides. J Neurochem 99(2):537–548.
314. Brenner M, Johnson AB, Boespflug-Tanguy O, Rodriguez D, Goldman JE, Messing A (2001) Mutations in GFAP, encoding glial fibrillary acidic protein, are associated with Alexander disease. Nat Genet 27(1):117–120.
315. Rodriguez D, Gauthier F, Bertini E, Bugiani M, Brenner M, N'Guyen S, Goizet C, Gelot A, Surtees R, Pedespan JM et al (2001) Infantile Alexander disease: spectrum of GFAP mutations and genotype-phenotype correlation. Am J Hum Genet 69(5):1134–1140.
316. Borrett D, Becker LE (1985) Alexander's disease. A disease of astrocytes. Brain 108 (Pt 2):367–385.
317. Head MW, Corbin E, Goldman JE (1993) Overexpression and abnormal modification of the stress proteins alpha B-crystallin and HSP27 in Alexander disease. Am J Pathol 143(6):1743–1753.
318. Li R, Messing A, Goldman JE, Brenner M (2002) GFAP mutations in Alexander disease. Int J Dev Neurosci 20(3–5):259–268.
319. Mignot C, Boespflug-Tanguy O, Gelot A, Dautigny A, Pham-Dinh D, Rodriguez D (2004) Alexander disease: putative mechanisms of an astrocytic encephalopathy. Cell Mol Life Sci 61(3):369–385.
320. Dahl D (1976) Glial fibrillary acidic protein from bovine and rat brain. Degradation in tissues and homogenates. Biochim Biophys Acta 420(1):142–154.
321. Dahl D (1976) Isolation and initial characterization of glial fibrillary acidic protein from chicken, turtle, frog and fish central nervous systems. Biochim Biophys Acta 446(1): 41–50.
322. Alpers BJ (1931) Diffuse progressive degeneration of the gray matter of the cerebrum. Arch Neurol Psychiatry 25:469–505.
323. Huttenlocher PR, Solitare GB, Adams G (1976) Infantile diffuse cerebral degeneration with hepatic cirrhosis. Arch Neurol 33(3):186–192.
324. Harding BN (1990) Progressive neuronal degeneration of childhood with liver disease (Alpers-Huttenlocher syndrome): a personal review. J Child Neurol 5(4):273–287.
325. Harding BN, Egger J, Portmann B, Erdohazi M (1986) Progressive neuronal degeneration of childhood with liver disease. A pathological study. Brain 109 (Pt 1):181–206.

326. Narkewicz MR, Sokol RJ, Beckwith B, Sondheimer J, Silverman A (1991) Liver involvement in Alpers disease. J Pediatr 119(2):260–267.
327. Naviaux RK, Nguyen KV (2004) POLG mutations associated with Alpers' syndrome and mitochondrial DNA depletion. Ann Neurol 55(5):706–712.
328. Kaguni LS (2004) DNA polymerase gamma, the mitochondrial replicase. Annu Rev Biochem 73:293–320.
329. Ferrari G, Lamantea E, Donati A, Filosto M, Briem E, Carrara F, Parini R, Simonati A, Santer R, Zeviani M (2005) Infantile hepatocerebral syndromes associated with mutations in the mitochondrial DNA polymerase-gammaA. Brain 128(Pt 4):723–731.
330. Luoma PT, Luo N, Loscher WN, Farr CL, Horvath R, Wanschitz J, Kiechl S, Kaguni LS, Suomalainen A (2005) Functional defects due to spacer-region mutations of human mitochondrial DNA polymerase in a family with an ataxia-myopathy syndrome. Hum Mol Genet 14(14):1907–1920.
331. Hayashi M, Imanaka-Yoshida K, Yoshida T, Wood M, Fearns C, Tatake RJ, Lee JD (2006) A crucial role of mitochondrial Hsp40 in preventing dilated cardiomyopathy. Nat Med 12(1):128–132.
332. Nance MA, Berry SA (1992) Cockayne syndrome: review of 140 cases. Am J Med Genet 42(1):68–84.
333. Friedberg EC (1996) Cockayne syndrome–a primary defect in DNA repair, transcription, both or neither? Bioessays 18(9):731–738.
334. Licht CL, Stevnsner T, Bohr VA (2003) Cockayne syndrome group B cellular and biochemical functions. Am J Hum Genet 73(6):1217–1239.
335. Troelstra C, van Gool A, de Wit J, Vermeulen W, Bootsma D, Hoeijmakers JH (1992) ERCC6, a member of a subfamily of putative helicases, is involved in Cockayne's syndrome and preferential repair of active genes. Cell 71(6):939–953.
336. Henning KA, Li L, Iyer N, McDaniel LD, Reagan MS, Legerski R, Schultz RA, Stefanini M, Lehmann AR, Mayne LV et al (1995) The Cockayne syndrome group A gene encodes a WD repeat protein that interacts with CSB protein and a subunit of RNA polymerase II TFIIH. Cell 82(4):555–564.
337. Selzer RR, Nyaga S, Tuo J, May A, Muftuoglu M, Christiansen M, Citterio E, Brosh RM, Jr., Bohr VA (2002) Differential requirement for the ATPase domain of the Cockayne syndrome group B gene in the processing of UV-induced DNA damage and 8-oxoguanine lesions in human cells. Nucleic Acids Res 30(3):782–793.
338. Melki J (1997) Spinal muscular atrophy. Curr Opin Neurol 10(5):381–385.
339. Lefebvre S, Burglen L, Reboullet S, Clermont O, Burlet P, Viollet L, Benichou B, Cruaud C, Millasseau P, Zeviani M et al (1995) Identification and characterization of a spinal muscular atrophy-determining gene. Cell 80(1):155–165.
340. Fischer U, Liu Q, Dreyfuss G (1997) The SMN-SIP1 complex has an essential role in spliceosomal snRNP biogenesis. Cell 90(6):1023–1029.
341. Pellizzoni L, Charroux B, Dreyfuss G (1999) SMN mutants of spinal muscular atrophy patients are defective in binding to snRNP proteins. Proc Natl Acad Sci U S A 96(20):11167–11172.
342. Buhler D, Raker V, Luhrmann R, Fischer U (1999) Essential role for the tudor domain of SMN in spliceosomal U snRNP assembly: implications for spinal muscular atrophy. Hum Mol Genet 8(13):2351–2357.
343. Selenko P, Sprangers R, Stier G, Buhler D, Fischer U, Sattler M (2001) SMN tudor domain structure and its interaction with the Sm proteins. Nat Struct Biol 8(1):27–31.
344. Brahms H, Raymackers J, Union A, de Keyser F, Meheus L, Luhrmann R (2000) The C-terminal RG dipeptide repeats of the spliceosomal Sm proteins D1 and D3 contain symmetrical dimethylarginines, which form a major B-cell epitope for anti-Sm autoantibodies. J Biol Chem 275(22):17122–17129.
345. Brahms H, Meheus L, de Brabandere V, Fischer U, Luhrmann R (2001) Symmetrical dimethylation of arginine residues in spliceosomal Sm protein B/B' and the Sm-like protein LSm4, and their interaction with the SMN protein. RNA 7(11):1531–1542.

346. Friesen WJ, Massenet S, Paushkin S, Wyce A, Dreyfuss G (2001) SMN, the product of the spinal muscular atrophy gene, binds preferentially to dimethylarginine-containing protein targets. Mol Cell 7(5):1111–1117.
347. Paushkin S, Gubitz AK, Massenet S, Dreyfuss G (2002) The SMN complex, an assemblyosome of ribonucleoproteins. Curr Opin Cell Biol 14(3):305–312.
348. Terns MP, Terns RM (2001) Macromolecular complexes: SMN–the master assembler. Curr Biol 11(21):R862–R864.
349. Sprangers R, Groves MR, Sinning I, Sattler M (2003) High-resolution X-ray and NMR structures of the SMN Tudor domain: conformational variation in the binding site for symmetrically dimethylated arginine residues. J Mol Biol 327(2):507–520.
350. Sprangers R, Selenko P, Sattler M, Sinning I, Groves MR (2003) Definition of domain boundaries and crystallization of the SMN Tudor domain. Acta Crystallogr D Biol Crystallogr 59(Pt 2):366–368.
351. Kerr DA, Nery JP, Traystman RJ, Chau BN, Hardwick JM (2000) Survival motor neuron protein modulates neuron-specific apoptosis. Proc Natl Acad Sci U S A 97(24):13312–13317.

Chapter 3
Dynamic Role of Ubiquitination in the Management of Misfolded Proteins Associated with Neurodegenerative Diseases

Esther S.P. Wong, Jeanne M.M. Tan and Kah-Leong Lim

Abstract Protein aggregation as a result of misfolding is a common theme underlying neurodegenerative diseases. Although a subject of intense research, how misfolded proteins bypass sophisticated protein quality control measures in the cell to be deposited as ubiquitin-enriched inclusion bodies remains poorly understood. Whilst proteasome dysfunction could account for this phenomenon, emerging evidence suggests otherwise. We have previously hypothesized that under conditions of proteolytic stress, the cell may switch to a non-proteolytic form of ubiquitination to help divert misfolded proteins away from an overloaded proteasome. In this way, the cell could preserve its proteasome function over prolonged periods of stress and recover thereafter. Supporting this, we recently found that non-proteolytic lysine (K) 63-linked ubiquitin modification promotes the formation of protein inclusions associated with several major neurodegenerative diseases. Importantly, we further found that K63-linked polyubiquitin selectively facilitates the subsequent clearance of inclusions via autophagy. In this chapter, we will discuss the apparent dynamic role of ubiquitination in the management of misfolded proteins.

3.1 Protein Misfolding and the Ubiquitin-Proteasome System

Protein misfolding is very much a part of life for most, if not all, living organisms. It has been estimated that as many as one-third of all newly synthesized proteins in eukaryotes might not be properly folded into their native conformations [1], presumably even with the assistance of molecular chaperones. Furthermore, misfolding of proteins could also arise from various post-synthetic modifications such as oxidation, glycation and nitrosylation, as well as from genetic mutations [2]. Likewise, prokaryotes are not spared from protein folding errors [3]. To ensure that the quality of intracellular proteins at the end of the production line is of functional

K.L. Lim (✉)
Neurodegeneration Research Lab, National Neuroscience Institute, Singapore and Duke-NUS Graduate Medical School, Singapore
e-mail: Kah_Leong_Lim@nni.com.sg

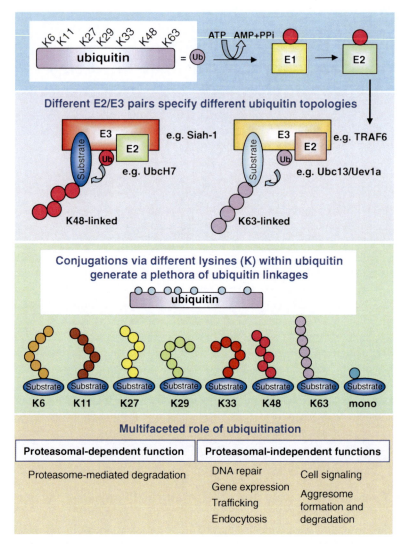

Fig. 3.1 The ubiquitin cascade. Ubiquitin attachment on a protein involves a linear reaction catalyzed by ubiquitin activating (E1), conjugating (E2) and ligating (E3) enzymes. Repeated actions of this multi-enzyme-mediated cascade lead to the formation of a polyubiquitin chain in which the terminal residue (G76) of one ubiquitin molecule is linked through an iso-peptide bond to a lysine (K) residue within another. The ubiquitin sequence contains seven lysine residues (at positions 6, 11, 27, 29, 33, 48 and 63) that can support the assembly of polyubiquitin of different chain topologies. The plethora of ubiquitin linkages is specified by different E2-E3 combinations. For example, the E3 TRAF6 interacts with the E2 Ubc13 to mediate K63-linked polyubiquitin chain assembly. Because of this versatility, protein ubiquitination serves diverse roles in the cell; many of these are notably not coupled to the proteasome. We have featured a non-canonical role for protein ubiquitination in the formation and clearance of protein inclusions associated with neurodegenerative diseases in this chapter

standard, both prokaryotic and eukaryotic cells had evolved complex intracellular proteolytic machineries to destroy faulty proteins rapidly, which in the process of doing so, also serve to prevent the accumulation of misfolded proteins that would otherwise compromise normal cell function and survivability in eukaryotes and prokaryotes alike.

The selection of proteins to be degraded is a highly regulated process. In eukaryotes, this process involves labeling target proteins with ubiquitin, a highly conserved 76 amino acid residue protein that is encoded by multiple genes and expressed ubiquitously [4]. The majority of the cytosolic proteins destined for degradation in a eukaryotic cell are first covalently tagged with a polymeric ubiquitin chain in which the terminal residue (G76) of one ubiquitin molecule is linked through an iso-peptide bond to a lysine (K) residue (most commonly K48) within another [5]. This ligation reaction is elaborate and requires the sequential and repetitive actions of ubiquitin-activating (E1), -conjugating (E2) and -ligating (E3) enzymes (Fig. 3.1). The G76-K48-linked polyubiquitinated substrate is then targeted for degradation by the 26S proteasome, a large protease complex consisting of a barrel-shaped 20S proteolytic core in association with two 19S (PA700) regulatory caps, one on each side of the barrel's openings [6]. The components of the 19S cap play vital roles in the initial steps of substrate proteolysis, including the recognition, unfolding and translocation of substrate proteins into the lumen of the proteolytic core [7–9]. Individual ubiquitin monomers are regenerated in the process by the actions of deubiquitinating enzymes (DUBs) in a manner analogous to the actions of phosphatases on phosphorylated amino acid residues [10]. It is important to highlight that substantial metabolic energy is required to drive the ubiquitin-proteasome system (UPS) machinery. ATP is needed both at the start of the ubiquitination reaction by E1 to charge the ubiquitin molecule and towards the end of the process by the 19S regulatory cap to open the channel and unfold ubiquitinated substrate for subsequent translocation into the 20S proteolytic chamber. Notably, the 19S cap contains not one but six different ATPase subunits to fulfill its tasks [11]. This energy requirement by the UPS distinguishes it from lysosomal proteases, which degrades proteins in an exergonic manner. Furthermore, ATP is also required to assemble the proteasome complex [6]. Conceivably, conditions that promote mitochondrial dysfunction and thereby energy depletion are likely to interfere with the efficiency of proteasome-mediated degradation, which in turn could trigger the accumulation of misfolded proteins and compromise cellular survivability. Indeed, mitochondrial impairment is intimately associated with UPS dysfunction in a wide variety of neurodegenerative diseases [12].

3.2 Protein Misfolding, UPS Disruption and Neurodegeneration

An almost invariant hallmark of human neurodegenerative disorders is the presence of insoluble proteinaceous deposits in surviving populations of affected neurons. These include neurofibrillary tangles in Alzheimer's disease (AD) and

Frontotemporal Dementia (FTD), Lewy bodies (LBs) in Parkinson's disease (PD) and Bunina bodies in Amyotrophic Lateral Sclerosis (ALS) [13]. Although a diverse array of pathogenic proteins bearing little structural and functional similarities with each other has been identified as key contributors to the formation of protein inclusion bodies in these varied diseases, they all share the tendency to become misfolded and thereby aggregation-prone. This suggests a common mechanism underlying the biogenesis of protein inclusions associated with neurodegenerative diseases. Since the UPS represents a major intracellular pathway responsible for the management of misfolded proteins, its involvement would seem obvious here. Indeed, UPS dysfunction is widely thought to precipitate the accumulation of misfolded proteins and their subsequent aggregation into inclusion bodies [12–16]. Consistent with this, neurodegenerative disease-associated protein deposits are almost always enriched with ubiquitin and components of the UPS [2, 13, 16]. Furthermore, post-mortem analysis of brain samples from individuals afflicted with AD or PD revealed structural and/or functional impairments of the UPS in affected areas [17–20] (although one cannot exclude the participation of glial cells within the sample preparations as a confounding factor in these analyses). Moreover, when cells are cultured in the presence of proteasome inhibitors, they tend to generate ubiquitin-positive juxtanuclear inclusions within one to two days in culture. Likewise, delivery of proteasomal inhibitors into adult rats appears sufficient to promote the formation of neuronal inclusions [21, 22]. Although these findings collectively suggest a role for UPS in neurodegeneration, whether UPS dysfunction plays a primary or secondary role in the pathogenesis of neurodegenerative diseases is less clear, as inhibition of UPS activity conceivably could be a consequence of protein aggregation. One scenario would be that misfolded protein aggregates tagged with ubiquitin resist full entry into the narrow catalytic axial pore of the barrel-shaped proteasome complex and cause steric occlusions, thereby denying itself and other proteins destined for degradation from efficient proteasomal clearance. This appears to be the case with disease-associated proteins containing expanded polyglutamine [14, 23, 24].

Perhaps the strongest evidence to date directly linking the UPS with neurodegeneration is the association of genetic mutations in the *parkin* gene with autosomal recessive parkinsonism [25]. Parkin is a member of the E3 family of enzymes and disease-causing mutations of parkin have been demonstrated by several groups to compromise its ubiquitination activity [26–28]. A working hypothesis that ensues is that loss of parkin function leads to a neurotoxic accumulation of one or several of its substrates. Indeed, numerous reported substrates of parkin were later found to accumulate in the brains of PD patients carrying parkin mutations [29–32]. This hypothesis is further fuelled by the discovery of a missense mutation (I93M) in UCHL1, a DUB, in a pair of German siblings with inherited PD [33]. Notably, the I93M UCHL1 mutant has markedly reduced ubiquitin hydrolase activity *in vitro* [33], suggesting that impaired polyubiquitin hydrolysis leading to a shortage of free ubiquitin might also promote the accumulation of toxic proteins and contribute to neuronal death. The involvement of UCHL1 I93M mutation in PD pathogenesis has however become contentious in recent years as its occurrence to date is restricted to the pair of German siblings [34]. Notwithstanding this, it is clear that the ubiquitin molecule, through its functional coupling with the proteasome, plays an important role in maintaining protein homeostasis and thereby survivability of post-mitotic neurons. The relationship between UPS disruption and neurodegenerative disorders is therefore unlikely a trivial one. Furthermore, the model also provides a persuasive explanation to how ubiquitin-tagged aggregation-prone proteins are able to escape proteasome-mediated degradation. Otherwise, the proteolytic role of ubiquitin tagging and the non-proteolytic role of inclusions formation would appear to be contradictory. Given this, it may be rather surprising to note that a number of recent reports found little or no evidence of proteasome dysfunction associated with spinocerebellar ataxia type 1 (SCA1) or PD even in the most vulnerable neurons despite the presence of ubiquitin-positive inclusions and neuropathology [35, 36]. These findings are in apparent contradiction to earlier ones by others that proteasome impairment accompanies neurodegeneration [17–20]. Importantly, they highlighted the capacity of ubiquitinated proteins to evade degradation even in the presence of functional proteasomes. While this may appear paradoxical, we now know that protein ubiquitination is a highly versatile modification that is not obligatorily linked to the proteasome (see Sect. 3.3). Accordingly, proteins modified by non-proteolytic forms of ubiquitination could potentially save themselves from destruction by the proteasome.

3.3 Diversity of Ubiquitin Modifications

Although ubiquitin tagging of proteins was originally identified as a signal for proteasome-mediated degradation, it is currently recognized that the reaction is highly versatile and can support alternative forms of ubiquitin modification that are non-proteolytic [37]. As the ubiquitin molecule contains seven lysine residues (at positions 6, 11, 27, 29, 33, 48 and 63), seven types of ubiquitin chain topologies are in theory possible depending on which of these lysines are used for the isopeptide formation during the assembly of polyubiquitin. Unconventional ubiquitin linkages that has been reported include those mediated by K6 [38], K11 [38], K29 [39] and K63 [39]. Although ubiquitin linkages involving K27 and K33 appear to be comparatively rarer, all the seven lysines on the ubiquitin molecule, including K27 and K33, were found to be modified in a recent proteomic-based study on ubiquitin conjugates from *Saccharomyces cerevisiae* [40]. Collectively, these findings provide support for an unexpected diversity in polyubiquitin chain topology *in vivo*. Adding to this layer of complexity is the recognition that the different types of ubiquitin chains on a protein do not represent static tags, but instead are dynamic signals that could be edited by DUBs and remodeled [41, 42]. The remodeling of ubiquitin tags on a substrate has been shown to influence it's commitment towards being degraded by the proteasome [41]. DUBs are therefore not mere housekeepers at the terminal end of the UPS but rather should be regarded as active players in the ubiquitination cascade as these enzymes could directly modify the chain length and thereby the fate of polyubiquitinated proteins. This property of DUBs is suitably exemplified by HAUSP (Herpes virus-associated ubiquitin specific protease), a DUB that deubiquitinates p53 in the presence of excess Mdm2 (the cognate E3 of p53) and concomitantly stabilizes p53 levels [43]. Finally, proteins can also be monoubiquitinated and this form of ubiquitin modification can occur at single or multiple sites on a protein [44]. Notably, both K63-linked polyubiquitination and monoubiquitination of proteins are not typically associated with their degradation [37, 44]. Taken together, it is apparent that a plethora of ubiquitin modification exists within the cell and this versatility is expected to offer the cell extraordinary flexibility in modulating the fate and function of its resident proteins under different conditions. Indeed, protein ubiquitination rivals protein phosphorylation as a major post-translational event in the cell. Little wonder that an enormous number of cellular proteins (>1000), comprising of various E1s, E2s, E3s, DUBs and other related members, are involved in protein ubiquitination. This would include a whole collection of ubiquitin-binding protein containing structurally diverse domains that bind non-convalently, and sometimes specifically, to monomeric ubiquitin or polymeric ubiquitin of different chain topologies [44]. Ubiquitin-binding proteins function as important downstream mediators of ubiquitin-mediated events, by helping to interpret and transmit information encoded by the ubiquitin tag [44]. Not surprising, protein ubiquitination serves a multitude of cellular roles including regulation of gene expression, cell division and differentiation, signal transduction, protein trafficking, endocytosis and DNA repair [45]. This list of services provided by ubiquitination is likely to be continually expanded as novel functions underscored especially by the less-studied ubiquitin linkages will invariably be uncovered with time. Here, we shall discuss a recent discovery by our laboratory regarding the role of K63-linked polyubiquitin as a key regulator of the biogenesis and clearance of protein inclusions associated with neurodegenerative diseases.

3.4 Non-Proteolytic Ubiquitination and Protein Inclusion Biogenesis

For whatever reasons the proteasome becomes compromised in its function, it is difficult to imagine that the cell will continue to burden the machinery under such conditions with an endless stream of cargo proteins to be degraded. We have developed a hypothesis about two years ago that non-proteolytic ubiquitination of proteins may help divert proteins originally destined for proteasomal degradation away from the system when it becomes overwhelmed under conditions of proteolytic stress [46]. The diverted ubiquitin-enriched proteins are then sequestered into inclusion bodies following their accumulation. In this way, the cell could preserve its proteasome function over prolonged periods of proteolytic stress and recover thereafter.

Support for the above hypothesis first came from our work on parkin. An interesting feature of parkin-related PD cases is the general absence of LBs [47–49], which suggests that parkin-mediated ubiquitination is a key promoter of LB biogenesis.

Consistent with this, we have previously demonstrated that LB-like inclusions formed in cultured cells via the co-expression of the PD-linked genes, α-synuclein and synphilin-1, is significantly enhanced in the presence of wild type but not mutant parkin overexpression [50]. However, as synphilin-1 is a substrate of parkin, it did seem paradoxical to us at that time that parkin-mediated ubiquitination promotes synphilin-1 accumulation and subsequent aggregation into LB-like inclusions rather than facilitates its clearance from the cell. This is unlike the action of Siah-1, another E3 that targets synphilin-1 specifically for degradation [51]. This conundrum was finally resolved when we and others subsequently found that parkin is a multi-functional E3 enzyme that is capable of catalyzing both degradation-associated and non-proteolytic forms of protein ubiquitination under different cellular conditions [51–54]. We demonstrated that parkin-mediated polyubiquitination of synphilin-1 normally occurs via K63-linked chains and this non-proteolytic form of ubiquitin modification of synphilin-1 by parkin promotes its accumulation and subsequent aggregation into inclusion bodies [51]. This observation has provided us with the first evidence that K63-linked ubiquitination of a protein could precipitate its deposition into an inclusion body. Interestingly, others have reported that the ubiquitin hydrolase UCHL1 also possesses a second catalytic activity but one that apparently opposes its hydrolytic role [55]. This unusual ligase-like activity of UCHL1 has similarly been shown to promote K63-linked ubiquitination. A substrate of UCHL1 ligase activity that has been identified in the study is α-synuclein, which expectedly accumulates in the cell following UCHL1-mediated K63-linked ubiquitination [55]. Based on these results, it is tempting to speculate that K63-linked ubiquitin chains may represent a common denominator underlying inclusions biogenesis. Supporting this, we recently found that K63-linked ubiquitination also facilitates the formation of inclusions mediated by FTD-associated tau mutant as well as ALS-associated superoxide dismutase1 (SOD1) mutant [56]. Importantly, we further found that protein inclusions formation in the presence of K63-linked ubiquitination can occur in the apparent absence of 20S proteasome impairment, although we cannot exclude the possibility that this form of ubiquitin modification may exert inhibitory effects through interactions with the regulatory components present within the 19S cap [56]. Together, our data collectively support a role for K63-linked ubiquitination in regulating the biogenesis of protein inclusions. Nonetheless, a formal concern regarding our studies above is that they were done in the setting of ectopic overexpression of ubiquitin mutants that support the formation of various chain topologies. For example, the K48 and K63 ubiquitin mutants used in our studies contain arginine substitutions on all their lysine residues except the one at position 48 and 63 respectively, and are thus expected to promote the proteasome-linked G76-K48 and the proteasome-independent G76-K63 ubiquitin linkages respectively. On the other hand, the K0 ubiquitin mutant is a lysineless ubiquitin theoretically only capable of mediating monoubiquitination. Because the number of mutations on these various ubiquitin mutants occupies almost 10% of the protein sequence of a wild type ubiquitin molecule, it is legitimate to express doubt on whether ubiquitin topologies mediated by these mutants are identical to those assembled endogenously. To address this concern, we have subsequently examined the effects of promoting endogenous K63-linked ubiquitination on inclusion dynamics. This is achieved by taking advantage of the specificity of a heterodimeric E2 pair, namely Ubc13 and Uev1a, in directing K63-linked ubiquitin chain assembly [57]. Supporting our findings derived from ubiquitin mutants, we found that Ubc13/Uev1a-mediated K63-linked ubiquitination enhances the formation of ubiquitin-positive protein inclusions to a similar extent to those brought about by K63 ubiquitin mutant over expression, which are significantly higher than those generated in the presence of exogenously-introduced UbcH7 and UbcH8, E2s for which no specificity for K63-linked chain has been reported [58]. Thus, the promotion of endogenous K63-linked ubiquitination alone appears to be sufficient to precipitate protein inclusions formation.

Another limitation that has however persisted in our investigations is the inability to detect for K63-linked polyubiquitin directly in protein inclusions due to the current lack of ubiquitin linkage-specific antibodies. To circumvent this, we have characterized the utility of ubiquitin binding proteins as indirect markers for K63-linked ubiquitin chains. The sequestosome p62 is one such protein that has been previously proposed by Seibenhener et al to have specificity for K63-linked polyubiquitinated proteins *in vivo* [59], although the isolated ubiquitin-associated (UBA) domain apparently also binds with low affinity to monoubiquitin and K48-linked chains [60]. Consistent with this, a recent NMR-based study revealed that p62-UBA has a higher affinity for single ubiquitin molecule and may explain its preference for the more opened and extended K63-linked chains [61]. In our hands, we found that p62 has a preference to interact with long ($>$ Ub7) K63-linked chains over corresponding K48-linked chains, although the protein does not appear to distinguish between short chains containing either forms of ubiquitin linkages [58]. Notably, p62 is a prominent component of cytoplasmic inclusions in several protein aggregation disorders including AD and PD, suggesting indirectly that K63-linked polyubiquitination may be a physiologically-relevant player in the biogenesis of disease-associated inclusions. This suggestion is corroborated, in part, by a recent mass spectrometry-based study from Ron Kopito's group who demonstrated the presence and accumulation of K63-linked polyubiquitin in a mouse model of Huntington's disease [15]. Further, we also found an enrichment of K63-linked ubiquitin chains in aggregate-laden fractions of post-mortem AD brains using similar methods (unpublished results). Taken together, the findings above by our group and others support a role for non-proteolytic K63-linked ubiquitin modification as a physiologically meaningful manager of aggregation-prone protein.

3.5 Aggresome Formation and Clearance

An interesting feature of cytosolic inclusion bodies generated by aggregation-prone proteins is that they are often located juxtaposed to the nucleus. Johnston and Kopito have coined the term "aggresomes" to describe these peri-nucleus structures and proposed that aggresome formation is a general response of cells which occurs when the capacity of the proteasome is exceeded by the production of aggregation-prone misfolded proteins [62]. It is currently recognized that aggresomes are formed via a microtubule-driven process by which non-degradable protein aggregates are trafficked along microtubules to the microtubule organizing center (MTOC) juxtaposed to the nucleus. Interestingly, protein inclusions associated with several neurodegenerative diseases bear striking resemblance to aggresomes [63–65]. As aggresomes are seemingly inert structures, their formation could be considered as a proactive way by which the cell deals with its non-disposable proteins, and thereby a protective response by the cell in times of proteolytic stress. Consistent with this, several studies supporting a neuroprotective role for inclusions formation have emerged recently [35, 66–68]. These studies collectively suggest that the accumulation of disease-associated proteins in a diffuse non-aggregated form is more toxic than when they are sequestered into inclusion bodies. Notwithstanding this, inclusion bodies and aggresomes alike are space-filling entities that could physically disrupt cellular function if their size were to exceed a certain threshold. Conceivably, the growth of an aggresome/inclusion body is a regulated process. Emerging evidence implicates macroautophagy (hereafter referred to as autophagy), a lysosome-mediated bulk degradation system, as a key regulator of inclusion dynamics. Autophagy is an evolutionarily conserved process for subcellular degradation and serves as a cellular survival mechanism in times of starvation. Excellent reviews regarding this pathway have been written elsewhere (for example, see [69]). Briefly, cytosolic materials targeted by autophagic degradation are first sequestered by a double-membrane vesicle known as the autophagosome. Following the fusion of the outer membrane of this structure with a lysosome, the cargo-containing inner membrane is then lysed and its contents degraded by lysosomal proteases. Morphological evidence of autophagy is present in several neurodegenerative disorders, including AD, PD and Huntington's disease [70, 71]. Notably, two independent groups who analyzed the consequence of ablating autophagy function specifically in the neural cells of mice clearly indicated that autophagy dysfunction is a key precipitator of protein inclusions in neurons and that impaired clearance of neural inclusions is intimately associated with neurodegeneration [72, 73]. Further, several other studies demonstrated that pharmacological activation of autophagy promotes the degradation of aggregation-prone proteins, including huntingtin and tau mutants, and concomitantly improve cellular survivability [74–77]. Relevant to this is the proposal that aggresome formation facilitates the delivery of dispersed protein aggregates to the autophagic pathway [78]. However, whether autophagy is clearing monomeric and oligomeric precursors of aggregates, or aggresomes is currently controversial. For example, work from Rubinsztein's laboratory suggests that autophagy can efficiently target mutant species of α-synuclein that do not form aggregates large enough to be seen with light microscopy [79]. On the other hand, Fortun et al demonstrated that autophagy inhibition retards the clearance of aggresome clearance formed by mutant peripheral myelin protein PMP22 [80]. Similarly, Iwata et al showed that aggresomes generated by mutant huntingtin stain positively for several autophagy-linked proteins and are amenable to clearance by autophagy [81]. More recently, a live cell imaging study by a German group suggests that the sequestration of α–synuclein into aggresomes facilitates its clearance from the cell [82]. It is thus tempting to think that autophagy could clear proteins of all conformations, particularly under conditions of proteasome impairment. However, unlike the UPS where the mechanism of cargo selection is well defined, an important mystery to be solved is how cargo is selected for autophagic degradation. At least with aggresomes, our recent studies, as discussed below, have illuminated the possibility that K63-linked polyubiquitin might also act as a cargo recognition signal for the autophagic system.

3.6 K63-Linked Polyubiquitination – A Novel Cargo Recognition Signal for Autophagy-Mediated Degradation

An important clue pointing to the participation of protein ubiquitination in the clearance of aggresomes by autophagy is that these structures are frequently enriched with p62. We have discussed earlier about the ubiquitin binding properties of p62, i.e. that it preferentially binds long K63-linked ubiquitin chains. Importantly, p62 has been demonstrated recently by Johansen's group to enhance the autophagic clearance of aggresome-inclusions via its direct interaction with the autophagosome marker, Light Chain 3 (LC3) [83, 84], suggesting that p62 facilitates the recruitment of the autophagy machinery to aggresomes. Consistent with this, loss of p62 in mice leads to the hyper-accumulation of K63-linked chains in the insoluble fraction of their brains, where protein aggregates usually reside [85]. In view of the relationship between p62 and K63-linked polyubiquitin, we wondered whether K63-linked polyubiquitin could potentially partner with p62, thereby acting as a cargo selection signal for autophagic degradation. We found that ubiquitin-positive, aggresome-like inclusions formed in cultured cells ectopically expressing either K63 ubiquitin mutant or the heterodimeric Ubc13/Uev1a E2 pair are rapidly cleared when autophagy is induced [56, 58]. In contrast, those generated in cells ectopically expressing either K63R ubiquitin mutant (which prevents K63-linked ubiquitin chain assembly), UbcH7 or UbcH8 resisted autophagy-mediated removal [56, 58]. It thus appears that K63-linked polyubiquitin could help provide the bridge between the autophagy machinery and aggresomes. Further supporting this, we also found that K63 ubiquitin-positive inclusions exhibit a predominant tendency (∼80%) to colocalize with the lysosomal membrane protein LAMP-1 even in the absence of autophagy induction, whereas only a fraction (∼20%) of K63R-ubiquitin-positive inclusions stains positively with LAMP-1 under the same condition [56]. This suggests an efficient recruitment of lysosomal structures to K63 polyubiquitinated proteins during aggresome formation, which concomitantly primes the structure for autophagic clearance. It is thus tempting to suggest that p62 and K63-linked polyubiquitin form a partnership to promote the clearance of aggresome-like inclusions by autophagy.

It is noteworthy to highlight that besides p62, the microtubule-associated deacetylase HDAC6 has also been proposed to connect aggresomes with the autophagy machinery via the ubiquitin molecule [86, 87]. Like p62, HDAC6 interacts with polyubiquitinated misfolded proteins but binds additionally to dynein, a minus end-directed microtubule motor that moves cargo from the cell periphery towards its center [88]. Dynein mutations have previously been demonstrated to impair autophagic clearance of aggregation-prone proteins [88]. By virtue of its dual interaction with dynein and ubiquitinated proteins, HDAC6 not only helps to transport aggregates to the MTOC, but also appears to facilitate the delivery of autophagy apparatus to aggresomes [86, 87]. Indeed, cells deficient in HDAC6 cannot form aggresomes properly and fail to clear misfolded proteins from the cytoplasm [87]. Because HDAC6 has the ability to bind polyubiquitinated misfolded proteins, the obvious question here is whether different ubiquitin chain topologies would influence their interaction. The answer in part, has been provided by a recent report by Olzmann et al, who identified parkin-mediated K63-linked ubiquitination as a signal that couples misfolded proteins to the dynein complex via HDAC6 and thereby promoting the sequestration of proteins into aggresomes and subsequent clearance by autophagy [89]. This finding by Olzmann lends further support to our demonstration that K63-linked polyubiquitination is a key regulator of aggresomes formation and clearance.

3.7 A Model of Inclusion Biogenesis and Clearance

Taking all the findings that we have discussed above together, we could envisage a model of inclusion biogenesis and clearance whereby the nature of ubiquitin topology would play a deterministic role in deciding the fate of a misfolded protein (Fig. 3.2). In this model, we embrace the widely accepted notion that the first cellular quality control system a nascent misfolded protein would encounter is the chaperone system (*more information on this system is provided in Sect. 8.4*). Failure of the chaperone system to refold the malfolded protein would then trigger its ubiquitination via proteasome-associated K48-linked chains, although the precise mechanism governing this cellular triage decision is not entirely clear at this moment. Under normal conditions, the K48-linked polyubiquitinated substrate would be targeted for degradation by the proteasome. However, under conditions whereby the proteasome becomes overwhelmed or otherwise impaired, misfolded proteins to be modified by ubiquitination will receive one that is uncoupled from the proteasome. Although our results favor the participation of K63-linked polyubiquitination under these circumstances, it is possible that monoubiquitination might also have a role here. Furthermore, a very recent study demonstrated that monoubiquitination of α-synuclein promotes its aggregation into inclusion bodies [90]. Alternatively, K48-linked polyubiquitinated substrate could be remodeled, firstly by a DUB and subsequently by an E2/E3 pair that specify K63 ubiquitin linkages. In this scenario, K48 ubiquitin chains on proteins that are unable to be degraded by the proteasome are hydrolyzed by DUB to a monoubiquitin, from which K63-linked chains could be subsequently extended. Indeed, evidence exists for a specific DUB enzyme that

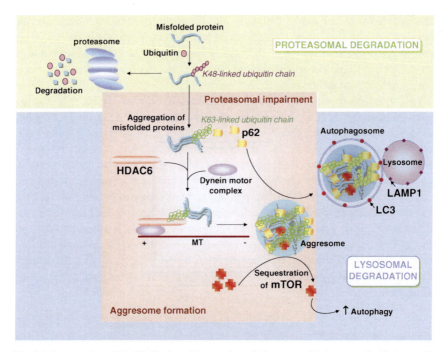

Fig. 3.2 Proposed model of inclusions biogenesis and clearance. Under normal cellular conditions, proteins destined for degradation by the proteasome are tagged with a chain of K48-linked ubiquitin. In times of proteolytic stress, the cell switches to K63-linked ubiquitination to divert the protein load originally targeted for proteasomal degradation away from the otherwise overloaded machinery. Alternatively, K48-linked polyubiquitinated substrate could be remodeled under conditions of stress, firstly by a DUB and subsequently by an E2/E3 pair that specify K63 ubiquitin linkages. K63-linked polyubiquitin, through its interactions with HDAC6 and p62, helps sequestering the diverted protein load into an aggresome. At the same time, K63-linked polyubiquitin also acts as a cargo recognition signal, presumably in partnership with p62, for the autophagy machinery thereby facilitating the clearance of aggresomes *via* lysosomal-mediated degradation

could readily disassemble polyubiquitin chains to leave behind a monoubiquitin remnant at the site of the original chain [42]. Being non-proteolytic, K63-linked polyubiquitination of misfolded proteins originally destined for proteasomal degradation would serve to divert the substrates away from an otherwise overloaded machinery. However, in doing so, the population of ubiquitinated misfolded proteins will invariably increase. By virtue of its association with HDAC6 and also p62, K63-linked polyubiquitin would have the ability to re-route these protein cargoes towards the MTOC to facilitate the formation of an aggresome, a move that can be considered as cytoprotective. From this transit station, the team comprising of K63-linked polyubiquitin, p62, HDAC6 then acts to ensure that the protein cargoes make a final exit from the cell by linking them to the autophagic machinery. Again, the precise mechanism regarding how this tripartite partnership works remains to be elucidated. From our proposed model, it is apparent that the dynamism of ubiquitin modification would enable this simple molecule to govern the fate of misfolded proteins at different stages of their life cycle in the cell. Without this dynamism, the cell would be denied of the flexibility to manage its misfolded protein load differently under different circumstances.

3.8 E2/E3 Pairs – Triage Officers?

An important gap that is apparent in our proposed model is the current lack of information regarding the existence and the nature of the molecular switch that mediates a cell's decision to degrade its misfolded proteins either via the proteasomal or lysosomal pathway. While the identity of this triage decision maker remains to be elucidated, we believe that the executors almost certainly would involve E2s and E3s. Because Ubc13, in association with either of the E2 variants, Uev1a or Mms2, is the only E2 known to date to mediate the formation of K63-linked ubiquitin chains [57], its role in executing the triage between proteasomal or lysosomal degradation is immediately implied. Indeed, we have demonstrated that over expression of heterodimeric Ubc13/Uev1a pair alone is sufficient to promote inclusions formation and their subsequent clearance by autophagy [58]. Accordingly, E3 enzymes identified thus far that partners with Ubc13 to perform K63-linked polyubiquitination, including TRAF6 [91], CHIP [92] and parkin [52], are attractive candidates for triage officers. Interestingly, both CHIP and parkin are also capable of assembling ubiquitin chains alternative to those linked via K63. This dual function is dependent on the E2 they interact with. For example, parkin can bind to UbcH7 or H8 to mediate K48-linked ubiquitin chains [27–29] or to Ubc13/UeV1a to mediate K63-linked ubiquitin chains [52, 89]. Likewise, CHIP is also capable of forming K48-linked or K63-linked ubiquitin chains by switching between its E2 partners, UbcH5a and Ubc13/UeV1a [92]. Certainly, it would be important to elucidate the factors governing the choice of E2 an E3 chooses to partner with, which remains elusive at this moment. Given their multifunctional ligase activity, and their intimate relationship to protein aggregation diseases, both parkin and CHIP would offer themselves as strong contenders to undertake the role of triaging misfolded proteins. Indeed, a recent study suggest that CHIP mediates α-synuclein degradation decisions between proteasomal and lysosomal pathways [93]. Apparently, the decision to degrade α-synuclein via autophagy is dependent on the U-box domain of CHIP, which contains the E3 ligase activity. The authors suggest, albeit implicitly, that ubiquitin ligation via K63 is involved in this decision. As for parkin, we like to think that it may have an analogous role to CHIP for the following reasons: 1) Parkin is a multifunctional E3 ligase capable of mediating both degradation-associated and non-proteolytic forms of ubiquitination (presumably under different conditions), 2) Parkin-mediated K63-linked ubiquitination facilitates the formation of aggresomes and their subsequent clearance by autophagy, 3) LBs associated with PD are frequently absent in cases involving parkin mutations, suggesting that parkin-mediated ubiquitination is needed for inclusions biogenesis, and finally 4) parkin interacts with CHIP directly and this interaction has been shown to enhance parkin-mediated ubiquitination [26]. A role for parkin in the triage of misfolded proteins between proteasomal and lysosomal degradation thus appears compelling to us.

3.9 Conclusions

Mechanistically, the involvement of non-proteolytic modes of ubiquitination in protein inclusion formation is attractive as it provides a potential mechanism by which ubiquitinated protein aggregates evade proteasomal degradation and conceivably, a protective route by which misfolded or aggregated proteins could be diverted away from an overloaded proteasome to be subsequently handled by the autophagic machinery. Although the elegance of a proposal is never to be taken as a proof of concept, current evidences from our laboratory and others strongly support a non-trivial role for K63-linked ubiquitination in regulating the dynamics of protein aggregation. The participation of this unconventional ubiquitin modification as a key manager of protein quality control in the cell that is otherwise performed by the classic K48-linked ubiquitin chain is a testimony to the versatility of ubiquitination. It is indeed remarkable that a humble ubiquitin molecule could have such an extraordinary kaleidoscopic personality.

Acknowledgments The work in my laboratory covered in this chapter is funded by A*STAR Biomedical Research Council and National Medical Research Council. E.W and J.T. are recipients of postdoctoral fellowship awarded by the Parkinson's disease Foundation (USA) and the Singapore Millennium Foundation, respectively.

Abbreviations

AD, Alzheimer's disease; ALS, Amyotrophic Lateral Sclerosis; DUB, deubiquitinating enzyme; FTD, Frontotemporal Dementia; LB, Lewy body; PD, Parkinson's disease; MTOC, microtubule organizing center; UPS, ubiquitin-proteasome system.

References

1. Schubert U, Anton LC, Gibbs J, Norbury CC, Yewdell JW, Bennink JR (2000) Rapid degradation of a large fraction of newly synthesized proteins by proteasomes. Nature 404:770–774
2. Goldberg AL (2003) Protein degradation and protection against misfolded or damaged proteins. Nature 426:895–899
3. Goldberg AL (1972) Degradation of abnormal proteins in *Escherichia coli* (protein breakdown-protein structure-mistranslation-amino acid analogs-puromycin). Proc Natl Acad Sci USA 69:422–426
4. Hershko A, Ciechanover A (1998) The ubiquitin system. Annu Rev Biochem 67:425–479
5. Chau V, Tobias JW, Bachmair A, Marriott D, Ecker DJ, Gonda DK, Varshavsky A (1989) A multiubiquitin chain is confined to specific lysine in a targeted short-lived protein. Science 243:1576–1583

6. Pickart CM, Cohen RE (2004) Proteasomes and their kin: proteases in the machine age. Nat Rev Mol Cell Biol 5:177–187
7. Braun BC, Glickman M, Kraft R, Dahlmann B, Kloetzel PM, Finley D, Schmidt M (1999) The base of the proteasome regulatory particle exhibits chaperone-like activity. Nat Cell Biol 1:221–226
8. Glickman MH, Rubin DM, Coux O, Wefes I, Pfeifer G, Cjeka Z, Baumeister W, Fried VA, Finley D (1998) A subcomplex of the proteasome regulatory particle required for ubiquitin-conjugate degradation and related to the COP9-signalosome and eIF3. Cell 94:615–623
9. Navon A, Goldberg AL (2001) Proteins are unfolded on the surface of the ATPase ring before transport into the proteasome. Mol Cell 8:1339–1349
10. Wilkinson KD (1997) Regulation of ubiquitin-dependent processes by deubiquitinating enzymes. Faseb J 11:1245–1256
11. Smith DM, Benaroudj N, Goldberg A (2006) Proteasomes and their associated ATPases: a destructive combination. J Struct Biol 156:72–83
12. Forman MS, Trojanowski JQ, Lee VM (2004) Neurodegenerative diseases: a decade of discoveries paves the way for therapeutic breakthroughs. Nat Med 10:1055–1063
13. Ross CA, Poirier MA (2004) Protein aggregation and neurodegenerative disease. Nat Med 10 (Suppl):S10–S17
14. Bence NF, Sampat RM, Kopito RR (2001) Impairment of the ubiquitin-proteasome system by protein aggregation. Science 292:1552–1555
15. Bennett EJ, Shaler TA, Woodman B, Ryu KY, Zaitseva TS, Becker CH, Bates GP, Schulman H, Kopito RR (2007) Global changes to the ubiquitin system in Huntington's disease. Nature 448:704–708
16. Sherman MY, Goldberg AL (2001) Cellular defenses against unfolded proteins: a cell biologist thinks about neurodegenerative diseases. Neuron 29:15–32
17. McNaught KS, Belizaire R, Isacson O, Jenner P, Olanow CW (2003) Altered proteasomal function in sporadic Parkinson's disease. Exp Neurol 179:38–46
18. McNaught KS, Jenner P (2001) Proteasomal function is impaired in substantia nigra in Parkinson's disease. Neurosci Lett 297:191–194
19. Lam YA, Pickart CM, Alban A, Landon M, Jamieson C, Ramage R, Mayer RJ, Layfield R (2000) Inhibition of the ubiquitin-proteasome system in Alzheimer's disease. Proc Natl Acad Sci USA 97:9902–9906
20. Keller JN, Hanni KB, Markesbery WR (2000) Impaired proteasome function in Alzheimer's disease. J Neurochem 75:436–439
21. McNaught KS, Perl DP, Brownell AL, Olanow CW (2004) Systemic exposure to proteasome inhibitors causes a progressive model of Parkinson's disease. Ann Neurol 56:149–162
22. Miwa H, Kubo T, Suzuki A, Nishi K, Kondo T (2005) Retrograde dopaminergic neuron degeneration following intrastriatal proteasome inhibition. Neurosci Lett 380:93–98
23. Venkatraman P, Wetzel R, Tanaka M, Nukina N, Goldberg AL (2004) Eukaryotic proteasomes cannot digest polyglutamine sequences and release them during degradation of polyglutamine-containing proteins. Mol Cell 14:95–104
24. Verhoef LG, Lindsten K, Masucci MG, Dantuma NP (2002) Aggregate formation inhibits proteasomal degradation of polyglutamine proteins. Hum Mol Genet 11:2689–2700
25. Kitada T, Asakawa S, Hattori N, Matsumine H, Yamamura Y, Minoshima S, Yokochi M, Mizuno Y, Shimizu N (1998) Mutations in the parkin gene cause autosomal recessive juvenile parkinsonism. Nature 392:605–608
26. Imai Y, Soda M, Takahashi R (2000) Parkin suppresses unfolded protein stress-induced cell death through its E3 ubiquitin–protein ligase activity. J Biol Chem 275:35661–35664
27. Shimura H, Hattori N, Kubo S, Mizuno Y, Asakawa S, Minoshima S, Shimizu N, Iwai K, Chiba T, Tanaka K et al. (2000) Familial Parkinson disease gene product, parkin, is a ubiquitin–protein ligase. Nat Genet 25:302–305
28. Zhang Y, Gao J, Chung KK, Huang H, Dawson VL, Dawson TM (2000) Parkin functions as an E2-dependent ubiquitin–protein ligase and promotes the degradation of the synaptic vesicle-associated protein, CDCrel-1. Proc Natl Acad Sci USA 97:13354–13359

29. Imai Y, Soda M, Inoue H, Hattori N, Mizuno Y, Takahashi R (2001) An unfolded putative transmembrane polypeptide, which can lead to endoplasmic reticulum stress, is a substrate of Parkin. Cell 105:891–902
30. Ko HS, Kim SW, Sriram SR, Dawson VL, Dawson TM (2006) Identification of far upstream element-binding protein-1 as an authentic Parkin substrate. J Biol Chem 281:16193–16196
31. Ko HS, von Coelln R, Sriram SR, Kim SW, Chung KK, Pletnikova O, Troncoso J, Johnson B, Saffary R, Goh EL et al. (2005) Accumulation of the authentic parkin substrate aminoacyl-tRNA synthetase cofactor, p38/JTV-1, leads to catecholaminergic cell death. J Neurosci 25:7968–7978
32. Staropoli JF, McDermott C, Martinat C, Schulman B, Demireva E, Abeliovich A (2003) Parkin is a component of an SCF-like ubiquitin ligase complex and protects postmitotic neurons from kainate excitotoxicity. Neuron 37:735–749
33. Leroy E, Boyer R, Auburger G, Leube B, Ulm G, Mezey E, Harta G, Brownstein MJ, Jonnalagada S, Chernova T et al. (1998) The ubiquitin pathway in Parkinson's disease. Nature 395:451–452
34. Healy DG, Abou-Sleiman PM, Wood NW (2004) Genetic causes of Parkinson's disease: UCHL-1. Cell Tissue Res 318:189–194
35. Bowman AB, Yoo SY, Dantuma NP, Zoghbi HY (2005) Neuronal dysfunction in a polyglutamine disease model occurs in the absence of ubiquitin-proteasome system impairment and inversely correlates with the degree of nuclear inclusion formation. Hum Mol Genet 14:679–691
36. Tofaris GK, Razzaq A, Ghetti B, Lilley KS, Spillantini MG (2003) Ubiquitination of alpha-synuclein in Lewy bodies is a pathological event not associated with impairment of proteasome function. J Biol Chem 278:44405–44411
37. Pickart CM (2000) Ubiquitin in chains. Trends Biochem Sci 25:544–548
38. Baboshina OV, Haas AL (1996) Novel multiubiquitin chain linkages catalyzed by the conjugating enzymes E2EPF and RAD6 are recognized by 26S proteasome subunit 5. J Biol Chem 271:2823–2831
39. Arnason T, Ellison MJ (1994) Stress resistance in Saccharomyces cerevisiae is strongly correlated with assembly of a novel type of multiubiquitin chain. Mol Cell Biol 14:7876–7883
40. Peng J, Schwartz D, Elias JE, Thoreen CC, Cheng D, Marsischky G, Roelofs J, Finley D, Gygi SP (2003) A proteomics approach to understanding protein ubiquitination. Nat Biotechnol 21:921–926
41. Crosas B, Hanna J, Kirkpatrick DS, Zhang DP, Tone Y, Hathaway NA, Buecker C, Leggett DS, Schmidt M, King RW et al. (2006) Ubiquitin chains are remodeled at the proteasome by opposing ubiquitin ligase and deubiquitinating activities. Cell 127:1401–13
42. Hanna J, Hathaway NA, Tone Y, Crosas B, Elsasser S, Kirkpatrick DS, Leggett DS, Gygi SP, King RW, Finley D (2006) Deubiquitinating enzyme Ubp6 functions noncatalytically to delay proteasomal degradation. Cell 127:99–111
43. Li M, Chen D, Shiloh A, Luo J, Nikolaev AY, Qin J, Gu W (2002) Deubiquitination of p53 by HAUSP is an important pathway for p53 stabilization. Nature 416:648–653
44. Hicke L, Schubert HL, Hill CP (2005) Ubiquitin-binding domains. Nat Rev Mol Cell Biol 6:610–621
45. Aguilar RC, Wendland B (2003) Ubiquitin: not just for proteasomes anymore. Curr Opin Cell Biol 15:184–190
46. Lim KL, Dawson VL, Dawson TM (2006) Parkin-mediated lysine 63-linked polyubiquitination: a link to protein inclusions formation in Parkinson's and other conformational diseases? Neurobiol Aging 27:524–529
47. Hayashi S, Wakabayashi K, Ishikawa A, Nagai H, Saito M, Maruyama M, Takahashi T, Ozawa T, Tsuji S, Takahashi H (2000) An autopsy case of autosomal-recessive juvenile parkinsonism with a homozygous exon 4 deletion in the parkin gene. Mov Disord 15:884–888

48. Mori H, Kondo T, Yokochi M, Matsumine H, Nakagawa-Hattori Y, Miyake T, Suda K, Mizuno Y (1998) Pathologic and biochemical studies of juvenile parkinsonism linked to chromosome 6q. Neurology 51:890–892
49. Takahashi H, Ohama E, Suzuki S, Horikawa Y, Ishikawa A, Morita T, Tsuji S, Ikuta F (1994) Familial juvenile parkinsonism: clinical and pathologic study in a family. Neurology 44: 437–441
50. Chung KK, Zhang Y, Lim KL, Tanaka Y, Huang H, Gao J, Ross CA, Dawson VL, Dawson TM (2001) Parkin ubiquitinates the alpha-synuclein-interacting protein, synphilin-1: implications for Lewy-body formation in Parkinson disease. Nat Med 7:1144–1150
51. Lim KL, Chew KC, Tan JM, Wang C, Chung KK, Zhang Y, Tanaka Y, Smith W, Engelender S, Ross CA et al. (2005) Parkin mediates nonclassical, proteasomal-independent ubiquitination of synphilin-1: implications for Lewy body formation. J Neurosci 25:2002–2009
52. Doss-Pepe EW, Chen L, Madura K (2005) Alpha-synuclein and parkin contribute to the assembly of ubiquitin lysine 63-linked multiubiquitin chains. J Biol Chem 280:16619–16624
53. Hampe C, Ardila-Osorio H, Fournier M, Brice A, Corti O (2006) Biochemical analysis of Parkinson's disease-causing variants of Parkin, an E3 ubiquitin–protein ligase with monoubiquitylation capacity. Hum Mol Genet 15:2059–2075
54. Matsuda N, Kitami T, Suzuki T, Mizuno Y, Hattori N, Tanaka K (2006) Diverse effects of pathogenic mutations of Parkin that catalyze multiple monoubiquitylation in vitro. J Biol Chem 281:3204–3209
55. Liu Y, Fallon L, Lashuel HA, Liu Z, Lansbury PT Jr. (2002) The UCH-L1 gene encodes two opposing enzymatic activities that affect alpha-synuclein degradation and Parkinson's disease susceptibility. Cell 111:209–218
56. Tan JM, Wong ES, Kirkpatrick DS, Pletnikova O, Ko HS, Tay SP, Ho MW, Troncoso J, Gygi SP, Lee MK et al. (2008) Lysine 63-linked ubiquitination promotes the formation and autophagic clearance of protein inclusions associated with neurodegenerative diseases. Hum Mol Genet 17:431–439
57. Hofmann RM, Pickart CM (1999) Noncanonical MMS2-encoded ubiquitin-conjugating enzyme functions in assembly of novel polyubiquitin chains for DNA repair. Cell 96: 645–653
58. Tan JM, Wong ES, Dawson VL, Dawson TM, Lim KL (2008) Lysine 63-linked polyubiquitin potentially partners with p62 to promote the clearance of protein inclusions by autophagy. Autophagy 4:251–253
59. Seibenhener ML, Babu JR, Geetha T, Wong HC, Krishna NR, Wooten MW (2004) Sequestosome 1/p62 is a polyubiquitin chain binding protein involved in ubiquitin proteasome degradation. Mol Cell Biol 24:8055–8068
60. Raasi S, Varadan R, Fushman D, Pickart CM (2005) Diverse polyubiquitin interaction properties of ubiquitin-associated domains. Nat Struct Mol Biol 12:708–714
61. Long J, Gallagher TR, Cavey JR, Sheppard PW, Ralston SH, Layfield R, Searle MS (2008) Ubiquitin recognition by the ubiquitin-associated domain of p62 involves a novel conformational switch. J Biol Chem 283:5427–5440
62. Johnston JA, Ward CL, Kopito RR (1998) Aggresomes: a cellular response to misfolded proteins. J Cell Biol 143:1883–1898
63. Mishra RS, Bose S, Gu Y, Li R, Singh N (2003) Aggresome formation by mutant prion proteins: the unfolding role of proteasomes in familial prion disorders. J Alzheimers Dis 5: 15–23
64. Olanow CW, Perl DP, DeMartino GN, McNaught KS (2004) Lewy-body formation is an aggresome-related process: a hypothesis. Lancet Neurol 3:496–503
65. Waelter S, Boeddrich A, Lurz R, Scherzinger E, Lueder G, Lehrach H, Wanker EE (2001) Accumulation of mutant huntingtin fragments in aggresome-like inclusion bodies as a result of insufficient protein degradation. Mol Biol Cell 12:1393–1407
66. Arrasate M, Mitra S, Schweitzer ES, Segal MR, Finkbeiner S (2004) Inclusion body formation reduces levels of mutant huntingtin and the risk of neuronal death. Nature 431:805–810

67. Cummings CJ, Reinstein E, Sun Y, Antalffy B, Jiang Y, Ciechanover A, Orr HT, Beaudet AL, Zoghbi HY (1999) Mutation of the E6-AP ubiquitin ligase reduces nuclear inclusion frequency while accelerating polyglutamine-induced pathology in SCA1 mice. Neuron 24:879–892
68. Klement IA, Skinner PJ, Kaytor MD, Yi H, Hersch SM, Clark HB, Zoghbi HY, Orr HT (1998) Ataxin-1 nuclear localization and aggregation: role in polyglutamine-induced disease in SCA1 transgenic mice. Cell 95:41–53
69. Klionsky DJ (2007) Autophagy: from phenomenology to molecular understanding in less than a decade. Nat Rev Mol Cell Biol 8:931–937
70. Cuervo AM (2006) Autophagy in neurons: it is not all about food. Trends Mol Med 12: 461–464
71. Rubinsztein DC, Difiglia M, Heintz N, Nixon RA, Qin ZH, Ravikumar B, Stefanis L, Tolkovsky A (2005) Autophagy and its possible roles in nervous system diseases, damage and repair. Autophagy 1:11–22
72. Hara T, Nakamura K, Matsui M, Yamamoto A, Nakahara Y, Suzuki-Migishima R, Yokoyama M, Mishima K, Saito I, Okano H et al. (2006) Suppression of basal autophagy in neural cells causes neurodegenerative disease in mice. Nature 441:885–889
73. Komatsu M, Waguri S, Chiba T, Murata S, Iwata J, Tanida I, Ueno T, Koike M, Uchiyama Y, Kominami E et al. (2006) Loss of autophagy in the central nervous system causes neurodegeneration in mice. Nature 441:880–884
74. Berger Z, Ravikumar B, Menzies FM, Oroz LG, Underwood BR, Pangalos MN, Schmitt I, Wullner U, Evert BO, O'Kane CJ et al. (2006) Rapamycin alleviates toxicity of different aggregate-prone proteins. Hum Mol Genet 15:433–442
75. Ravikumar B, Berger Z, Vacher C, O'Kane CJ, Rubinsztein DC (2006) Rapamycin pre-treatment protects against apoptosis. Hum Mol Genet 15:1209–1216
76. Ravikumar B, Vacher C, Berger Z, Davies JE, Luo S, Oroz LG, Scaravilli F, Easton DF, Duden R, O'Kane CJ et al. (2004) Inhibition of mTOR induces autophagy and reduces toxicity of polyglutamine expansions in fly and mouse models of Huntington disease. Nat Genet 36: 585–595
77. Sarkar S, Davies JE, Huang Z, Tunnacliffe A, Rubinsztein DC (2007) Trehalose, a novel mTOR-independent autophagy enhancer, accelerates the clearance of mutant huntingtin and alpha-synuclein. J Biol Chem 282:5641–5652
78. Kopito RR (2000) Aggresomes, inclusion bodies and protein aggregation. Trends Cell Biol 10:524–530
79. Webb JL, Ravikumar B, Atkins J, Skepper JN, Rubinsztein DC (2003) Alpha-Synuclein is degraded by both autophagy and the proteasome. J Biol Chem 278:25009–25013
80. Fortun J, Dunn WA Jr., Joy S, Li J, Notterpek L (2003) Emerging role for autophagy in the removal of aggresomes in Schwann cells. J Neurosci 23:10672–10680
81. Iwata A, Christianson JC, Bucci M, Ellerby LM, Nukina N, Forno LS, Kopito RR (2005) Increased susceptibility of cytoplasmic over nuclear polyglutamine aggregates to autophagic degradation. Proc Natl Acad Sci USA 102:13135–13140
82. Opazo F, Krenz A, Heermann S, Schulz JB, Falkenburger BH (2008) Accumulation and clearance of alpha-synuclein aggregates demonstrated by time-lapse imaging. J Neurochem. 106:529–540
83. Bjorkoy G, Lamark T, Brech A, Outzen H, Perander M, Overvatn A, Stenmark H, Johansen T (2005) p62/SQSTM1 forms protein aggregates degraded by autophagy and has a protective effect on huntingtin-induced cell death. J Cell Biol 171:603–614
84. Pankiv S, Clausen TH, Lamark T, Brech A, Bruun JA, Outzen H, Overvatn A, Bjorkoy G, Johansen T (2007) p62/SQSTM1 Binds Directly to Atg8/LC3 to Facilitate Degradation of Ubiquitinated Protein Aggregates by Autophagy. J Biol Chem 282:24131–24145
85. Wooten MW, Geetha T, Babu JR, Seibenhener ML, Peng J, Cox N, Diaz-Meco MT, Moscat J (2008) Essential role of sequestosome 1/p62 in regulating accumulation of Lys63-ubiquitinated proteins. J Biol Chem 283:6783–6789

86. Iwata A, Riley BE, Johnston JA, Kopito RR (2005) HDAC6 and microtubules are required for autophagic degradation of aggregated huntingtin. J Biol Chem 280:40282–40292
87. Kawaguchi Y, Kovacs JJ, McLaurin A, Vance JM, Ito A, Yao TP (2003) The deacetylase HDAC6 regulates aggresome formation and cell viability in response to misfolded protein stress. Cell 115:727–738
88. Ravikumar B, Acevedo-Arozena A, Imarisio S, Berger Z, Vacher C, O'Kane CJ, Brown SD, Rubinsztein DC (2005) Dynein mutations impair autophagic clearance of aggregate-prone proteins. Nat Genet 37:771–776
89. Olzmann JA, Li L, Chudaev MV, Chen J, Perez FA, Palmiter RD, Chin LS (2007) Parkin-mediated K63-linked polyubiquitination targets misfolded DJ-1 to aggresomes via binding to HDAC6. J Cell Biol 178:1025–1038
90. Rott R, Szargel R, Haskin J, Shani V, Shainskaya A, Manov I, Liani E, Avraham E, Engelender S (2008) Monoubiquitylation of alpha-synuclein by seven in absentia homolog (SIAH) promotes its aggregation in dopaminergic cells. J Biol Chem 283:3316–3328
91. Deng L, Wang C, Spencer E, Yang L, Braun A, You J, Slaughter C, Pickart C, Chen ZJ (2000) Activation of the IkappaB kinase complex by TRAF6 requires a dimeric ubiquitin-conjugating enzyme complex and a unique polyubiquitin chain. Cell 103:351–361
92. Zhang M, Windheim M, Roe SM, Peggie M, Cohen P, Prodromou C, Pearl LH (2005) Chaperoned ubiquitylation–crystal structures of the CHIP U box E3 ubiquitin ligase and a CHIP-Ubc13-Uev1a complex. Mol Cell 20:525–538
93. Shin Y, Klucken J, Patterson C, Hyman BT, McLean PJ (2005) The co-chaperone carboxyl terminus of Hsp70-interacting protein (CHIP) mediates alpha-synuclein degradation decisions between proteasomal and lysosomal pathways. J Biol Chem 280:23727–23734

Chapter 4
Protein Misfolding and Axonal Protection in Neurodegenerative Diseases

Haruhisa Inoue, Takayuki Kondo, Ling Lin, Sha Mi, Ole Isacson and Ryosuke Takahashi

Abstract Genetically engineered mouse model studies show that neuronal dysfunction caused by protein aggregation/misfolding are reversible, indicating that injured neurons are alive even under disease states. Protein misfolding/aggregation in axons and distal dominant axonal degeneration are observed in a subgroup of degenerative diseases and in certain experimental conditions. Moreover, therapeutic approaches towards axonal protection are effective in neurodegenerative disease mouse models; (a) axonal regeneration, (b) anti-Wallerian degeneration, (c) autophagy enhancement, and (d) stabilization of microtubules. These studies demonstrate that axonal protection/functional repair of axons can be general therapeutic interventions for neurodegenerative diseases.

4.1 Neuronal Dysfunction in Neurodegeneration is a Reversible Process

It had been believed that neurodegeneration is not reversible. However, recent studies of transgenic mouse models, which express abnormal proteins associated with Alzheimer's disease, diffuse Lewy body disease, Parkinson's disease (PD), Huntington's disease (HD) and tauopathies such as frontotemporal dementia develop distinct disease-related neurological impairments, elegantly show that some neurological deficits of neurodegenerative cascades can be prevented or reversed by removing abnormal proteins, without obvious alteration of the number of neuronal cell bodies [1]. Thus, neurological impairments that are associated with neurodegenerative conditions might be caused by neuronal dysfunction to some extent rather than neuronal loss. These studies also demonstrate that symptoms arise from neuronal dysfunction which precedes neuronal death [1]. In HD mice model, as in most of the other triplet repeat diseases, the mutant huntingtin proteins form misfolded nuclear aggregates,

---

H. Inoue (✉)
Department of Neurology, Kyoto University Graduate School of Medicine, 54 Kawahara-cho Shogoin, Sakyo-ku, Kyoto 606-8507, Japan
e-mail: haruhisa@kuhp.kyoto-u.ac.jp

which are highly insoluble. The double mutant huntingtin transgenic mice, in which the bidirectional transgene expression is activated by the removal of doxycycline at birth, express high levels of both mutant huntingtin and lacZ in the striatum, cortex, and hippocampus [2, 3]. Most of striatal neurons are stained with an anti-huntingtin antibody, showing diffuse nuclear aggregates. By 8 weeks of age, striatal morphological alterations in the mutant huntingtin transgenic mice include a reduced size, reactive gliosis, and a decrease in D1 receptors (a feature seen in HD patients). All of the mice at this age also show a behavioral abnormality common to mouse models of HD: when suspended by their tails, they clasp their limbs. This behavioral phenotype was aggravated over time. Neuropathological examination demonstrated the colocalization of various molecular chaperones, ubiquitin, and proteasome subunits with the aggregated proteins. Surprisingly, abolishing the expression of mutant huntingtin by Cre-loxP system in mutant huntingtin transgenic mice with neurodegenerative phenotype results in either a halt of the disease progression or a full recovery from the disease phenotype including pathological changes [2, 3]. This observation indicates that irreversible changes that commit the neurons to persistent dysfunction or death do not necessarily take place in the neurodegenerative process. These observations suggest that therapeutic approaches aiming at elimination of misfolded proteins might be effective in treating neuronal dysfunction. Furthermore, the recovery from motor disturbances indicates that plastic changes can occur when the toxic insult ceases [3]. Recent studies of mutant huntingtin transgenic mice show that the neuronal dysfunction may be caused by misfolded mutant huntingtin protein, at synaptosomal proteasome and mitochondria, which seem to trigger vicious cycles of aberrant neuronal activity [4].

4.2 Neuronal Dysfunction Is Not Treatable by Anti-Cell Death Therapy

Although an important role of apoptosis is implicated in neurodegenerative diseases, data from both humans and animal models indicate that neurodegeneration is often a long-lasting process that finish with cell death only after a prolonged period of disease state.

In PD model mice study, although peptide inhibitors of caspases block 1-methyl-4-phenylpyridinium (MPP+)-induced dopaminergic neuronal death, dopaminergic neuronal terminals are not rescued [5]. Similarly, adenovirus-mediated transgene expression of X-linked inhibitor of apoptosis protein (XIAP) blocks death of dopaminergic neurons in a N-methyl-4-phenyl-1,2,3,6-tetrahydropyridine (MPTP)-induced PD mouse model, but does not prevent the decrease of dopaminergic terminal markers in the striatum [5]. Moreover the resistance of the dopaminergic neurons in the pro-apoptotic Bax protein knockout mice against MPTP toxicity is accompanied by a significant, although less prominent, sparing of striatal dopamine contents [6]. In the superoxide dismutase 1 (*SOD1*) transgenic mouse models of amyotrophic lateral sclerosis (ALS) study, we have also shown that overexpression of XIAP in spinal motor neurons rescues cell bodies of motor neuron without inhibition of neuronal dysfunction [7]. Similarly, removal of *Bax* gene resulted in complete rescue of cell bodies of motor neurons in ALS model mice, but denervation and axonal degeneration still occurred [8, 9]. Moreover with Bcl-2 transgenic mice crossed with *pmn* mice to block cellular apoptosis, motor neurons were completely rescued, but motor axons degenerate to the same extent as in *pmn* mice with normal levels of Bcl-2, and there is no change in muscle strength or life span [9, 10]. Consistent with these findings, although an important role for synaptic caspase activation and apoptosis has been proposed, axonal degeneration after withdrawal of trophic support occurs without activation of caspases in contrast to cell death of the cell body [9, 11]. These studies support the idea that axonal degeneration/dysfunction may proceed independently from the molecular events regulating cell death, and that apoptosis plays a critical role in neuronal cell body death, and neuronal dysfunction is not treatable by anti-cell death therapy. Therefore, anti-dysfunction therapies which target axonal degeneration/dysfunction are promising for treatment of neurodegenerative diseases.

4.3 Morphological Aspects of Neuronal Dysfunction Caused by Protein Aggregation/Misfolding in Human Neurodegenerative Disorder

It is hard to morphologically evaluate the neuronal dysfunction caused by protein aggregation/misfolding in the central nervous system, because neurons possess intricate three-dimensional structure, and are embedded deep in the brain which prevents accurate observation of cell shape. In contrast, morphological evaluation of the peripheral nervous system shows that degeneration of the cardiac sympathetic nerve occurs in PD and diffuse Lewy body disease, both of which are caused by accumulation of misfolded α-synuclein, and that degeneration of their distal axons precedes loss of their neuronal cell bodies in the paravertebral sympathetic ganglia [12]. This interesting observation suggests that distal dominant axonal degeneration precedes cell death not only in peripheral sympathetic, but central dopaminergic neurons of PD. Moreover, it is implicated that the centripetal degeneration may represent the common pathological process underlying various neurodegenerative disorders.

In ALS study, there are data supporting the hypothesis that the pathology of ALS starts with distal axonal degeneration [9]. Neuropathological studies, by quantitative morphometry, demonstrate a distal-to-proximal gradient of axonal pathology in phrenic nerves from ALS patients [9, 13]. Moreover, an autopsy case of an ALS patient, who died unexpectedly during a minor surgical procedure, revealed severe denervation and reinnervation changes demonstrated by electromyography, but there were no detectable changes in the corresponding spinal motor neuronal cell bodies [9, 14]. Threshold tracking, which measures axonal excitability, is an alternative electrophysiological technique that demonstrates early abnormalities in

ALS patients. An apparent increase in persistent Na+ current and a decrease in K+ conductance is observed in two ALS patients [9, 15, 16], and these changes are more prominent distally than proximally [9, 17]. Genetical studies of ALS also provide further evidence for the potential importance of axonal pathology in ALS [9, 18].

4.4 Protein Misfolding and Axonal Degeneration in Experimental Animal Models

Recent studies demonstrate that genetically engineered mice with misfolded protein accumulation display axonal degeneration phenotype [19].

One of the excellent examples is the knockout mouse of an essential autophagy gene, *Atg7*, whose alterations have also been observed in several neurodegenerative diseases [20, 21]. Ablation of panneuronal autophagy causes ubiquitin-p62 positive aggregation in neuronal cell body [20, 21]. Conditional knockout of *Atg7* in Purkinje cells initially causes cell-autonomous, progressive dystrophy (manifested by axonal swellings) and degeneration of the axon terminals [22]. Consistent with suppression of autophagy, no autophagosomes are observed in these dystrophic swellings [22]. Axonal dystrophy of mutant Purkinje cells proceeds with little sign of dendritic or spine atrophy, indicating that axon terminals are much more vulnerable to autophagy impairment than dendrites. This early pathological event in the axons is followed by Purkinje cell death. Furthermore, ultrastructural analyses of mutant Purkinje cells reveal an accumulation of aberrant membrane structures in the axonal dystrophic swellings, indicating that the autophagic machinery component Atg7 is required for membrane trafficking and turnover in the axons, and that impairment of axonal autophagy as a possible mechanism for axonal degeneration associated with neurodegeneration [22]. Accordingly, significant accumulation of ubiquitinated proteins is noted in Atg7-deficient brain, but their levels, especially insoluble ubiquitinated proteins, are lower than in Atg7-deficient liver, and formation of the inclusion is found in restricted groups of neurons. Several ubiquitin-positive aggregates are recognized in Atg7-deficient brain regions in the presence of mild neuronal loss [22]. Direct degradation of aggregates/misfolded protein by autophagy is contradictory to the recent hypothesis that the generation of protein aggregates represents a protective mechanism [23]. However, the primary targets of autophagy are likely to be diffuse cytosolic proteins, not inclusion bodies themselves, suggesting that inclusion body formation in autophagy-deficient cells is an event secondary to impaired general protein turnover [23]. However, it is still possible that misfolded proteins in soluble or oligomeric states could be preferentially recognized by autophagosomal membranes, which might also be mediated by ubiquitin–p62–LC3 interactions [23, 24].

A recent study also showed that axonal degeneration is relevant to autophagy caused by protein mislocalization [25]. Adaptor protein-4 (AP-4) is a member of the adaptor protein complexes, which control vesicular trafficking of membrane proteins. Although AP-4 has been suggested to contribute to basolateral sorting in epithelial cells, its function in neurons is unknown. A recent study showed that disruption of the gene encoding the β subunit of AP-4 resulted in increased accumulation of axonal autophagosomes, which contained alpha-amino-3-hydroxy-5-methyl-4-isoxazolepropionic acid (AMPA) receptors and transmembrane AMPA receptor regulatory proteins (TARPs), in axons of hippocampal neurons and cerebellar Purkinje cells both *in vitro* and *in vivo* [25]. AP-4 indirectly associates with the AMPA receptor via TARPs, and the specific disruption of the interaction between AP-4 and TARPs causes the mislocalization of endogenous AMPA receptors in axons of wild-type neurons. These results indicate that AP-4 may regulate proper somatodendritic-specific distribution of its cargo proteins, including AMPA receptor-TARP complexes and that protein mislocalization may disturb the autophagic pathway(s) in neurons [25].

4.5 Therapeutic Approaches to Treat Neuronal Dysfunction by Axonal Protection

4.5.1 Axonal Regeneration

From previous studies showing that neuronal dysfunction, which may morphologically reflect axonal degeneration by misfolded proteins, precedes neuronal cell death, we hypothesized that axonal regeneration may protect axons from degeneration and have therapeutic effects against neuronal dysfunction in neurodegeneration [26]. We have tested this hypothesis using anti-LINGO-1 antagonists in experimental PD models induced by either oxidative (6-hydroxydopamine) or mitochondrial (MPTP) toxicity [25]. LINGO-1 is the nervous system-specific leucine-rich repeat Ig-containing protein, and associated with the Nogo-66 receptor (NgR) complex and is endowed with a canonical EGF receptor (EGFR)-like tyrosine phosphorylation site, playing a critical role as an inhibitor of axonal regeneration (Fig. 4.1) [27, 28]. LINGO-1 antagonists, which block signal transduction of LINGO-1 complex (Fig. 4.2) [28], include decoy protein LINGO-1-Fc, Lenti-virus-dominant negative LINGO-1, and anti-LINGO-1 blocking antibody. We examined the role of LINGO-1 in cell damage responses of dopaminergic neurons. In LINGO-1 knockout mice, dopaminergic neuronal survival is increased and behavioral abnormalities are reduced compared with wild-type ones. This neuroprotection is accompanied by increased Akt phosphorylation [26]. Similar *in vivo* neuroprotective effects on midbrain dopaminergic neurons are obtained in wild-type mice by blocking LINGO-1 activity using LINGO-1-Fc protein which inhibit LINGO-1 function. Neuroprotection and enhanced neurite growth are also demonstrated for midbrain dopaminergic neurons *in vitro* [26]. LINGO-1 antagonists improve dopaminergic neuronal survival in response to MPTP in part by mechanisms that involve activation of the EGFR/Akt signaling pathway through a direct inhibition of the binding LINGO-1 to EGFR (Fig. 4.3) [26]. LINGO-1 is also upregulated in compromised, probably dysfunctional, neurons in spinal cord injury [29] or kainic acid injection

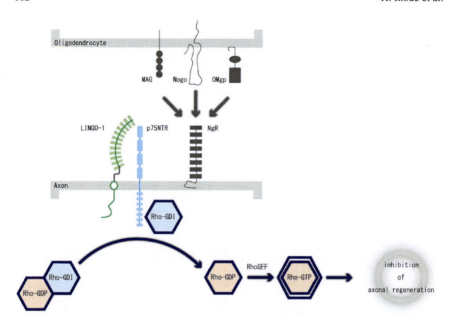

Fig. 4.1 Molecular signaling of Nogo-66 receptor (NgR) complex. The potential role of LINGO-1 is revealed as a component of NgR complex, which is comprised of NgR and p75 neurotrophin receptor (p75NTR) or an orphan TNF receptor Taj/Troy [27, 28]. Activated p75NTR binds the RhoA-GTP dissociation inhibitor (Rho-GDI), thus enabling RhoA activation via the exchange of GDP for GTP, and inhibits axonal regeneration upon binding to inhibitory molecules such as myelin-associated glycoprotein (MAG), oligodendrocyte myelin glycoprotein (OMgp), and Nogo-66 (Nogo) expressed in oligodendrocytes [40, 41]

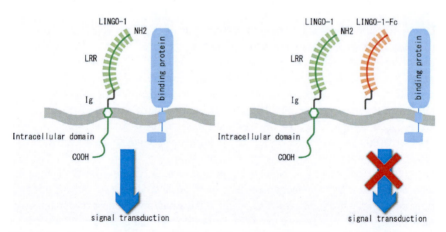

Fig. 4.2 Functional mechanisms of LINGO-1 antagonist(s). LINGO-1-Fc, one of LINGO-1 antagonists, is the soluble, truncated form of LINGO-1, and inhibits LINGO-1 modulating signaling transduction by inhibiting LINGO-1 to bind its binding protein(s) [26, 28]

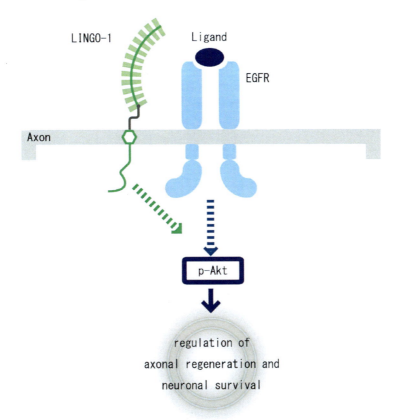

Fig. 4.3 LINGO-1 effect on EGF receptor (EGFR). LINGO-1 binds EGFR, and regulates EGFR expression level, leading to control axonal regeneration and neuronal survival via phosphorylation of Akt [26, 28]

[30]. We found that LINGO-1 expression is elevated in compromised, dysfunctional neurons including in the substantia nigra of PD patients compared with age-matched controls and in animal models of PD after neurotoxic lesions [25]. These results show that inhibitory agents of LINGO-1 activity can protect dopaminergic neurons from degeneration caused by PD. It is necessary to test whether LINGO-1 inhibition of function has protective effects on genetic PD models and/or other neurodegenerative disease models in the future.

4.5.2 Anti-Wallerian Degeneration

Axonal degeneration in "dying back" disorders seems to be different from Wallerian degeneration which is triggered by focal lesion (Fig. 4.4) [19]. However, apparent differences in the directionality of degeneration have been controversial [19].

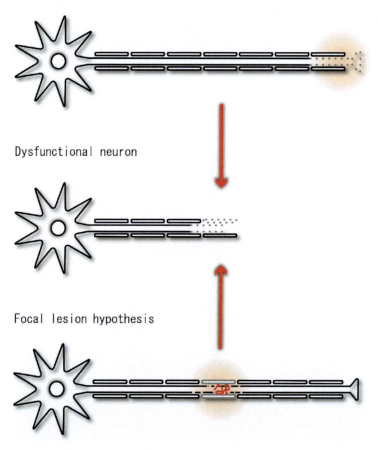

Fig. 4.4 "Dying back" and focal lesion models of axonal degeneration/dysfunction. The centripetal axonal degeneration in neurodegenerative disease may be caused either by the "dying back" process or by repetitive Wallerian degeneration from focal lesion(s) [19]

Wallerian degeneration is a simple experimental model of axonal degeneration, in which the distal stump of an injured axon degenerates rapidly after a reproducible latent phase [19]. In Wallerian degeneration slow (*WldS*) mice, Wallerian degeneration in response to axonal injury is delayed because of a mutation that results in overexpression of a chimeric protein (WldS) composed of the ubiquitin assembly protein Ufd2a and the nicotinamide adenine dinucleotide (NAD) biosynthetic enzyme Nmnat1 [31]. With the discovery of the *WldS* mouse, the hypothesis could be tested. In *WldS* mice, injury-induced Wallerian degeneration is delayed ~tenfold (for 2–3 weeks) by a dominant mutation that acts intrinsically in neurons. In crossbreeding with progressive motor neuronopathy (*pmn*) mice and myelin protein zero (P0) null mutants, a model of Charcot-Marie-Tooth disease, WldS significantly delayed axonal degeneration [19]. In the central nervous system, WldS also protects against both genetic and toxic insults. Some nigrostriatal axons, which degenerate in PD, are spared and remain functional after 6-hydroxydopamine lesions in *WldS* mice [19, 32]. Axonal spheroids, which are presumably composed by misfolded protein(s) accumulation, are reduced in number in the gracile tract of mice with gracile axonal dystrophy (*gad*), deficient in ubiquitin carboxyterminal hydrolase L1 (UCHL1) crossed with *WldS* mice [19, 33]. Not all axonal degeneration is delayed by WldS. WldS have modest effect on the ALS model mice with protection of the terminal axon only at its early stage [34, 35]. The failure to protect axons under certain circumstances indicates the existence of multiple axonal degeneration mechanisms. WldS may have protective effect(s) in rapidly degenerative or acute disorders.

4.5.3 Autophagy Enhancement

It is reasonable to assume that autophagy could represent a therapeutic target for axonal degeneration because the deletion of essential components of autophagy causes axonal degeneration, and relevance(s) of autophagy with degeneration are observed in several neurodegenerative diseases [23]. Autophagy enhancement by the regulatory protein kinase complex Target of Rapamycin (TOR) inhibitors such as rapamycin and its analogue CCI-779 protects against neurodegeneration seen in polyglutamine disease models in Drosophila and mice [23, 36]. A screened small molecule enhancers of rapamycin improve the clearance of mutant huntingtin and α-synuclein, and protect against neurodegeneration in a fruit-fly HD model [23, 37]. These results provide us with the evidence supporting autophagy enhancement as a therapeutic strategy against the toxicity of misfolded proteins in neurodegenerative diseases.

Tsc2, also known as tuberin, is a GTPase activating protein that regulates the G protein Rheb, an activator of mTOR (mammalian Target of Rapamycin) [38]. Tuberous sclerosis is a single-gene disorder caused by heterozygous mutations in the *TSC1* or *TSC2* genes and is frequently associated with mental retardation, autism and epilepsy [38]. Even individuals with tuberous sclerosis and a normal intelligence quotient are commonly affected with specific neuropsychological problems, including long-term and working memory deficits [38]. Mice heterozygous for the deletion of the *Tsc2* gene in $Tsc2(+/-)$ mice show deficits in learning and memory [38]. A recent study showed that hyperactive hippocampal signaling led to abnormal long-term potentiation in the CA1 region of the hippocampus and consequently to deficits in hippocampal-dependent learning in TSC mice [38]. Moreover, a brief treatment with the mTOR inhibitor rapamycin in adult mice rescues not only the synaptic plasticity, but also the behavioral deficits in this animal model of tuberous sclerosis, demonstrating that treatment with mTOR antagonists ameliorates cognitive dysfunction in the TSC mice model [38]. Autophagy may be included in axon and/or synaptic dysfunction in degeneration via the mTOR signaling pathway(s).

4.5.4 Stabilization of Microtubules

In tauopathy model(s), misfolded mutant tau protein causes axonal dysfunction/degeneration [39]. Microtubule-binding drugs can be therapeutically beneficial in tauopathy models by functionally substituting for the microtubule-binding protein tau, which is sequestered into inclusions of human tauopathies and transgenic mouse models [39]. Mutant tau transgenic mice treated with paclitaxel (Paxceed) showed that fast axonal transport in spinal axons is restored, and that microtubule numbers and stable tubulins are increased compared with sham treatment [39]. Moreover, Paxceed ameliorated motor impairments in tau transgenic mice [39]. Thus, microtubule-stabilizing drugs have therapeutic potential for axonal dysfunction/degeneration, in tauopathies by offsetting losses of tau function that result from the sequestration of this microtubule-stabilizing protein into filamentous misfolded inclusions [39].

4.6 Concluding Remarks

Accumulating experimental evidence suggests that protection/functional repair of axons (Fig. 4.5) can be a general therapeutic strategy for neuronal dysfunction caused by misfolded protein(s) deposition in neurodegenerative disease(s). The translation of these results into human disease therapies should be done in the near future.

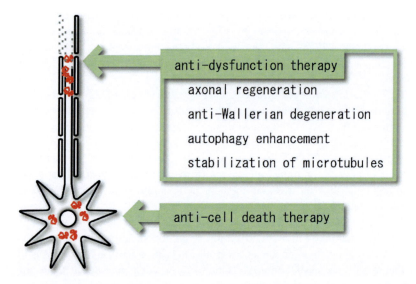

Fig. 4.5 Anti-dysfunction therapy for axonal protection and anti-cell death therapy for inhibiting neuronal loss. Neuronal dysfunction in neurodegeneration is reversible. Axonal regeneration, anti-Wallerian degeneration, autophagy enhancement, and/or stabilization of microtubules may be effective as anti-dysfunction therapies by protecting axon from neurodegenerative diseases, although anti-cell death therapies may inhibit only neuronal cell body loss

Abbreviations

amyotrophic lateral sclerosis (ALS)
alpha-amino-3-hydroxy-5-methyl-4-isoxazolepropionic acid (AMPA)
adaptor protein-4 (AP-4)
EGF receptor (EGFR)
Huntington's disease (HD)
N-methyl-4-phenyl-1,2,3,6-tetrahydropyridine (MPTP)
mammalian target of rapamycin (mTOR)
Parkinson's disease (PD)
progressive motor neuronopathy (*pmn*)
transmembrane AMPA receptor regulatory protein (TARP)
leucine-rich repeat Ig-containing protein (LINGO-1)
Wallerian degeneration slow (*Wlds*)

References

1. Palop JJ, Chin J, Mucke L (2006) A network dysfunction perspective on neurodegenerative diseases. Nature 443:768–773
2. Yamamoto A, Lucas JJ, Hen R (2000) Reversal of neuropathology and motor dysfunction in a conditional model of Huntington's disease. Cell 101:57–66
3. Orr HT, Zoghbi HY. (2000) Reversing neurodegeneration: a promise unfolds. Cell 101:1–4
4. Wang J, Wang C-E, Orr A, Tydlacka S, Li S-H, Li X-J (2008) Impaired ubiquitin – proteasome system activity in the synapses of Huntington's disease mice. J Cell Biol 180:1177–1189
5. Eberhardt O, Coelln RV, Kugler S, Lindenau J, Rathke-Hartlieb S, Gerhardt E, Haid S, Isenmann S, Gravel C, Srinivasan A et al. (2000) Protection by synergistic effects of adenovirus-mediated X chromosome-linked inhibitor of apoptosis and glial cell line-derived neurotrophic factor gene transfer in the 1-methyl-4-phenyl-1,2,3,6-tetrahydropyridine model of Parkinson's disease. J Neurosci 20:9126–9134
6. Vila M, Jackson-Lewis V, Vukosavic S, Djaldetti R, Liberatore G, Offen D, Korsmeyer SJ, Przedborski S (2001) Bax ablation prevents dopaminergic neurodegeneration in the 1-methyl-4-phenyl-1,2,3,6-tetrahydropyridine mouse model of Parkinson's disease. Proc Natl Acad Sci USA 98:2837–2842
7. Inoue H, Tsukita K, Iwasato T, Suzuki Y, Tomioka M, Tateno M, Nagao M, Kawata A, Saido TC, Miura M et al. (2003) The crucial role of caspase-9 in the disease progression of a transgenic ALS mouse model. EMBO J 22:6665–6674
8. Gould TW, Buss RR, Vinsant S, Prevette D, Sun W, Knudson CM, Milligan CE, Oppenheim (2006) Complete dissociation of motor neuron death from motor dysfunction by Bax deletion in a mouse model of ALS. J Neurosci 26:8774–8786
9. Fischer LR, Glass JD (2007) Axonal degeneration in motor neuron disease. Neurodegener Dis 4:431–442
10. Sagot Y, Vejsada R, Kato A (1997) Clinical and molecular aspects of motoneurone diseases: animal models, neurotrophic factors and Bcl-2 oncoprotein. Trends Pharmacol Sci 18:330–337
11. Finn JT, Weil M, Archer F, Siman R, Srinivasan A, Raff MC (2000) Evidence that Wallerian degeneration and localized axon degeneration induced by local neurotrophin deprivation do not involve caspases. J Neurosci 20:1333–1341
12. Orimo S, Uchihara T, Nakamura A, Mori F, Kakita A, Wakabayashi K, Takahashi H (2008) Axonal α-synuclein aggregates herald centripetal degeneration of cardiac sympathetic nerve in Parkinson's disease. Brain 131:642–650

13. Bradley WG, Good P, Rasool CG, Adelman LS (1983) Morphometric and biochemical studies of peripheral nerves in amyotrophic lateral sclerosis. Ann Neurol 14:267–277
14. Fischer LR, Culver DG, Tennant P, Davis AA, Wang M, Castellano-Sanchez A, Khan J, Polak MA, Glass JD (2004) Amyotrophic lateral sclerosis is a distal axonopathy: evidence in mice and man. Exp Neurol 185:232–240
15. Kanai K, Kuwabara S, Misawa S, Tamura N, Ogawara K, Nakata M, Sawai S, Hattori T, Bostock H (2006) Altered axonal excitability properties in amyotrophic lateral sclerosis: impaired potassium channel function related to disease stage. Brain 129:953–962
16. Vucic S, Kiernan MC (2006) Axonal excitability properties in amyotrophic lateral sclerosis. Clin Neurophysiol 117:1458–1466
17. Nakata M, Kuwabara S, Kanai K, Misawa S, Tamura N, Sawai S, Hattori T, Bostock H (2006) Distal excitability changes in motor axons in amyotrophic lateral sclerosis. Clin Neurophysiol 117:1444–1448
18. Pasinelli P, Brown RH (2006) Molecular biology of amyotrophic lateral sclerosis: insights from genetics. Nat Rev Neurosci 7:710–723
19. Coleman M (2005) Axon degeneration mechanisms: commonality amid diversity. Nat Rev Neurosci 6:889–898
20. Hara T, Nakamura K, Matsui M, Yamamoto A, Nakahara Y, Suzuki-Migishima R, Yokoyama M, Mishima K, Saito I, Okano H et al. (2006) Suppression of basal autophagy in neural cells causes neurodegenerative disease in mice. Nature 441:885–889
21. Komatsu M, Waguri S, Chiba T, Murata S, Iwata J, Tanida I, Ueno T, Koike M, Uchiyama Y, Kominami E et al. (2006) Loss of autophagy in the central nervous system causes neurodegeneration in mice. Nature 441:880–884
22. Komatsu M, Wang QJ, Holstein GR, Friedrich VL Jr, Iwata J, Kominami E, Chait BT, Tanaka K, Yue Z. (2007) Essential role for autophagy protein Atg7 in the maintenance of axonal homeostasis and the prevention of axonal degeneration. Proc Natl Acad Sci USA 104: 14489–14494
23. Mizushima N, Levine B, Cuervo AM, Klionsky DJ (2008) Autophagy fights disease through cellular self-digestion. Nature 451:1069–1075
24. Komatsu M, Waguri S, Koike M, Sou Y, Ueno T, Hara T, Mizushima N, Iwata J, Ezaki J, Murata S et al. (2007) Homeostatic levels of p62 control cytoplasmic inclusion body formation in autophagy-deficient mice. Cell 131:1149–1163
25. Matsuda S, Miura E, Matsuda K, Kakegawa W, Kohda K, Watanabe M, Yuzaki M (2008) Accumulation of AMPA receptors in autophagosomes in neuronal axons lacking adaptor protein AP-4. Neuron 57:730–745
26. Inoue H, Lin L, Lee X, Shao Z, Mendes S, Snodgrass-Belt P, Sweigard H, Engber T, Pepinsky B, Yang L et al. (2007) Inhibition of the leucine-rich repeat protein LINGO-1 enhances survival, structure, and function of dopaminergic neurons in Parkinson's disease models. Proc Natl Acad Sci USA 104:14430–14435
27. Bandtlow C, Dechant G (2004) From cell death to neuronal regeneration, effects of the p75 neurotrophin receptor depend on interactions with partner subunits. Sci STKE 235:pe24
28. Mi S, Sandrock A, Miller RH (2008) LINGO-1 and its role in CNS repair. Int J Biochem Cell Biol 40:1971–1978
29. Wang J, So K-F, McCoy JM, Pepinsky RB, Mi S, Relton JK (2006) LINGO-1 antagonist promotes functional recovery and axonal sprouting after spinal cord injury. Mol Cell Neurosci 33:311–320
30. Trifunovski A, Josephson A, Ringman A, Brene S, Spenger C, Olson L (2004) Neuronal activity-induced regulation of Lingo-1. Neuroreport 15:2397–2400
31. Araki T, Sasaki Y, Milbrandt J (2004) Increased nuclear NAD biosynthesis and SIRT1 activation prevent axonal degeneration. Science 305:1010–1013
32. Sajadi A, Schneider BL, Aebischer P (2004) WldS-mediated protection of dopaminergic fibers in an animal model of Parkinson disease. Curr Biol 14:326–330
33. Mi W, Beirowski B, Gillingwater TH, Adalbert R, Wagner D, Grumme D, Osaka H, Conforti L, Arnhold S, Addicks K et al. (2005) The slow Wallerian degeneration gene, *Wld*S, inhibits axonal spheroid pathology in gracile axonal dystrophy mice. Brain 128:405–416

34. Fischer LR, Culver DG, Davis AA, Tennant P, Wang M, Coleman M, Asress S, Adalbert R, Alexander GM, Glass JD (2005) The *WldS* gene modestly prolongs survival in the SOD1G93A fALS mouse. Neurobiol Dis 19:293–300
35. Vande Velde C, Garcia ML, Yin X, Trapp BD, Cleveland DW (2004) The neuroprotective factor WldS does not attenuate mutant SOD1-mediated motor neuron disease. Neuromolecular Med 5:193–204
36. Ravikumar B, Vacher C, Berger Z, Davies JE, Luo S, Oroz LG, Scaravilli F, Easton DF, Duden R, O'Kane CJ et al. (2004) Inhibition of mTOR induces autophagy and reduces toxicity of polyglutamine expansions in fly and mouse models of Huntington disease. Nat Genet 36: 585–595
37. Sarkar S, Perlstein EO, Imarisio S, Pineau S, Cordenier A, Maglathlin RL, Webster JA, Lewis TA, O'Kane CJ, Schreiber SL et al. (2007) Small molecules enhance autophagy and reduce toxicity in Huntington's disease models. Nat Chem Biol 3:331–338
38. Ehninger D, Han S, Shilyansky C, Zhou Y, Li W, Kwiatkowski DJ, Ramesh V, Silva AJ (2008) Reversal of learning deficits in a Tsc2 + /− mouse model of tuberous sclerosis. Nat Med. 14:843–848
39. Zhang B, Maiti A, Shively S, Lakhani F, McDonald-Jones G, Bruce J, Lee E B, Xie S X, Joyce S, Li C et al. (2005) Microtubule-binding drugs offset tau sequestration by stabilizing microtubules and reversing fast axonal transport deficits in a tauopathy model. Proc Natl Acad Sci USA 102:227–231
40. Yamashita T, Tohyama M (2003) The p75 receptor acts as a displacement factor that release Rho from Rho-GDI. Nat Neurosci. 6:461–467
41. Hasegawa Y, Yamagishi S, Fujitani M, Yamashita T (2004) p75 neurotrophin receptor signaling in the nervous system. Biotechnol Annu Rev.10:123–149

Chapter 5
Endoplasmic Reticulum Stress in Neurodegeneration

Jeroen J.M. Hoozemans and Wiep Scheper

Abstract Accumulation of misfolded proteins in the endoplasmic reticulum triggers a cellular stress response called the unfolded protein response (UPR) that protects the cell against the toxic buildup of misfolded proteins. Neurodegenerative disorders like Alzheimer's disease, Parkinson's disease, prion disease, Huntington's disease, frontotemporal dementia, and amyotrophic lateral sclerosis are characterized by the accumulation and aggregation of misfolded proteins. In this chapter we will discuss the different levels of protein quality control systems in the endoplasmic reticulum. The role of these systems and especially the UPR will be reviewed in view of current data about the expression and role of the UPR markers in the pathology of neurodegenerative disorders.

5.1 Introduction

Neurodegenerative disorders are often characterized by the aggregation and accumulation of misfolded proteins. Aggregated proteins are very toxic to cells in culture, and both *in vitro* and *in vivo* there is overwhelming evidence that these aberrant proteins are responsible for neurodegeneration. Protein quality control is a cellular defense mechanism against misfolded proteins and prevents aggregate formation under physiological condition. The presence of accumulated aggregates of misfolded proteins in many neurodegenerative disorders, suggests that protein quality control failed to restore homeostasis, indicating the importance of proper quality control and suggesting that it may provide a target for therapeutic intervention. In this chapter we will focus on the protein quality control systems in the endoplasmic reticulum (ER), where all non-cytosolic proteins are synthesized. The involvement of ER quality control in neurodegenerative diseases will be addressed.

J.J.M. Hoozemans (✉)
Department of Pathology, VU University Medical Center, P.O. Box 7057
1007 MB Amsterdam, The Netherlands
e-mail: jjm.hoozemans@vumc.nl

5.2 Protein Quality Control in the Endoplasmic Reticulum

In mammalian cells, proteins are co-translationally imported in the ER, where they are kept in a folding competent state by chaperone proteins, among which the ER heath shock protein (Hsp)70 family member BiP (heavy chain binding protein or glucose-regulated protein 78, Grp78). The immature N-glycans are transferred, probably co-translationally and initial processing of N-glycans occurs in the ER. Also intermolecular disulfide bridges are formed, facilitated by the oxidizing environment. Misfolding of polypeptide chains in the ER may occur in case of overexpression, mutations or aberrant post-translational processing. To ensure the fidelity of protein synthesis and folding, quality control systems are in place. When misfolding occurs, the first line of defense involves refolding, mediated by chaperone proteins. If this fails the misfolded protein will be targeted for degradation. In case the refolding/degradation machinery is impaired or insufficient to deal with the protein load it is presented with, a stress response is activated. This involves upregulation of refolding and degradation factors, but if the misfolded protein stress is too severe, a cell death program will be activated. In the next sections we will discuss the sequential steps of protein quality control in the ER: Folding (Chaperones, 2.1), triage (ERAD, 2.2), degradation (2.3 Ubiquitin proteasome system and autophagy), stress response (Unfolded protein response, 2.4) and cell death (2.5).

5.2.1 Folding: Chaperones

The first line of defense against misfolded proteins is provided by chaperones that bind to unfolded stretches in proteins and keep them in a folding-competent state while preventing aggregation. Several cycles of binding and release are usually required to obtain a folded protein. With these properties they also assist to dissolve aggregates and target misfolded proteins for degradation. Many chaperones are also called heatshock proteins (Hsp) because their expression is regulated by heat and other forms of stress [1]. They are classified according to their molecular weight (e.g. Hsp100, Hsp70, small Hsp [2]). Chaperones can be involved in folding of specific proteins, bind to specific features in proteins (e.g. the lectins calnexin and calreticulin that bind carbohydrate modifications) or have specific activities (e.g. isomerases, that catalyze the cis-trans isomerisation of peptide bonds, or Ero1 an oxidoreductase involved in disulfide bond formation). Of particular interest for the prevention of aggregate formation is the Hsp70 class of chaperones [3]. In the ER, the Hsp70 chaperone BiP (or Grp78) plays an important role in quality control. BiP binds to small hydrophobic stretches in proteins, in cooperation with a co-chaperone of the Hsp40 family. Binding requires ATP hydrolysis and with the help of an exchange factor the ADP is exchanged for a new ATP, which will be hydrolyzed in order to release the substrate, after which BiP may bind again. BiP is a key protein in ER protein quality control and its proper function is essential to prevent aggregation. This is illustrated by the observation that mutations in SIL1, an exchange factor for BiP, cause aggregate formation and neurodegeneration in a spontaneous mouse model as well as in a human disease, Marinesco-Sjögren syndrome [4, 5]. This is to our knowledge the only example of a mutation in the ER quality control system leading to a hereditary neurodegenerative disorder. In contrast to other disruptions of the ER quality control system, disruption of SIL1 in yeast is not lethal [6], which may explain the rarity of such mutations.

5.2.2 Triage: ERAD

Secretory and transmembrane proteins are synthesized and folded in the ER. Misfolded proteins are exported from the ER and degraded by proteasomes in the cytosol (a process called ERAD, ER associated degradation). The recognition of misfolded substrates and targeting to the export channel in the membrane involves different ER chaperones [7].

Several checkpoints are in place where a decision is made to let a protein pass, repair it or target it for degradation (reviewed in [8]). There are specific molecular mechanisms underlying this triage that can be different for different classes of proteins, and depends on their localization (membrane associated or luminal, [9, 10]). The best studied class is quality control of glycoproteins, which we will use to illustrate the intricate machinery involved in ER quality control (reviewed in [11]). The N-glycan structure in glycoproteins binds to calnexin and calreticulin ("the calnexin/calreticulin cycle"). Trimming of the terminal glucose by glucosidase II releases the protein from calnexin/calreticulin. Misfolded N-glycoproteins are re-glucosylated by UDP-glucose-glycoprotein glycosyltransferase (UGT1) resulting in retention in the ER and re-entry in the calnexin/calreticulin cycle. This calnexin/calreticulin cycle functions as a timer for folding, from which proteins can still obtain their native fold. Prolonged interaction with calnexin/calreticulin permits demannosylation. Improperly folded glycoproteins are transferred to ER degradation-enhancing alpha-mannosidase-like proteins (EDEM1-3), that trim the mannosidase tree and are keyfactors in targeting the misfolded glycoproteins to the retrotranslocon [12]. The identity of this translocon is still not completely resolved, but there is data to suggest that the Sec61 complex that is involved in import into the ER is also involved in export [13, 14]. More recently, another channel with the protein Derlin-1 was shown to be involved in retrotranslocation as well [15, 16]. After export to cytoplasm the proteins are degraded by the ubiquitin proteasome system (UPS; see below). In order to be degraded, deglycosylation by peptide: N-glycanase has to take place, which is probably associated with the ER membrane via Derlin-1 [17]. The AAA-ATPase P97/cdc48 is thought to play a role in unfolding the proteins before feeding into the catalytic chamber of the proteasome [8, 18]. Reports indicating association of proteasomes with the ER membrane and more in particular with the Sec61 channel [19, 20] suggest that retrotranslocated proteins may be fed into the proteasome directly, without entering the cytosol. Recently, P58(IPK) has been identified as a co-chaperone that functions to target proteins that are stalled during translocation to the proteasome [21]. This is an important mechanism during ER stress, and again indicates a very direct coupling between the translocation channel and the UPS.

5.2.3 Degradation: Ubiquitin Proteasome System and Autophagy

5.2.3.1 The Ubiquitin Proteasome System

The ubiquitin proteasome system (UPS) is the main cellular protein degradation system that tags and targets aberrant and misfolded proteins for destruction [22, 23]. This cytosolic system therefore plays an essential role in ER protein quality control [24]. A first step is the tagging of the substrate proteins by the covalent attachment of a poly-ubiquitin chain to the protein, which is a three step process [25]. This is initiated by the activation of ubiquitin by E1 enzymes, which is subsequently transferred to an E2 conjugating enzyme, which facilitates the attachment of ubiquitin to a lysine residue in the target protein by an E3 ubiquitin ligase. The substrate specificity lies within the E3 enzyme, which therefore encompasses the largest class of ubiquitinating enzymes [26]. Hrd1 for example is a ubiquitin ligase specifically involved in ERAD of a broad range of substrates [10]. Parkin is a ubiquitin ligase that mediates the degradation of a limited number of substrates. It is found mutated in autosomal-recessive juvenile parkinsonism (AR-JP, see also Section 3.2 below) and this leads to the accumulation of specific proteins among which the Pael receptor [27].

The polyubiquitinated substrates are targeted for destruction to the 26S proteasome [28]. The 20S core is a chamber that consists of two central rings containing the catalytic β subunits, and two distal α rings. The 19S lid is a regulatory complex required for binding of polyubiquitinated substrates and the site for deubiquitination. The exact mechanism of the delivery of polyubiquitinated cargo to the 26S proteasome is not fully elucidated yet, but it is clear that a number of "escort-factors" like Rad23, Dsk2 and Cdc48 are involved [29, 30]. The target protein is unfolded and fed into the 20S core of the proteasome, where the protein is degraded by three enzymatic activities residing in the β-rings: Trypsin-like, chymotrypsin-like and PGPH-like [31].

5.2.3.2 Autophagy

Autophagy is a lysosomal pathway for the turn-over of organelles and long-lived proteins and has been identified initially as a cellular survival response to starvation [32]. This involves the wrapping of an isolation membrane around cell organelles or other structures, like proteins aggregates. This initial phagophore is elongated and finally ends up as autophagosome when the edges of the membrane fuse resulting in a double membrane structure. The autophagy process is regulated and executed by autophagy related proteins (Atg's), such as Atg 5 and Atg 16. Although the exact function of these factors is largely unknown, the conjugation of the ubiquitin-like protein Atg 12 to Atg 5 appears to be an important step in enlargement of the phagophore membrane [32]. During formation of the autophagosome, the microtubule-associated protein light chain 3 protein (LC3) is recruited and modified with phosphatidylethanolamine. This conversion from LC3-I to LC3-II is a widely used marker of autophagy [33]. Subsequently, the outer membrane of the autophagosome fuses with a lysosome to form an autophagolysosome, where the sequestered material is degraded [34]. For long, autophagy was thought of as a mechanism that is only used for degradation of proteins under extreme conditions (i.e. amino acid starvation). However, recent studies of mice that are incapable to activate autophagy due to a deficiency in Atg5 [35] or Atg7 [36], revealed accumulation of protein aggregates. This occured most notably in neurons and the mice presented with neurodegeneration. This shows that autophagy plays an important role in protein quality control under normal conditions [37]. There is a growing body of evidence that autophagy is triggered by proteasome impairment, indicating that the two degradational pathways used by the ER are tightly coupled [38, 39]. ER activated autophagy has even been termed ERADII [40]. More importantly, autophagy protects against ER stress toxicity [41, 42].

5.2.4 Stress Response: The Unfolded Protein Response

Accumulation of misfolded proteins in the ER leads to induction of the unfolded protein response (UPR) [43]. The UPR is initiated by the binding of the ER Hsp70 chaperone BiP to the misfolded proteins. This releases BiP from three sensors at the ER membrane that are subsequently activated: The RNA-activated protein kinase R (PKR)-like ER kinase (PERK), activating transcription factor 6 (ATF6) and inositol requiring enzyme 1 (Ire1). The activation of PERK is initiated by autophosphorylation, which results in translational attenuation through phosphorylation of eukaryotic initiation factor2α (eIF2α) [44]. This prevents the further build-up of protein load in the ER. Paradoxically, this leads to selective translation of activating transcription factor 4 (ATF4). ATF4 contains short open reading frames upstream in its 5′ UTR that inhibit translation of ATF4 under normal conditions [45]. When eIF2α is phosphorylated and becomes limiting (e.g. during activation of the UPR), this allows "leaky" scanning of the ribosome so that the ATF4 startcodon can be used. ATF4 activates transcription of pro-survival genes, but also induction of the pro-apoptotic gene C/EBP homologous protein (CHOP) is highly dependent on ATF4 (see below). Dephosphorylation of eIF2α is facilitated by growth arrest- and DNA damage-inducible gene (GADD) 34 and relieves the translation block [46]. Release of BiP from ATF6 allows its transport to the Golgi, where it is processed by site1 and site2 proteases [47]. This releases ATF6 from the membrane after which it translocates to the nucleus, where it activates transcription of ER stress response genes. Ire1 is also activated by autophosphorylation upon BiP release. The phosphorylation activates the endonuclease activity of Ire1, which leads to the non-conventional cytoplasmic splicing of X-Box binding protein 1 (XBP1) mRNA [48].

The splicing excises a 26 nt intron, shifting the reading frame and resulting in a larger product the active transcription factor XBP1, that also activates transcription of ER stress responsive genes. The shorter protein synthesized from the unspliced XBP1 mRNA functions as a suppressor of the UPR, by sequestering the spliced XBP1 and targeting it for degradation [49]. The transcriptional activation of ER stress responsive genes is mainly mediated by ATF6 and XBP1 [50], resulting in the increased expression of ER chaperones and ERAD components. Recently it has become clear that different classes of genes differ in their requirement for either ATF6 or XBP1 [51].

5.2.5 ER-Stress-Induced Cell Death

The UPR is aimed to restore homeostasis when misfolded proteins accumulate, but prolonged stress will activate an apoptotic program (reviewed in [46]). This may involve specific caspases; it has been suggested that caspase 12 in rodents and caspase 4 in humans are responsible for ER stress mediated apoptosis [52, 53]. CHOP, also known as GADD153 [54] is an ER stress responsive gene, activated by all UPR transducers (see above) and therefore an important protein in ER stress mediated apoptosis [46]. CHOP deficient fibroblasts are more resistant to cell death induced by ER stress [55]. One of the genes induced by CHOP is GADD34. The resulting restoration of translation (see above) is thought to facilitate synthesis of pro-apoptotic factors, in fact, inhibition of GADD34 inhibits ER stress toxicity [56]. In addition, ER stress activates apoptosis signal regulating kinase (ASK 1) through formation of an Ire1-TRAF2-ASK1 complex, subsequently leading to activation of apoptosis via the Jun Kinase pathway [57].

5.3 ER Stress in Neurodegenerative Disorders

Neurons are particularly vulnerable to the toxic effects of mutant or misfolded proteins. Because they are *post mitotic* cells they rely on protein quality control and stress responses like the UPR to adapt to and survive in a constantly changing environment. Neurodegenerative disorders like Alzheimer's disease (AD), Parkinson's disease (PD), prion disease, Huntington's disease (HD), frontotemporal dementia (FTD), and amyotrophic lateral sclerosis (ALS) are characterized by the aggregation and accumulation of misfolded proteins. Neurodegenerative diseases may also be caused by specific mutant proteins that accumulate as misfolded proteins and escape degradation, but more often this is not the case. Since age is an important risk factor for several neurodegenerative disorders, age-related decrease in proteasome activity and the increase of reactive oxygen species in the aging brain could contribute to decreased removal and the increasing build-up of misfolded and unfolded proteins [58–60]. The UPR- and ER stress-induced cell death are therefore likely to be involved in the pathogenesis of several neurodegenerative disorders (for overview see Table 5.1).

Table 5.1 UPR markers in neuropathology

Neurodegenerative disorders characterized by protein aggregation	Toxic protein	UPR markers described in pathology	References
Alzheimer's disease	Aβ, tau	BiP, pPERK, peIF2α, hHRD1	[62–64, 67]
Parkinson's disease	α-synuclein	BiP, pPERK, peIF2α	[89]
Prion disease	PrP^{Sc}	BiP, caspase 12	[64, 93]
Tauopathy	tau	pPERK, peIF2α	[64]
Polyglutamine disease	polyQ	not determined	–
Amyotrophic lateral sclerosis	SOD1	not determined	–
White matter disorders			
Multiple sclerosis	–	CHOP, BiP, XBP1	[113]
Vanishing White Matter disease	–	pPERK, peIF2α, ATF4, CHOP, XPB1, GADD34, ATF6	[118, 119]

5.3.1 Alzheimer's Disease

AD is the most prevalent neurodegenerative disease and the most common form of dementia. Deposits of aggregated proteins are a prominent neuropathological hallmark of AD: intracellular aggregates of tau in the neurofibrillary tangles (NFTs) and extracellular aggregates of β-amyloid (Aβ) in the senile plaques. AD thus represents a prime example of a protein folding disease [61].

Several reports of our group and others, have indicated that UPR activation is increased in AD brain [62–64]. This is in contrast to other reports that either showed a decrease of BiP in AD patients [65] or no differences at all between controls, sporadic or presenilin 1 (PS1) mediated familial AD patients [66]. These differences may be explained by the use of different antibodies (a generic anti-KDEL antibody and BiP specific antibodies in the other studies), or analysis of different brain areas. Compared to nondemented control cases, increased protein expression levels of the ER chaperone BiP, which is indicative for UPR activation, are found in the temporal cortex and the hippocampus of AD cases [63]. Comparing the expression of BiP in the different pathological stages of AD pathology (Braak stage) suggests that UPR activation occurs relatively early in AD. It appears that BiP immunoreactivity is primarily observed in healthy neurons [62, 63]. More direct evidence that the UPR is activated in AD neurons is provided by the detection of phosphorylated (activated) PERK and phosphorylated eIF2α [63, 67] (Fig. 5.1a,b). When activated, PERK phosphorylates eIF2α, which subsequently inhibits the assembly of the 80S ribosome and thereby facilitates the translational block of the UPR. Although eIF2α can also be phosphorylated by protein kinase R, the similar distribution and colocalization of pPERK and peIF2α in AD brain (Fig. 5.1c,d) suggests that eIF2α is directly phosphorylated by PERK upon induction of ER stress in AD neurons. Double-labeling experiments on AD brain tissue show that neurons with increased levels of BiP or pPERK do not contain NFTs as observed by immunoreactivity for

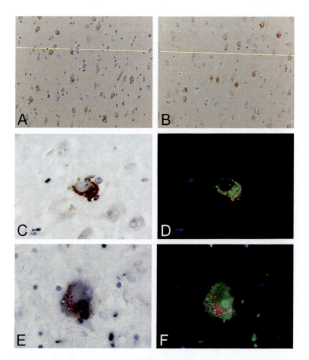

Fig. 5.1 Immunohistochemical detection of UPR activation markers in Alzheimer's and Parkinson's disease. (**A**) AD cases show increased neuronal expression of pPERK (DAB, *brown*) in the CA1 region and subiculum of the hippocampus. Nuclei are stained with haematoxylin. (**B**) Also peIF2α (DAB, *brown*) expression in neurons in the hippocampus of AD cases is indicative of UPR activation in AD. (**C**) pPERK (DAB, *brown*) and peIF2α (AEC, *red*) colocalize in AD neurons in the hippocampus. (**D**) Double immunohistochemistry as shown in (C) analyzed using a spectral imaging system [89]. The different reaction products were spectrally unmixed: green, pPERK; *red*, peIF2α; *blue*, haematoxilin. (**E**) Double immunohistochemistry for pPERK (*red*) and α-synuclein (*blue*) shows double-labeling of melanin containing dopaminergic neurons in the substantia nigra in a PD case. (**F**) Staining as shown in (E) unmixed by spectral imaging: *red*, pPERK; *green*, α-synuclein; *blue*, haematoxilin; *white*, neuromelanin

phosphorylated tau. However, pPERK immunoreactivity can be observed in neurons that show diffuse labeling for phosphorylated tau. This suggests that an increase in UPR precedes the formation of NFTs. Interestingly, in our studies we find that Rab6, a protein involved in retrograde trafficking from the Golgi to the ER is upregulated in AD in a manner that correlates strongly with the levels of ER stress [68]. Our data indicate that the Rab6 pathway is a post-ER protein quality mechanism, and therefore it appears that an array of protein quality control in the secretory pathway is activated during AD pathogenesis [69].

Taken together, these data suggest that UPR activation is initiated as a protective response in the early stages of AD pathology to increase ER associated protein degradation. For instance, hHRD1, a recently identified ubiquitin ligase involved in the ER-associated protein degradation [70], has been found to be expressed in pretangle neurons in AD hippocampus [71].

Alternatively, ER stress may also aggravate the severity and/or accelerate the onset of the disease by favoring amyloidogenic processing of the amyloid precursor protein (APP) [72]. Although ER stress has been shown to be involved in AD pathogenesis, the aggregated proteins, i.e. tangles and plaques, are not found in the ER. This raises the question about the source of the ER stress. First indications that suggested that protein quality control mechanisms are involved in AD came from studies showing increased ubiquitin immunoreactivity. Ubiquitin colocalizes with NFTs, neuritic plaques and neuropil threads [73] and it has been suggested that dysfunction in ubiquitin-mediated proteolysis and the resulting accumulation of ubiquitin-conjugated proteins may contribute to the origination of dystrophic neurites and NFTs. However, it is unlikely that the UPR activation is secondary to decreased proteasomal activity, as the latter is observed only in the final stages of AD [74].

In vitro studies have demonstrated that $A\beta_{1-42}$ is generated in the ER of neuronal cells [75, 76] and although it is unclear how extracellular $A\beta$ signals to the ER, it has been shown that extracellular $A\beta$ can induce the UPR [77–79]. $A\beta$ induces Ca^{2+} release from the ER [80] and inhibition of Ca^{2+} release, which concomitantly attenuates the upregulation of BiP, partly relieves $A\beta$ toxicity [77]. In contrast, upregulation of BiP is also suggested to be protective in $A\beta$ toxicity [78]. These apparently contradicting observations may relate to the balance between protective and destructive responses of the UPR in AD pathology [69]. A direct correlation between intracellular $A\beta$ and activation of the UPR during AD pathology needs to be addressed in future studies.

5.3.2 Parkinson's Disease

The typical clinical features of PD are caused by loss of dopaminergic neurons from the substantia nigra pars compacta (SN). Pathologically, PD is characterized by the accumulation of misfolded proteins [61, 81]. PD neurons show increased presence of ubiquitinated protein deposits in the neuronal cytoplasm, called Lewy bodies, and protein inclusions in the neurites. Lewy bodies are composed largely of α-synuclein, a small presynaptic protein. The function of α-synuclein is still unclear, but its involvement in PD pathogenesis is further indicated by a subset of familial cases of PD that carry either a missense mutation in the α-synuclein gene or have a duplication of the α-synuclein locus [82–85].

The involvement of the UPR in PD has, primarily, been shown in cellular models using the PD mimetics 6-hydroxydopamine (6-OHDA), 1-methyl-4-phenyl-pyridinium (MPP^+) and rotenone [86]. More substantial evidence for a role of the UPR in PD pathogenesis comes from a juvenile onset autosomal recessive form of PD (AR-JP) that is caused by mutation of the parkin gene, which compromises the ubiquitin ligase function of the protein [87]. This loss of activity results in accumulation of a substrate of parkin in the ER of SN neurons, leading to ER stress and cell death [27]. In addition, overexpression of wild-type or mutant α-synuclein induces

UPR activation in yeast [88]. Although the involvement of the UPR in neuronal cell death in PD pathogenesis is widely suggested, there is hardly any data on UPR activation from *post mortem* studies on PD cases.

In a recent study we investigated the immunohistochemical expression and localization of pPERK and peIF2α in the substantia nigra of PD and control cases [89]. Immunoreactivity for pPERK and peIF2α was absent in control cases and in PD cases observed in neuromelanin containing neurons in the SN. Neurons in PD cases positive for pPERK contained neuromelanin and increased immunoreactivity for α-synuclein localized in the cytoplasm and/or α-synuclein positive Lewy bodies (Fig. 5.1e,f). These data suggest a strong relation between α-synuclein and ER stress in dopaminergic neurons. α-Synuclein could induce ER stress in different ways. Overexpression of mutant forms of α-synuclein in neuronal cells leads to formation of cytoplasmic aggregates and disruption of the UPS [90, 91]. Decreasing protein degradation could induce the protein load in the ER, making the cell more vulnerable for ER stress. Recently it was shown in different cellular models for PD that α-synuclein induces ER stress by disrupting ER-Golgi vesicular trafficking [88].

The initial activation of the UPR in PD pathogenesis might have a neuroprotective role, in an attempt to remove the neurotoxic unfolded proteins. However, on the long term ER stress and UPR activation play a critical role in neuronal cell death in PD pathogenesis. *In vitro* studies indicate that ER stress is involved in neuronal cell death promoted by parkinsonism-mimicking agents 6-OHDA, MPP$^+$ and rotenone [86, 92]. Also α-synuclein aggregation *in vitro* results in protein accumulation in the ER and activation of the UPR, resulting in cell death [88]. Removal of this burden on the ER by stimulating protein trafficking out of the ER protects dopaminergic neurons from α-synuclein induced cell death. These data indicate that prolonged endurance of ER stress could determine the final outcome towards neurodegeneration in PD.

5.3.3 Prion Disease

Prion disease or transmissible spongiform encephalopathies (TSEs) are fatal neurodegenerative disorders that occur in humans (eg. Creutzfeldt-Jakob (CJD), Gerstmann-Sträussler-Schenker disease (GSS), fatal familial insomnia (FFI), and Kuru). These disorders are characterized by rapidly progressed neuronal loss and by extracellular accumulation of the prion protein scrapie (PrP^{Sc}), a pathological isoform of the normal cellular prion protein (PrP).

In prion-diseased patients that show no neurofibrillary pathology, no immunoreactivity for pPERK or peIF2α is observed [64]. Investigation of pPERK in brains of scrapie-inoculated mice only showed low activation of PERK. These data suggest no significant role for ER stress in neuronal death in prion diseases. Nonetheless, increased caspase 12 activation and increased expression levels of ER-stress markers Grp58, Grp78, Grp94 and Hsp70 in sporadic CJD and variant CJD human brain samples have been reported [93]. It has been suggested that mutant PrP molecules are delayed in their exit from the ER and may therefore activate the UPR [94, 95]. In addition, the ER stress induced chaperone BiP interacts with mutant PrP retained in the ER and mediates the degradation of PrP by the proteasomal pathway [95]. These data suggest that the BiP plays a role in maintaining the quality control in the PrP maturation pathway. Nonetheless, activation of the UPR in prion disease remains controversial. The pro-apoptotic transcription factor CHOP is not induced *in vitro* in neuronal cells treated with PrP^{Sc} as well as in prion infected mice [96]. ER chaperones could be involved in the ER-associated degradation of PrP, however, the absence of phosphorylated PERK, necessary for the induction of the proapoptotic transcription factor CHOP, does not support that UPR activation is involved in neuronal cell loss in prion disease [64, 96]. Because prion disease has a very rapid progression, it is possible that the degeneration is too fast and severe at the time of autopsy to allow detection of all UPR markers.

5.3.4 Tauopathies

Prominent filamentous tau inclusions and neurodegeneration in the absence of Aβ deposits are hallmarks of tauopathies including sporadic corticobasal degeneration (CBD), progressive supranuclear palsy (PSP), and Pick's disease (PiD), as well as hereditary FTD and parkinsonism linked to chromosome 17 (FTDP-17). Mutations in the tau gene result in FTDP-17 and tau polymorphisms are genetic risk factors for PSP and CBD. This indicates that tau abnormalities are linked directly to the etiology and pathogenesis of these neurodegenerative disorders.

Unterberger and colleagues have described UPR activation in two cases with tauopathy (CBD/PSP) [64]. In both cases, activated PERK was detected in the frontal cortex and hippocampus, which also contained abnormal tau protein. In addition to neurons, also in astrocytes and oligodendrocytes immunoreactivity for pPERK could be observed. The presence of tau-positive glial inclusions is a consistent feature in the brains of patients with PSP, CBD and PiD [97]. Remarkably, in the tauopathy cases immunoreactivity for peIF2α could be observed in neurons, similar to that seen in AD, but remained absent in the glia. Double labeling experiments showed that pPERK colocalizes with phospho-tau in the neurons [64]. These data together with the observation that pPERK is associated with phospho-tau containing neurons in AD and prion disease cases [63, 64] suggest that UPR activation and tau phosphorylation are closely linked during neurodegeneration. The link between UPR activation and tau phosphorylation is still elusive. Several reports have shown that ER stress results in increased activation of glycogen synthase kinase 3β (GSK-3β) [98, 99]. GSK-3β is a physiological kinase for tau and is a candidate protein kinase involved in the hyperphosphorylation of tau present in NFTs [100]. Alternatively, treating cells with nocodazole, which interferes with the polymerization of microtubules, induces phosphorylation of eIF2α and increases the expression of Grp94 [101]. Together these data suggest a connection between ER stress and phosphorylation of tau in neurodegeneration.

5.3.5 Polyglutamine Diseases

Expanded polyglutamine (polyQ) repeats found in different proteins can cause human inherited neurodegenerative diseases, such as HD, spinobulbar muscular atrophy, dentatorubral-pallidoluysian atrophy and spinocerebellar ataxia (SCA). These disorders are characterized by accumulation of intracellular protein aggregates and selective neuronal death. The role of the polyQ expansions in neuronal death is under debate. So far it is unclear whether the polyQ protein aggregates are deleterious for neurons or work neuroprotective. The induction of mitochondrial dysfunction and subsequent excitotoxic injury, oxidative stress and apoptosis have been suggested to be involved in the pathogenesis of polyglutamine diseases. It has also been suggested that ER stress and UPR activation are involved in these disorders. The cytoplasmic accumulation of polyQ triggers ER stress by inhibiting the UPS, and subsequent caspase 12-mediated apoptosis [57, 102, 103]. *Post mortem* brain samples from patients with HD show a decrease in ER-associated α-glucosidase and fucosyl-transferase activities, suggesting that these changes reflect highly specific alterations in glycoprotein synthesis and processing [104]. It remains to be studied whether ER stress and UPR activation markers are present in brain samples derived from patients afflicted by polyQ disorders.

5.3.6 Amyotrophic Lateral Sclerosis

ALS is characterized by the degeneration of motorneurons in the spinal cord, cortex and brain stem, leading to muscle atrophy and paralysis [105]. Motor neuron loss is accompanied by reactive gliosis, intracytoplasmic neurofilament abnormalities, and axonal spheroids. In end-stage disease, there is significant loss of large myelinated fibers in the corticospinal tracts and ventral roots as well as evidence of Wallerian degeneration and atrophy of the myelinated fibers [105]. Most ALS cases are sporadic, but about 10% of the patients have an inherited variant of the disease, called familial ALS (FALS). Some patients with FALS have mutations in the Cu/Zn superoxide dismutase (SOD1) enzyme that is involved in quenching reactive oxygen species.

The striking pathological and clinical similarity between familial and sporadic disease suggests that animal models based on mutant SOD1 might provide insight into mechanisms of both sporadic and familial disease. A feature common to all examples of SOD1 mutant-mediated disease in mice is prominent ubiquitin-positive, intracellular aggregates of SOD1 in motor neurons and astroctyes [106, 107]. Aggregated SOD1 may in conjunction with other proteins impair the function of the UPS or adversely affect some other functions related to protein quality control within motorneurons. Using biochemical and morphological methods it has been demonstrated that mutant SOD1 accumulates inside the ER, where it forms insoluble high molecular weight species and interacts with BiP [108]. In addition, pPERK and peIF2α as well as activated caspase 12 have been observed in the spinal cord of transgenic mice carrying the mutant SOD1 [109, 110]. Also other markers for

ER stress, such as BiP, were altered in the ALS mice model. These data suggest that ER stress contributes to FALS and motorneuron degeneration. Although there is accumulating evidence for ER stress in SOD1 models, it remains to be resolved whether UPR activation can be detected in motorneurons of FALS and sporadic ALS patients.

5.3.7 White Matter Disorders

5.3.7.1 Multiple Sclerosis

Multiple sclerosis (MS) is an inflammatory disease of the central nervous system (CNS) associated with the development of large plaques of demyelination, oligodendrocyte destruction, and axonal degeneration. Signaling pathways involved in particular forms of cellular stress have been shown to be involved in the formation of these plaques, for example oxidative stress and mitochondrial dysfunction [111] and excitotoxicity [112]. Increased expression of CHOP, BiP, and XBP1 is detected in multiple cell types, including oligodendrocytes, astrocytes, T cells, and microglia in active MS lesions [113]. Unlike other neurodegenerative disorders, MS can not directly be characterized as a disease where protein misfolding plays a central role in disease pathogenesis. Probably, ER stress markers are upregulated in MS as response to ischemic injury or occur in association with an inflammatory response.

The role of the UPR in MS pathogenesis could be dual. Although ER stress could be involved in selective loss of oligodendrocytes, there are also indications that (parts) of the UPR are involved in preventing demyelination. Interferon (IFN)-gamma, which activates the ER stress response in oligodendrocytes, is believed to play a critical role in regulating the immune response in MS and its mouse model, experimental autoimmune encephalomyelitis (EAE). It is reported that CNS delivery of IFN-gamma before EAE onset ameliorates the disease course and prevents demyelination, axonal damage, and oligodendrocyte loss [114]. The beneficial effects of IFN-gamma are accompanied by PERK activation in oligodendrocytes and are abrogated in PERK-deficient animals. These results suggest that activation of PERK, by IFN-gamma, might work beneficial in MS [114].

5.3.7.2 Leukoencephalopathy with Vanishing White Matter

Leukoencephalopathy with Vanishing White Matter (VWM), also termed Childhood Ataxia with CNS Hypomyelination, is an autosomal recessive white matter disorder [115]. Its clinical course is chronic and progressive with intermittent episodes of major and rapid neurological deterioration. These episodes are provoked by minor head trauma and particularly febrile infections. VWM is caused by mutations in any of the five genes encoding the subunits of the eIF2B [116]. eIF2B plays an important role in the regulation of protein synthesis. Mutant eIF2B may impair the ability of cells to regulate protein synthesis under normal conditions and in response to stress. For unknown reasons, the central white matter appears to be selectively affected in VWM and there is increasing evidence that glial cell dysfunction is central to the pathophysiology of VWM. It has been shown that oligodendrocytes are being subjected to conflicting cell death and survival signals [117].

Recent studies demonstrate activation of the UPR in glial cells in patients with VWM [118, 119]. At the immunohistochemical level pPERK was observed in cells with the morphological feature of oligodendrocytes as well as in dysmorphic astrocytes. In addition, pPERK was observed in oligodendrocytes with abundant or foamy cytoplasm, which is characteristic of VWM. Next to pPERK also peIF2α, ATF6, BiP, and XBP1 can be observed in astrocytes and oligodendrocytes in cerebral white matter of VWM patients by immunohistochemistry [118, 119]. In addition, mild immunoreactivity for ATF4 and nuclear staining of CHOP can be detected in glial cells and cortical neurons. In comparison to normal controls ATF4 and CHOP were reported to be increased in VWM cases in protein extracts of the white matter [118]. At the mRNA level, splicing of XBP1 mRNA was observed in gray and white matter of VWM brain and not in control brain [119]. In addition, a significant increase in BiP, CHOP and GADD34 mRNA levels was reported in VWM brain compared to control brain.

These data show that activation of the UPR in VWM is mainly restricted to the white matter. The enhanced expression of UPR markers is observed in the cells, i.e. oligodendrocytes and astrocytes that are known to be involved in the pathology of VWM. It is therefore suggested that inappropriate activation of the UPR may play a role in the pathophysiology of VWM [119].

5.4 Conclusions

The UPR is activated in several neurodegenerative disorders (Table 5.1). Exploiting the ER stress response, by stimulating protective pathways or inhibiting degenerative responses, may be beneficial for treatment of various neuronal injuries and brain disorders. Pharmacological induction of molecular chaperones may protect against aggregate toxicity [120]. The therapeutic potential of such a strategy is best illustrated by treatment of mutant SOD1 mice with arimoclomol, a co-inducer of heat shock proteins [121]. This treatment resulted in decreased neuronal loss and consequently improved motorfunction, as well as increased lifespan. Another way to employ protein quality control to tackle neurodegeneration is via the UPR. Recently, the compound salubrinal was identified as an inhibitor of eIF2α dephosphorylation and shown to protect against ER stress-induced cell death [56, 122]. Salubrinal is protective in a PD cell model [123], and in ischemia induced ER stress in mice [124], indicating its potential for the treatment of neurodegenerative diseases. A drawback is that constitutive phosphorylation of eIF2α is unlikely to present a long-term treatment opportunity [122], therefore more selective targeting should be investigated. Valproate, a drug widely prescribed in the treatment of bipolar disorder and epilepsy, has been shown to increase the levels of BiP and other ER chaperones [125, 126]. Although valproate increases the expression of ER chaperones it has been reported that ER stress induced cellular dysfunction is reduced by valproate through inhibition of GSK-3α/β [127]. Valproate could work protective in AD reducing ER stress and reducing tau phosphorylation by GSK-3β. Currently, clinical trails have been started investigating the effect of valproate on AD patients.

A very interesting finding was recently reported by Kondo and colleagues. They report that a transcription factor called OASIS specifically stimulates BiP upregulation (a positive response), but down-regulates the ER stress induced cell death pathways [128]. OASIS is specifically induced in astrocytes, but factors with similar properties may exist in neurons as well, and present an ideal therapeutic target by stimulating positive and down-regulating negative responses through a single protein.

Nonetheless, in view of future therapeutic possibilities directed at protein quality control in neurodegeneration it is important to know which role the UPR has in neurodegeneration and to know more about which part of the ER stress response to target. Despite the accumulating evidence for the presence of ER stress and UPR markers in neurodegenerative disorders (Table 5.1), most research on ER stress has mainly been performed on *in vitro* and *in vivo* models for neurodegenerative disorders. Future studies on UPR markers in human brain tissue will increase our knowledge on the role of ER stress in neurodegenerative diseases.

Acknowledgments The authors thank Elise van Haastert and Chris van der Loos for their technical assistance and assistance with the spectral imaging system. Work discussed in this chapter has been supported by the Netherlands Organisation for Scientific Research (NWO; Meervoud Grant (#836.05.060) to W.S. and VENI Grant (#916.76.013) to J.J.M.H.), the Internationale Stichting Alzheimer Onderzoek (ISAO grant 05508 and 07506 to W.S.), and the EU 6th Framework Program (EDAR).

Abbreviations

Aβ	β amyloid
AD	Alzheimer's disease
ALS	amyotrophic lateral sclerosis
AR-JP	autosomal-recessive juvenile parkinsonism
ATF	activating transcription factor
Atg	autophagy related protein
BiP	heavy chain binding protein
CBD	corticobasal degeneration
CHOP	C/EBP homologous protein
CJD	Creutzfeldt-Jakob disease
CNS	central nervous system
EAE	experimental autoimmune encephalomyelitis
ER	endoplasmic reticulum
ERAD	ER associated degradation
eIF	eukaryotic initiation factor
FALS	familial amyotrophic lateral sclerosis

FTD	Frontotemporal dementia
FTDP-17	FTD and parkinsomism linked to chromosome 17
GADD	growth arrest- and DNA damage-inducible gene
Grp	glucose-regulated protein
GSK	glycogen synthase kinase
HD	Huntington's disease
Hsp	heath shock protein
IFN	interferon
Ire1	inositol-requiring enzyme 1
LC3	light chain 3 protein
MPP+	1-methyl-4-phenyl-pyridinium
MS	multiple sclerosis
NFT	neurofibrillary tangle
6-OHDA	6-hydroxydopamine
PD	Parkinson's disease
PERK	protein kinase R (PKR)-like ER kinase
PiD	Pick's disease
polyQ	polyglutamine
PrP	prion protein
PSP	progressive supranuclear palsy
SN	substantia nigra pars compacta
SOD1	Cu/Zn superoxide dismutase
UPR	unfolded protein response
UPS	ubiquitin proteasome system
VWM	vanishing white matter
XBP1	X-Box binding protein 1

References

1. Richter-Landsberg C, Goldbaum O (2003) Stress proteins in neural cells: functional roles in health and disease. Cell Mol Life Sci 60:337–349
2. Lee S, Tsai FT (2005) Molecular chaperones in protein quality control. J Biochem Mol Biol 38:259–265
3. Mayer MP, Bukau B (2005) Hsp70 chaperones: cellular functions and molecular mechanism. Cell Mol Life Sci 62:670–684
4. Senderek J, Krieger M, Stendel C et al (2005) Mutations in SIL1 cause Marinesco-Sjogren syndrome, a cerebellar ataxia with cataract and myopathy. Nat Genet 37:1312–1314
5. Zhao L, Longo-Guess C, Harris BS et al (2005) Protein accumulation and neurodegeneration in the woozy mutant mouse is caused by disruption of SIL1, a cochaperone of BiP. Nat Genet 37:974–979
6. Tyson JR, Stirling CJ (2000) LHS1 and SIL1 provide a lumenal function that is essential for protein translocation into the endoplasmic reticulum. EMBO J 19:6440–6452
7. Meusser B, Hirsch C, Jarosch E et al (2005) ERAD: the long road to destruction. Nat Cell Biol 7:766–772
8. Nakatsukasa K, Brodsky JL (2008) The recognition and retrotranslocation of misfolded proteins from the endoplasmic reticulum. Traffic 9:861–870.
9. Denic V, Quan EM, Weissman JS (2006) A luminal surveillance complex that selects misfolded glycoproteins for ER-associated degradation. Cell 126:349–359

10. Carvalho P, Goder V, Rapoport TA (2006) Distinct ubiquitin-ligase complexes define convergent pathways for the degradation of ER proteins. Cell 126:361–373
11. Molinari M (2007) N-glycan structure dictates extension of protein folding or onset of disposal. Nat Chem Biol. 3:313–320
12. Molinari M, Calanca V, Galli C et al (2003) Role of EDEM in the release of misfolded glycoproteins from the calnexin cycle. Science 299:1397–1400
13. Rapoport TA (2007) Protein translocation across the eukaryotic endoplasmic reticulum and bacterial plasma membranes. Nature 450:663–669
14. Johnson AE, Haigh NG (2000) The ER translocon and retrotranslocation: is the shift into reverse manual or automatic? Cell 102:709–712
15. Lilley BN, Ploegh HL (2004) A membrane protein required for dislocation of misfolded proteins from the ER. Nature 429:834–840
16. Wahlman J, DeMartino GN, Skach WR et al (2007) Real-time fluorescence detection of ERAD substrate retrotranslocation in a mammalian in vitro system. Cell 129:943–955
17. Katiyar S, Joshi S, Lennarz WJ (2005) The retrotranslocation protein Derlin-1 binds peptide: N-glycanase to the endoplasmic reticulum. Mol Biol Cell 16:4584–4594
18. Rabinovich E, Kerem A, Frohlich KU et al (2002) AAA-ATPase p97/Cdc48p, a cytosolic chaperone required for endoplasmic reticulum-associated protein degradation. Mol Cell Biol 22:626–634
19. Kalies KU, Allan S, Sergeyenko T et al (2005) The protein translocation channel binds proteasomes to the endoplasmic reticulum membrane. EMBO J 24:2284–2293
20. Ng W, Sergeyenko T, Zeng N et al (2007) Characterization of the proteasome interaction with the Sec61 channel in the endoplasmic reticulum. J Cell Sci 120:682–691
21. Oyadomari S, Yun C, Fisher EA et al (2006) Cotranslocational degradation protects the stressed endoplasmic reticulum from protein overload. Cell 126:727–739
22. Goldberg AL (2003) Protein degradation and protection against misfolded or damaged proteins. Nature 426:895–899
23. Varshavsky A (2005) Regulated protein degradation. Trends Biochem Sci 30:283–286
24. Hol EM, Fischer DF, Ovaa H et al (2006) Ubiquitin proteasome system as a pharmacological target in neurodegeneration. Expert Rev Neurother 6:1337–1347
25. Gao M, Karin M (2005) Regulating the regulators: control of protein ubiquitination and ubiquitin-like modifications by extracellular stimuli. Mol Cell 19:581–593
26. Hicke L, Schubert HL, Hill CP (2005) Ubiquitin-binding domains. Nat Rev Mol Cell Biol 6:610–621
27. Imai Y, Soda M, Inoue H et al (2001) An unfolded putative transmembrane polypeptide, which can lead to endoplasmic reticulum stress, is a substrate of Parkin. Cell 105:891–902
28. Ciechanover A (2005) Intracellular protein degradation: from a vague idea thru the lysosome and the ubiquitin–proteasome system and onto human diseases and drug targeting. Cell Death Differ 12:1178–1190
29. Elsasser S, Finley D (2005) Delivery of ubiquitinated substrates to protein-unfolding machines. Nat Cell Biol 7:742–749
30. Richly H, Rape M, Braun S et al (2005) A series of ubiquitin binding factors connects CDC48/p97 to substrate multiubiquitylation and proteasomal targeting. Cell 120:73–84
31. Pickart CM, Cohen RE (2004) Proteasomes and their kin: proteases in the machine age. Nat Rev Mol Cell Biol 5:177–187
32. Reggiori F, Klionsky DJ (2005) Autophagosomes: biogenesis from scratch? Curr Opin Cell Biol 17:415–422
33. Mizushima N, Yoshimori T (2007) How to interpret LC3 immunoblotting. Autophagy 3:542–545
34. Levine B, Yuan J (2005) Autophagy in cell death: an innocent convict? J Clin Invest 115:2679–2688
35. Hara T, Nakamura K, Matsui M et al (2006) Suppression of basal autophagy in neural cells causes neurodegenerative disease in mice. Nature 441:885–889

36. Komatsu M, Waguri S, Chiba T et al (2006) Loss of autophagy in the central nervous system causes neurodegeneration in mice. Nature 441:880–884
37. Mizushima N, Hara T (2006) Intracellular quality control by autophagy: how does autophagy prevent neurodegeneration? Autophagy 2:302–304
38. Iwata A, Riley BE, Johnston JA et al (2005) HDAC6 and microtubules are required for autophagic degradation of aggregated huntingtin. J Biol Chem 280:40282–40292
39. Iwata A, Christianson JC, Bucci M et al (2005) Increased susceptibility of cytoplasmic over nuclear polyglutamine aggregates to autophagic degradation. Proc Natl Acad Sci USA 102:13135–13140
40. Ding WX, Yin XM (2008) Sorting, recognition and activation of the misfolded protein degradation pathways through macroautophagy and the proteasome. Autophagy 4:141–150
41. Ding WX, Ni HM, Gao W et al (2007) Differential effects of endoplasmic reticulum stress-induced autophagy on cell survival. J Biol Chem 282:4702–4710
42. Ding WX, Ni HM, Gao W et al (2007) Linking of autophagy to ubiquitin–proteasome system is important for the regulation of endoplasmic reticulum stress and cell viability. Am J Pathol 171:513–524
43. Schroder M, Kaufman RJ (2005) The mammalian unfolded protein response. Annu Rev Biochem 74:739–789
44. Scheuner D, Song B, McEwen E et al (2001) Translational control is required for the unfolded protein response and in vivo glucose homeostasis. Mol Cell 7:1165–1176
45. Wek RC, Cavener DR (2007) Translational control and the unfolded protein response. Antioxid Redox Signal 9:2357–2371
46. Szegezdi E, Logue SE, Gorman AM et al (2006) Mediators of endoplasmic reticulum stress-induced apoptosis. EMBO Rep 7:880–885
47. Lee K, Tirasophon W, Shen X et al (2002) IRE1-mediated unconventional mRNA splicing and S2P-mediated ATF6 cleavage merge to regulate XBP1 in signaling the unfolded protein response. Genes Dev 16:452–466
48. Yoshida H (2007) Unconventional splicing of XBP-1 mRNA in the unfolded protein response. Antioxid Redox Signal 9:2323–2333
49. Yoshida H, Oku M, Suzuki M et al (2006) pXBP1(U) encoded in XBP1 pre-mRNA negatively regulates unfolded protein response activator pXBP1(S) in mammalian ER stress response. J Cell Biol 172:565–575
50. Yoshida H, Matsui T, Hosokawa N et al (2003) A time-dependent phase shift in the mammalian unfolded protein response. Dev Cell 4:265–271
51. Yamamoto K, Sato T, Matsui T et al (2007) Transcriptional induction of mammalian ER quality control proteins is mediated by single or combined action of ATF6alpha and XBP1. Dev Cell 13:365–376
52. Hitomi J, Katayama T, Eguchi Y et al (2004) Involvement of caspase-4 in endoplasmic reticulum stress-induced apoptosis and Abeta-induced cell death. J Cell Biol 165:347–356
53. Nakagawa T, Zhu H, Morishima N et al (2000) Caspase-12 mediates endoplasmic-reticulum-specific apoptosis and cytotoxicity by amyloid-beta. Nature 403:98–103
54. Oyadomari S, Mori M (2004) Roles of CHOP/GADD153 in endoplasmic reticulum stress. Cell Death Differ 11:381–389
55. Zinszner H, Kuroda M, Wang X et al (1998) CHOP is implicated in programmed cell death in response to impaired function of the endoplasmic reticulum. Genes Dev 12:982–995
56. Boyce M, Bryant KF, Jousse C et al (2005) A selective inhibitor of eIF2alpha dephosphorylation protects cells from ER stress. Science 307:935–939
57. Nishitoh H, Matsuzawa A, Tobiume K et al (2002) ASK1 is essential for endoplasmic reticulum stress-induced neuronal cell death triggered by expanded polyglutamine repeats. Genes Dev 16:1345–1355
58. Dahlmann B (2007) Role of proteasomes in disease. BMC Biochem 8(Suppl 1):S3
59. Katzman R (1986) Alzheimer's disease. N Engl J Med 314:964–973

60. Butterfield DA, Kanski J (2001) Brain protein oxidation in age-related neurodegenerative disorders that are associated with aggregated proteins. Mech Ageing Dev 122:945–962
61. Taylor JP, Hardy J, Fischbeck KH (2002) Toxic proteins in neurodegenerative disease. Science 296:1991–5
62. Hamos JE, Oblas B, Pulaski-Salo D et al (1991) Expression of heat shock proteins in Alzheimer's disease. Neurology 41:345–50
63. Hoozemans JJ, Veerhuis R, Van Haastert ES et al (2005) The unfolded protein response is activated in Alzheimer's disease. Acta Neuropathol (Berl) 110:165–172
64. Unterberger U, Hoftberger R, Gelpi E et al (2006) Endoplasmic reticulum stress features are prominent in Alzheimer disease but not in prion diseases in vivo. J Neuropathol Exp Neurol 65:348–357
65. Katayama T, Imaizumi K, Sato N et al (1999) Presenilin-1 mutations downregulate the signalling pathway of the unfolded-protein response. Nat Cell Biol 1:479–485
66. Sato N, Urano F, Yoon Leem J et al (2000) Upregulation of BiP and CHOP by the unfolded-protein response is independent of presenilin expression. Nat Cell Biol 2:863–870
67. Chang RC, Wong AK, Ng HK et al (2002) Phosphorylation of eukaryotic initiation factor-2alpha (eIF2alpha) is associated with neuronal degeneration in Alzheimer's disease. Neuroreport 13:2429–2432
68. Scheper W, Hoozemans JJ, Hoogenraad CC et al (2007) Rab6 is increased in Alzheimer's disease brain and correlates with endoplasmic reticulum stress. Neuropathol Appl Neurobiol 33:523–532
69. Scheper W, Hol EM (2005) Protein quality control in Alzheimer's disease: a fatal saviour. Curr Drug Targets CNS Neurol Disord 4:283–292
70. Nadav E, Shmueli A, Barr H et al (2003) A novel mammalian endoplasmic reticulum ubiquitin ligase homologous to the yeast Hrd1. Biochem Biophys Res Commun 303:91–97
71. Hou HL, Shen YX, Zhu HY et al (2006) Alterations of hHrd1 expression are related to hyperphosphorylated tau in the hippocampus in Alzheimer's disease. J Neurosci Res 84:1862–1870
72. Piccini A, Fassio A, Pasqualetto E et al (2004) Fibroblasts from FAD-linked presenilin 1 mutations display a normal unfolded protein response but overproduce Abeta42 in response to tunicamycin. Neurobiol Dis 15:380–386
73. Perry G, Friedman R, Shaw G et al (1987) Ubiquitin is detected in neurofibrillary tangles and senile plaque neurites of Alzheimer disease brains. Proc Natl Acad Sci USA 84:3033–3036
74. van Leeuwen FW, de Kleijn DP, van den Hurk HH et al (1998) Frameshift mutants of beta amyloid precursor protein and ubiquitin-B in Alzheimer's and Down patients. Science 279:242–247
75. Cook DG, Forman MS, Sung JC et al (1997) Alzheimer's A beta(1-42) is generated in the endoplasmic reticulum/intermediate compartment of NT2N cells. Nat Med 3:1021–1023
76. Hartmann T, Bieger SC, Bruhl B et al (1997) Distinct sites of intracellular production for Alzheimer's disease A beta40/42 amyloid peptides. Nat Med 3:1016–1020
77. Suen KC, Lin KF, Elyaman W et al (2003) Reduction of calcium release from the endoplasmic reticulum could only provide partial neuroprotection against beta-amyloid peptide toxicity. J Neurochem 87:1413–26
78. Yu Z, Luo H, Fu W et al (1999) The endoplasmic reticulum stress-responsive protein GRP78 protects neurons against excitotoxicity and apoptosis: suppression of oxidative stress and stabilization of calcium homeostasis. Exp Neurol 155:302–314
79. Chafekar SM, Hoozemans JJ, Zwart R et al (2007) Abeta 1–42 induces mild endoplasmic reticulum stress in an aggregation state-dependent manner. Antioxid Redox Signal 9:2245–2254
80. Suen KC, Yu MS, So KF et al (2003) Upstream signaling pathways leading to the activation of double-stranded RNA-dependent serine/threonine protein kinase in beta-amyloid peptide neurotoxicity. J Biol Chem 278:49819–49827

81. Rao RV, Bredesen DE (2004) Misfolded proteins, endoplasmic reticulum stress and neurodegeneration. Curr Opin Cell Biol 16:653–662
82. Polymeropoulos MH, Lavedan C, Leroy E et al (1997) Mutation in the alpha-synuclein gene identified in families with Parkinson's disease. Science 276:2045–2047
83. Kruger R, Kuhn W, Muller T et al (1998) Ala30Pro mutation in the gene encoding alpha-synuclein in Parkinson's disease. Nat Genet 18:106–108
84. Chartier-Harlin MC, Kachergus J, Roumier C et al (2004) Alpha-synuclein locus duplication as a cause of familial Parkinson's disease. Lancet 364:1167–1169
85. Ibanez P, Bonnet AM, Debarges B et al (2004) Causal relation between alpha-synuclein gene duplication and familial Parkinson's disease. Lancet 364:1169–1171
86. Ryu EJ, Harding HP, Angelastro JM et al (2002) Endoplasmic reticulum stress and the unfolded protein response in cellular models of Parkinson's disease. J Neurosci 22: 10690–10698
87. Shimura H, Hattori N, Kubo S et al (2000) Familial Parkinson disease gene product, parkin, is a ubiquitin-protein ligase. Nat Genet 25:302–305
88. Cooper AA, Gitler AD, Cashikar A et al (2006) Alpha-synuclein blocks ER-Golgi traffic and Rab1 rescues neuron loss in Parkinson's models. Science 313:324–328
89. Hoozemans JJ, Van Haastert ES, Eikelenboom P et al (2007) Activation of the unfolded protein response in Parkinson's disease. Biochem Biophys Res Commun 354:707–711
90. Tanaka Y, Engelender S, Igarashi S et al (2001) Inducible expression of mutant alpha-synuclein decreases proteasome activity and increases sensitivity to mitochondria-dependent apoptosis. Hum Mol Genet 10:919–926
91. Stefanis L, Larsen KE, Rideout HJ et al (2001) Expression of A53T mutant but not wild-type alpha-synuclein in PC12 cells induces alterations of the ubiquitin-dependent degradation system, loss of dopamine release, and autophagic cell death. J Neurosci 21:9549–9560
92. Silva RM, Ries V, Oo TF et al (2005) CHOP/GADD153 is a mediator of apoptotic death in substantia nigra dopamine neurons in an in vivo neurotoxin model of parkinsonism. J Neurochem 95:974–986
93. Hetz C, Russelakis-Carneiro M, Maundrell K et al (2003) Caspase-12 and endoplasmic reticulum stress mediate neurotoxicity of pathological prion protein. EMBO J 22: 5435–5445
94. Drisaldi B, Stewart RS, Adles C et al (2003) Mutant PrP is delayed in its exit from the endoplasmic reticulum, but neither wild-type nor mutant PrP undergoes retrotranslocation prior to proteasomal degradation. J Biol Chem 278:21732–21743
95. Jin T, Gu Y, Zanusso G et al (2000) The chaperone protein BiP binds to a mutant prion protein and mediates its degradation by the proteasome. J Biol Chem 275:38699–38704
96. Hetz C, Russelakis-Carneiro M, Walchli S et al (2005) The disulfide isomerase Grp58 is a protective factor against prion neurotoxicity. J Neurosci 25:2793–2802
97. Komori T (1999) Tau-positive glial inclusions in progressive supranuclear palsy, corticobasal degeneration and Pick's disease. Brain Pathol 9:663–679
98. Song L, De Sarno P, Jope RS (2002) Central role of glycogen synthase kinase-3beta in endoplasmic reticulum stress-induced caspase-3 activation. J Biol Chem 277:44701–44708
99. Brewster JL, Linseman DA, Bouchard RJ et al (2006) Endoplasmic reticulum stress and trophic factor withdrawal activate distinct signaling cascades that induce glycogen synthase kinase-3 beta and a caspase-9-dependent apoptosis in cerebellar granule neurons. Mol Cell Neurosci 32:242–253
100. Jope RS, Johnson GV (2004) The glamour and gloom of glycogen synthase kinase-3. Trends Biochem Sci 29:95–102
101. Seyb KI, Ansar S, Bean J et al (2006) beta-Amyloid and endoplasmic reticulum stress responses in primary neurons: effects of drugs that interact with the cytoskeleton. J Mol Neurosci 28:111–123
102. Kouroku Y, Fujita E, Jimbo A et al (2002) Polyglutamine aggregates stimulate ER stress signals and caspase-12 activation. Hum Mol Genet 11:1505–1515

103. Bence NF, Sampat RM, Kopito RR (2001) Impairment of the ubiquitin–proteasome system by protein aggregation. Science 292:1552–1555
104. Cross AJ, Crow TJ, Johnson JA et al (1985) Loss of endoplasmic reticulum-associated enzymes in affected brain regions in Huntington's disease and Alzheimer-type dementia. J Neurol Sci 71:137–143
105. Bruijn LI, Miller TM, Cleveland DW (2004) Unraveling the mechanisms involved in motor neuron degeneration in ALS. Annu Rev Neurosci 27:723–749
106. Bruijn LI, Becher MW, Lee MK et al (1997) ALS-linked SOD1 mutant G85R mediates damage to astrocytes and promotes rapidly progressive disease with SOD1-containing inclusions. Neuron 18:327–338
107. Bruijn LI, Houseweart MK, Kato S et al (1998) Aggregation and motor neuron toxicity of an ALS-linked SOD1 mutant independent from wild-type SOD1. Science 281:1851–1854
108. Kikuchi H, Almer G, Yamashita S et al (2006) Spinal cord endoplasmic reticulum stress associated with a microsomal accumulation of mutant superoxide dismutase-1 in an ALS model. Proc Natl Acad Sci USA 103:6025–6030
109. Wootz H, Hansson I, Korhonen L et al (2004) Caspase-12 cleavage and increased oxidative stress during motoneuron degeneration in transgenic mouse model of ALS. Biochem Biophys Res Commun 322:281–286
110. Nagata T, Ilieva H, Murakami T et al (2007) Increased ER stress during motor neuron degeneration in a transgenic mouse model of amyotrophic lateral sclerosis. Neurol Res 29:767–771
111. Lu F, Selak M, O'Connor J et al (2000) Oxidative damage to mitochondrial DNA and activity of mitochondrial enzymes in chronic active lesions of multiple sclerosis. J Neurol Sci 177:95–103
112. Werner P, Pitt D, Raine CS (2001) Multiple sclerosis: altered glutamate homeostasis in lesions correlates with oligodendrocyte and axonal damage. Ann Neurol 50:169–180
113. Mhaille AN, McQuaid S, Windebank A et al (2008) Increased expression of endoplasmic reticulum stress-related signaling pathway molecules in multiple sclerosis lesions. J Neuropathol Exp Neurol 67:200–211
114. Lin W, Bailey SL, Ho H et al (2007) The integrated stress response prevents demyelination by protecting oligodendrocytes against immune-mediated damage. J Clin Invest 117:448–456
115. van der Knaap MS, Barth PG, Gabreels FJ et al (1997) A new leukoencephalopathy with vanishing white matter. Neurology 48:845–855
116. van der Knaap MS, Leegwater PA, Konst AA et al (2002) Mutations in each of the five subunits of translation initiation factor eIF2B can cause leukoencephalopathy with vanishing white matter. Ann Neurol 51:264–270
117. Van Haren K, van der Voorn JP, Peterson DR et al (2004) The life and death of oligodendrocytes in vanishing white matter disease. J Neuropathol Exp Neurol 63:618–630
118. van der Voorn JP, van Kollenburg B, Bertrand G et al (2005) The unfolded protein response in vanishing white matter disease. J Neuropathol Exp Neurol 64:770–775
119. van Kollenburg B, van Dijk J, Garbern J et al (2006) Glia-specific activation of all pathways of the unfolded protein response in vanishing white matter disease. J Neuropathol Exp Neurol 65:707–715
120. Soti C, Nagy E, Giricz Z et al (2005) Heat shock proteins as emerging therapeutic targets. Br J Pharmacol 146:769–780
121. Kieran D, Kalmar B, Dick JR et al (2004) Treatment with arimoclomol, a coinducer of heat shock proteins, delays disease progression in ALS mice. Nat Med 10:402–405
122. Wiseman RL, Balch WE (2005) A new pharmacology–drugging stressed folding pathways. Trends Mol Med 11:347–350
123. Smith WW, Jiang H, Pei Z et al (2005) Endoplasmic reticulum stress and mitochondrial cell death pathways mediate A53T mutant alpha-synuclein-induced toxicity. Hum Mol Genet 14:3801–3811

124. Zhu Y, Fenik P, Zhan G et al (2008) Eif-2a protects brainstem motoneurons in a murine model of sleep apnea. J Neurosci 28:2168–2178
125. Wang JF, Bown C, Young LT (1999) Differential display PCR reveals novel targets for the mood-stabilizing drug valproate including the molecular chaperone GRP78. Mol Pharmacol 55:521–527
126. Bown CD, Wang JF, Young LT (2000) Increased expression of endoplasmic reticulum stress proteins following chronic valproate treatment of rat C6 glioma cells. Neuropharmacology 39:2162–2169
127. Kim AJ, Shi Y, Austin RC et al (2005) Valproate protects cells from ER stress-induced lipid accumulation and apoptosis by inhibiting glycogen synthase kinase-3. J Cell Sci 118:89–99
128. Kondo S, Murakami T, Tatsumi K et al (2005) OASIS, a CREB/ATF-family member, modulates UPR signalling in astrocytes. Nat Cell Biol 7:186–194

Chapter 6
Involvement of Alpha-2 Domain in Prion Protein Conformationally-Induced Diseases

Luisa Ronga, Pasquale Palladino, Ettore Benedetti, Raffaele Ragone and Filomena Rossi

Abstract A large number of human disorders, ranging from type II diabetes to Parkinson's and Alzheimer's diseases, are associated with protein aggregation resulting from aberrant folding or processing events. Despite its fundamental biological importance, little is known about the molecular basis or specificity of the general phenomenon of protein aggregation. Transmissible spongiform encephalopathies, also known as prion diseases, belong to this class. They are all characterized by progressive neuronal degeneration. In almost all cases there is a marked extracellular accumulation of an amyloidogenic conformer of the normal cellular prion protein (PrP^C), referred to as the scrapie isoform (PrP^{Sc}), which is thought to be responsible for the disease symptoms.

PrP is an ubiquitous 231-amino acid glycoprotein whose physiological role is still elusive. Its structure exhibits an N-terminal unfolded region and a C-terminal globular domain characterized by the presence of three α-helices (α1, α2 and α3), two short β-strands and an interhelical disulphide bridge, which confers structural stability. Particularly fascinating is the notion that the protein possesses one or several 'spots' of intrinsic conformational weakness, which may lead the whole secondary and tertiary structure to succumb in favour of more stable, but aggregation-prone conformations, depending on pH, redox condition or glycosylation. The C-terminal side of helix 2, containing four adjacent threonines, is decidedly suspected to be one of such spots and, in this regard, has recently gained the attention of several investigations. As α-helix 2 possesses chameleon conformational behaviour, gathering several disease-associated point mutations, it can be toxic to neuronal cells and strongly fibrillogenic and therefore, it is a suitable model to investigate both structural determinants of PrP^C misfolding and rational structure-based drug design of compounds able to block or prevent prion diseases. Huge spectroscopical, computational and

---

F. Rossi (✉)
Dipartimento delle Scienze Biologiche-Sez. Biostructure, Università degli Studi "Federico II" di Napoli, Via Mezzocannone, 16, I-80134 Napoli, Italy
e-mail: filomena.rossi@unina.it

biological literature data underline the intriguing structural properties of the α-2 C-globular domain.

6.1 Conformational Diseases

An increasing family of neurodegenerative disorders such as Alzheimer's, Parkinson's and Huntington's diseases and cystic fibrosis are currently classified as conformational diseases, which is a family of disorders where cellular functions are compromised because of protein misfolding and aggregation [1, 2]. In this context, prion diseases are a group of transmissible neurodegenerative disorders that enclose scrapie in sheep, spongiform encephalopathy in cattle (BSE), Creutzfeldt–Jakob disease (CJD), fatal insomnia and Gerstmann–Sträussler–Scheinker (GSS) disease in humans [1].

Since all members of this family of diseases are linked to a mechanism of aberrant protein folding, knowledge of the three-dimensional structure of the proteins implicated, both in their healthy and pathological forms, is a prerequisite for understanding the mechanism of aggregate formation and, eventually, preventing it. Yet, only relatively limited structural information is currently available. It is believed that the pathogenesis of these diseases is to be ascribed to reduced or lacking efficiency of physiological quality control systems, which leads to the formation of toxic protein aggregates, possibly affecting cellular function and eventually causing neuronal death [2]. Evidence has been accumulated that these aggregates possess various supramolecular architectures and, in most cases, form insoluble fibrillar deposits with well-defined structure, called amyloids [3]. A causative link between aggregation and disease is not, however, universally acknowledged, because amyloid fibril formation might be simply the consequence of a pathogenetic mechanism that could reside in causes yet to be identified [2]. The most widely accepted explanation for aggregation and amyloid formation is that the native fold of a protein isomerises to an improperly folded conformation prior to a structural reorganization resulting in protein aggregation and deposition. Fibrils are not toxic in themselves, but the quick β-strand-bonding-driven autolinkage of polypeptide chains may easily cause further linkage that leads to insoluble macrostructures with inflammatory or, more in general, toxic properties [4].

The term amyloidosis applies when deposition of such macrostructures in the tissue is a dominant, histologically apparent feature [5]. Amyloidosis is characterized by the accumulation of abnormal proteinaceous deposits in cell compartments and/or within the extra-cellular matrix, in which amyloid fibrils share a cross-β core structure [3, 6]. Amyloid formation can also occur when the plasma concentration of normal proteins is persistently increased, as with acute-phase proteins and immunoglobulins in chronic inflammation [7].

Such structural rearrangements likely take place in a class of degenerative neurological disorders involving the host-encoded prion protein (PrP), which are usually identified as Transmissible Spongiform Encephalopathies (TSE) [8]. As the general features of prion diseases are common to other amyloid disorders [2], the prion protein is adopted as a basic model for providing a comprehensive evaluation of protein misfolding mechanisms.

6.2 Prion Biology

In the most accredited model of infectious prion formation and replication, it is proposed that the pathogenic PrP^{Sc} acts as a template for direct interaction with the endogenous PrP^C substrate, thus driving the formation of nascent infectious prions [8]. Characterizing the exact intracellular localization of PrP^C and PrP^{Sc} is important for identifying the compartment and mechanism that underlie prion formation. Several studies have reported that the intracellular localization of some inherited pathological PrP mutants is altered, thus supporting the view that PrP^C misfolding and/or malfunction may correlate with defects in its trafficking [9]. It is not yet clear, however, whether mislocalization is the cause or the effect [10].

After modification with simple N-linked oligosaccharides and the glycosylphosphatidylinositol (GPI)-anchor in the endoplasmic reticulum and further oligosaccharide addition in the Golgi apparatus, most PrP^C arrives at the cell surface, where it is largely located in detergent-resistant microdomains (DRM) known as rafts or caveolae [11]. Transfected-cell studies have clarified that wild-type PrP cyclically moves between the cell surface and an early endocytic compartment, by association with clathrin-coated pits [12], which however may be completely bypassed by migration to endosomes or lysosomes through caveolae-containing endocytic structures [11]. Such classic and non-classic variations in PrP^C endocytic trafficking could be a sign of the cell type where exogenous PrP was expressed. Interference in normal intracellular trafficking can cause retrograde PrP^C transport toward the endoplasmic reticulum, where PrP^{Sc} may abnormally accumulate. The exact localization of PrP^C to PrP^{Sc} conversion is not known, but DRM [12], the endosomal pathway [13], and the endoplasmic reticulum (especially in familial TSE) are good candidates [14]. Initial PrP^{Sc} propagation during intercellular spread could take place in DRM, because insertion of PrP^{Sc} into the cell seems to be a prerequisite for membrane-associated conversion. Indeed, cell-free conversion models strongly suggest that physical contiguity is needed when different membrane components house PrP^C and PrP^{Sc} [15]. Converging evidence suggests that prion disorders are the result of a delicate balance between PrP^C synthesis followed by degradation and cellular quality control mechanisms, since misfolded protein is not detectable under usual conditions. Skillful alteration of biosynthetic and decomposition pathways has identified possible mediators of the complex PrP-related toxic effects. When proteasome activity is compromised, wild-type PrP accumulates in the cytoplasm, which correlates with PrP^{Sc}–free toxicity and neurodegeneration [16, 17]. However, other studies suggest that cytoplasmic accumulation of PrP^C may indicate that transit of the nascent PrP peptide to the endoplasmic reticulum is absent when PrP is overexpressed [18]. Nonetheless, as a common effect on the trafficking of

PrP, several pathogenic mutations impair delivery of PrP to the cell surface and cause its accumulation in the endoplasmic reticulum and cytoplasm in the absence of proteasomal inhibition [14]. Cytosolic accumulation of PrP causes PrP^{Sc}-like conversion and aggregation, which is able to self-perpetuate despite only transient proteasome inhibition [16]. Thus, unlike other proteasomaly-degraded proteins, PrP has an inherent capacity to promote and sustain its own conformational conversion in mammalian cells. Importantly, *in vitro* toxic effects did not correlate with appearance of PrP^{Sc} [17].

Overall, this suggests that age-related neurodegenerative diseases are governed by a generic mechanism, which allows harmful accumulation of soluble conformers as a consequence of compromise of quality control of endoplasmic reticulum protein synthesis from whatever cause. Once present, post-translationally PrP^{C}-derived PrP^{Sc} seems to catalyze amplified conversion of PrP^{C} to the abnormal TSE-associated form, needing at least temporary specific PrP^{C}-PrP^{Sc} interactions. In a cell-free system composed of substantially purified constituents [18], this property of PrP^{Sc} has been the basis for reproducing *in vitro* many of the species and strain characteristics noted in TSE. Knowledge of the *in vivo* pathway that leads to the transformation of PrP^{C} into PrP^{Sc} needs further studies, although it is likely that this proceeds through folding intermediates, including molten globule forms, with stepwise acquisition of altered biophysical properties [12].

6.3 Approaches to TSE Therapy: Anti-Prion Compounds

Approaches to the therapy of TSE are studded with many difficulties. First of all, the nature of the infectious agent is understood only in outline, and its composition, structure, and mode of replication are still shrouded in mystery. In addition, the mechanism of pathogenesis is not well understood. And finally, because the disease is usually recognized only after onset of severe clinical symptoms, only the preclinical diagnosis of TSE would permit the prevention or delay of neurodegeneration [19].

On the basis of the present knowledge on prion diseases, potential therapeutic strategies comprise: (a) stabilizing the structure of PrP^{C} via the formation of a PrP^{C}-drug complex; (b) preventing the formation or induce the degradation of amyloid aggregates; (c) hindering the conversion process of PrP^{C} or its binding to PrP^{Sc}; and (d) destabilizing the PrP^{Sc} structure or interfering with the cellular uptake of PrP^{C}/PrP^{Sc}. In this context, pharmacological studies are mostly focused on molecules able to interfere with fibrillogenesis such as sulphated polyanions [20–22], acridine-based compounds [23, 24], tetrapyrroles [25], the sulphonated azo-dye Congo red and some of its synthesized derivatives [26, 27], antibiotics [28], branched polyamines [29, 30], and synthetic peptides [31]. However, the intrinsic cytotoxicity and pharmacokinetic properties of these compounds, as well as their limited ability to pass the blood/brain barrier, strongly restrict their use, and the development of adequate therapeutic strategies against prion diseases is still waiting for an effective treatment.

Amongst others, the antibiotic tetracycline (TC) was described as able to prevent the aggregation of PrP-derived peptides, to reduce the protease resistance and disruption of their aggregates, and to abolish the neurotoxicity and astroglial proliferation induced by them. The anti-prion ability of TC has been investigated by modelling *in vitro* [32] its interaction with PrP synthetic peptides homologous to the N-terminal and C-terminal regions [32–34]. In an *in vivo* study [35], TC significantly delayed the onset of clinical signs of disease and prolonged survival into Syrian hamsters treated with scrapie-infected brain homogenates.

6.4 Immune Intervention

Since therapeutic and/or prophylactic intervention has at present few options available, new therapeutic strategies have turned to the possibility of developing means for an active immunoprophylaxis against prion diseases. Although autotolerance is a main problem in active immunization strategies [36], polyclonal anti-PrP autoantibodies have been induced employing dimeric PrP as an immunogen and shown to be able to significantly inhibit PrP^{Sc} propagation in prion-infected cells [37]. In addition, vaccination approaches with recombinant mouse prion protein [38] or by passive immunization are also promising [39]. Indeed, formation of protease-resistant PrP in a cell-free system has been inhibited by anti-PrP antibodies [40], which were also shown to prevent scrapie infection of susceptible mouse neuroblastoma cells [41] and inhibit prion replication in infected cells [42–44]. It is not yet clear through which mechanism anti-PrP antibodies interfere with PrPSc replication, but either a disruption of the interaction between PrP^C and PrP^{Sc} [42] or a perturbation of PrP^C trafficking and degradation [43] by the antibodies have been hypothesized.

Although abundant data are now available on anti-PrP monoclonal antibodies, poor knowledge on the mechanisms of prion diseases and the roles played by PrP^C and PrP^{Sc} in the brain dysfunctions limits their application in TSE therapy or diagnosis. Therefore, progress in therapy strongly depends on the basic comprehension of TSE. As a first step, it could be of help identifying and structurally defining epitopes of antibodies that cross-react with PrP^C and PrP^{Sc} to obtain structural details directly derived from the infectious agent involved in the mechanisms of PrP^{Sc} formation and spreading in infected organisms.

Recently, Eghiaian and co-authors [45] have reported the X-ray structures of the complexes between the C-terminal domain of three scrapie-susceptible sheep PrP variants and a Fab fragment that cross-reacts with PrP^C and PrP^{Sc}. The antibody epitope basically consists of the last two turns of the sPrP helix 2, which is known to be structurally invariant in the human protein. This has provided structural information on the PrP^C to PrP^{Sc} conversion. Hopefully, further structural characterization of the PrP^C to PrP^{Sc} conversion, and the subsequent development of a new class of biomolecules with *anti* prion activity, will benefit from the availability of additional antibodies, Fab fragments and molecules that bind to PrP^C and/or PrP^{Sc}.

6.5 Prion Protein Structure

The mammalian PrP gene encodes the PrP^C protein as a 253 amino acid polypeptide chain. The first 22 amino acids (signal peptide) are cleaved shortly after translation commences (Fig. 6.1), and addition of a C-terminal GPI-anchor, which facilitates glycolipid linkage of PrP^C to the cell membrane, post-translationally modifies the protein at residue 230 [46]. Recent studies on an anchorless, secreted version of PrP^C, expressed in transgenic mice, have clarified that membrane anchoring is a crucial prerequisite for prion toxicity resulting in clinical TSE [47]. Two N-linked glycosylation sites are located at residues 181 and 197. A nonapeptide followed by four identical octapeptide repeats are normally located between residues 51 and 91.

As schematically depicted in Fig. 6.2, the structure of the ubiquitous benign cellular form of PrP consists of an unstructured tail encompassing residues 23–125 and a globular domain, stabilized by an intramolecular disulphide bond (Cys179-Cys214) [48], which comprises residues 126–231 organized as three α-helices and a short β-sheet.

In a series of studies, Wüthrich and co-authors have reported the NMR structure of the globular C-terminal domain of recombinant human PrP^C (hPrP^C), also investigating several recombinant prion proteins from other species [49–51]. The overall structural organization of these PrPs is very similar, with residues 128–131, 144–154, 161–164, 173–195 and 200–228 forming the β-strand 1, the α-helix 1, the β-strand 2, the α-helix 2 and the α-helix 3, respectively. Crystallographic studies lend support to this monomeric structure, but in the dimeric form, an unusual domain swapping of α-helix 3 is apparent, with creation of a novel short anti-parallel β-sheet segment at the molecular interface [52].

As a consequence of a post-translational process, PrP^C is converted into the aberrantly folded and disease-specific scrapie isomer, PrP^{Sc}, through a process whereby a portion of its predominantly α-helical structure is refolded into β-sheet [5]. PrP^{Sc} exhibits resistance to proteinase K digestion [53, 54]. It is also known that the conversion of PrP^C into PrP^{Sc}, whose high β-sheet content is an essential constituent of putatively infectious prions [5, 55, 56], can also intrinsically occur as a result of a genetic mutation, as in rare familial encephalopathies. More disturbingly, however, transgenetic studies argue that inoculated PrP^{Sc} can impart conformational

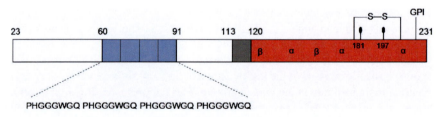

Fig. 6.1 Diagram of the prion protein sequence. The N-terminal segment, contains the octarepeat domain (*blue*). The C-terminal domain (*red*) contains three α-helical segments, two β-strands, a disulphide bond, and a glycosylphosphatidylinositol anchor

6 Involvement of Alpha-2 Domain in Prion Protein

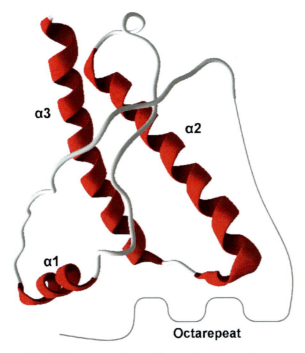

Fig. 6.2 Cartoon of the hPrP structure. The sketch was drawn according to current information [49–52]

variability to normal prions, thus triggering PrP^C refolding into a nascent PrP^{Sc} [5]. It can be concluded that the mechanisms that underlie pathological transitions remain unclear, despite attention paid to their understanding, because the highly aggregated state has hampered elucidation of the PrP^{Sc} structure at the atomic level.

6.6 Determinants of Prp^c Conversion: The N-Terminal Region

Partially protease-resistant forms of PrP are believed to mediate the neurodegeneration observed in spongiform encephalopathies. In fact, the neuropathological changes observed in prion disease are caused, at least in part, by the accumulation of proteinase K-resistant PrP^{Sc} [52–54]. This view is supported by the observation that the partially protease-resistant core of PrP^{Sc} displays a variety of pathogenic effects *in vitro*, including neurotoxicity and the ability to interact with plasma membrane, conferring an increased microviscosity. PrP^{Sc} accumulates in the central nervous system of affected individuals, and its partially protease-resistant core aggregates extracellularly into amyloid fibrils. The process is accompanied by nerve cell loss, whose pathogenesis and molecular basis are not well understood. Frankenfield and co-workers [57] compared the *in vitro* aggregation of a truncated portion (PrP[90–231]) and of a full-length version (PrP[23–231]) of the prion protein, which contains the largely unstructured N-terminal region in addition to the α-helical C-terminal one. They found that the full-length protein forms larger aggregates than the truncated protein, which indicates that the N-terminal region may mediate higher-order aggregation processes, possibly influencing the assembly state of PrP before aggregation begins. Other studies [58] have confirmed that the N-terminal region has a pivotal role in the development of prion misfolding and aggregation. Effects of aggregation have been observed with a short synthetic peptide fragment encompassing residues 106–126 of hPrP, which is toxic to cultured neurons depending on the expression of endogenous PrP. This synthetic peptide recapitulates several properties of PrP^{Sc}, including the propensity to form β-sheet-rich, insoluble and protease-resistant fibrils similar to those found in prion diseased brains [32]. Experimental data indicate that PrP[106–126] does not induce the formation of abnormal PrP species, suggesting, as an alternative explanation, that peptide toxicity depends on triggering alteration of a physiological function of PrP^C [59]. In fact, the N-terminal truncated PrP is toxic only to neurons that lack endogenous PrP, while PrP[106–126] is toxic only to neurons that express the endogenous protein. The structure of PrP[106–126] is modulated by pH, and its β-sheet content is higher at pH 5 than at pH 7. Furthermore, in the presence of lipids it acquires a predominantly β-sheet conformation. Extensive studies were performed to understand the relationships between toxicity and physicochemical properties of amyloid peptides. To determine the role of the hydrophobic palindromic sequence in PrP[106–126] toxicity, Jobling et al. [60] have generated a series of mutant PrP[106–126] peptides with hydrophilic substitutions in the hydrophobic core. The results of these studies correlate the neurotoxic action of PrP[106–126] to its secondary structure and subsequent fibril-forming propensity. The data suggest that the hydrophobic C-terminal valines and the palindromic region from Ala113 to Ala120 of PrP[106–126] are involved in the folding and/or stabilization of a β-sheet aggregate. These findings are similar to those described for amyloid β-peptide aggregates and strengthen the view of a common structure-function mechanism of amyloid generation in spongiform encephalopathies and Alzheimer's disease [2]. On the other hand, on consideration that, in infectious and familial prion disorders, neurodegeneration is often seen without deposits of PrP^{Sc}, Gu and co-workers [61] have shown that exposure of neuroblastoma cells to PrP[106–126] catalyses the aggregation of cellular prion protein to a weakly proteinase K-resistant form and induces the synthesis of transmembrane prion protein, suggesting that neurotoxicity is mediated by a complex pathway involving transmembrane prion protein and not only by deposits of aggregated and proteinase K-resistant PrP.

6.7 Determinants of Prp^c Conversion: The C-Globular Domain

In the X-ray structure of monomeric sheep PrP^C (sPrP^C), two potential loci of β-structure propagation were identified [62]. The former locus (residues 129–131) is involved in an intramolecular β-sheet with residues 161–163 and in lattice contacts about a crystal dyad to generate a four-stranded intermolecular β-sheet between neighbouring molecules. Modelling on the latter locus (residues 188–204) suggests that it is able to act as an α → β switch within the monomer. The α → β isomerization of PrP is most frequently observed *in vitro* in the pH range from 4 to 7 [63–67], and it has been postulated to be induced *in vivo* by the low pH of endosomal compartments [68]. A comparison between the C-*terminus* crystal structures of monomeric sPrP^C and dimeric hPrP^C showed that the dimer results from the swapping of the C-terminal α-helix 3 and rearrangement of the Cys179-Cys214 disulphide bond. An interchain two-stranded antiparallel β-sheet is formed at the dimer interface between the corresponding crystal-symmetry-related residues 190–194, which are located in α-helix 2 in the monomeric NMR structures [52]. The segment 188–201 (TVTTTTKGENFTET) is invariant across a wide variety of species [69] and, on the basis of its primary structure, several features emerge that might drive PrP^C reorganization. In particular, the seven threonine residues could confer the necessary conformational plasticity. Moreover, residues 188–201 in hPrP adopt an architecture that appears to be of lower stability as compared to the rest of the structure. The high intrinsic β-propensity of four adjacent threonines [70] makes this segment a good candidate to promote a local α → β transition, which, under suitable conditions, could lead to PrP^{Sc} formation, even independently of disulphide bridge rearrangement, since PrP^{Sc} monomers are not linked by intermolecular disulphide bonds. Furthermore, PrP^{Sc} can induce the conversion of the disulphide-intact form of the monomeric cellular prion protein to its protease-resistant form without the temporary breakage and subsequent re-formation of the disulphide bonds in cell-free reactions [71]. From the above studies, it emerges that quite small conformational adjustments can convert the monomeric PrP^C into a potentially oligomeric nucleating unit. It is likely that some conformational weaknesses converging on the sequence 190–195 or a shorter surrounding region are able to affect the whole protein architecture and promote the non-covalent association of misfolded monomers. It has been also proposed [62] that the synergic propagation of β-sheet association involving the whole molecule mediates protein oligomerization. Thompson and co-authors [4] have investigated the conformational and aggregation behaviour of synthetic peptides corresponding to PrP^C helices in aqueous solutions. The fragment corresponding to α-helix 1 exhibited a random coil CD spectrum at any pH value from 3 to 12, whereas in TFE the peptide was 20% helical and did not aggregate over time neither did it form amyloid fibrils. However, it has been also shown that α-helix 1 possesses a remarkably high intrinsic α-helical propensity [72] and retains significant helicity under a wide range of conditions, such as high salt, pH variation, and presence of organic co-solvents [73]. Because of its high stability against environmental changes, helix 1 is unlikely to be involved in the initial steps of the pathogenic conformational change and it could unfold in the late stage of the structural transition as a consequence of global conformational rearrangements occurring in other parts of the prion protein [73]. The fragment corresponding to α-helix 2 underwent a time dependent β-sheet rearrangement with formation of aggregates over time. However, electron microscopy showed that aggregates taken from CD samples were organized in fibrils, which were small at pH 7.2, but longer and more distinct at higher pH values. The fragment corresponding to α-helix 3 also underwent pH dependent β-sheet formation. The CD curve exhibited random organization at pH 6.0 and 7.2, and β-sheet at pH 3, with an aggregation dependent intensity decrease after 24 h. The precipitate did not show fibril formation indicating that this peptide is not truly amyloidogenic under the conditions studied. Therefore, it can be concluded from these observations, that the relationship between amyloidogenicity and neurotoxicity remains unclear, because the fact that a peptide is somehow prone to aggregate and readily form a β-sheet structure does not necessarily imply that it forms amyloid fibrils. Additional evidence indicates that an intermediate along the pathway to fibril formation could cause toxicity, whereas large fibrils may not be toxic in themselves. Gallo and co-workers [74] have recently reported that the conserved capping box (Thr199-Glu200-Thr201-Asp202) and, in part, the ionic bond formed between Glu200 and Lys204 render the PrP^C segment corresponding to the α-helix 3 structurally autonomous, in contrast to α-helix 1 and α-helix 2 peptides. In fact, the D202N capping mutation associated to the GSS disease almost completely destabilizes the isolated α-helix 3 peptide, thus possibly initiating the PrP^C pathogenic process associated with this substitution. Moreover, cell culture data based on the NMR structure of mouse PrP^C suggest that the highly conserved hydrophobic side chain at residue 204 of α-helix 3 is required for folding and maturation of PrP^C, providing an essential stabilization of α-helix 1 structure by interacting with Phe140, Glu145, Tyr148, and Tyr149. Disruption of α-helix 1 prevented attachment of the GPI anchor and the formation of complex N-linked glycans. In the absence of a C-terminal membrane anchor, however, α-helix 1 induced the formation of deglycosylated and partially protease-resistant PrP aggregates [75]. This result is confirmed by molecular dynamics simulations, in which disturbances of the folding and maturation process of PrP^C have been interpreted as consequences of mutation-induced structural changes in PrP, involving α-helix 1 and its attachment to α-helix 3 [76]. A number of results on cellular toxicity [4], fibrillization capabilities [48], and metal binding properties [77] of synthetic variants of the α-helix 2 point to an important contribution of this region to the overall biological behaviour of the prion protein. In fact, perturbations leading to structural rearrangements that may strongly affect the stability of the α-helix 2 could involve deglycosylation of Asn181 [48] and/or copper binding to His187 [77]. Rearrangements of the α-helix 2 could promote β-sheet-mediated protein association leading to a further α → β transition and subsequently to aggregation. In a novel thermodynamic study, the α → β conversion of the N- and C-*termini* blocked peptide corresponding to the α-helix 2, PrP[173–195], has been characterized by measuring α-helical and β-structure formation propensities in the temperature interval from 280 K to 350 K [78]. The scheme reported in Fig. 6.3 shows that the two ordered conformations were found to be separated by 5–8 kJ/mol, with an entropic advantage of 0.04 kJ mol^{-1} K^{-1} favouring the α-helical organization. This subtle free energy difference was interpreted as denoting the chameleon-like character of PrP[173–195], which could be governed, in the protein, by the cellular microenvironment, according to the finding that slight conditional changes may cause chameleon sequences to fold into either α- or β-structure [79]. In this context, it is worth noting that, in the

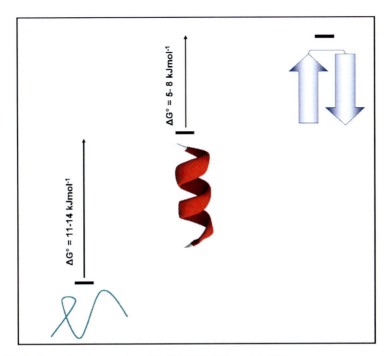

Fig. 6.3 Conformational energetics of the PrP(173–195) peptide [78]

whole PrP^C, the close packing of the first three turns of the α-helix 2 against the α-helix 3 generates a complementary interface [49, 52] that strongly stabilizes the helix up to around residue 188, with the glycosyl moiety bound to Asn181 providing further stabilization [48]. Conversely, the region spanning residues 190–195 is rather apart from the α-helix 3, which is well characterized only up to Thr219 in the NMR structures of mouse PrP^C [80] and found in β-conformation in the dimeric crystal structure of hPrP^C [52], suggesting that this site is one of the most prone to structural rearrangements upon suitable perturbation. Thus, the short C-terminal end of the full length α-helix 2 could be involved in the nucleation process of prion misfolding and oligomerization, possibly in cooperation with the N-terminal fragment 82–146, whose intrinsic properties are dependent upon the integrity of the C-terminal region [81].

6.8 Prion-Metal Ion Binding

Research over the past few years clarified that PrP^C can exist in a Cu-metalloprotein form *in vivo* [82] and displays high selectivity for Cu^{2+} [83]. Screens against other divalent cations, such as Ca^{2+}, Co^{2+}, Mg^{2+}, Mn^{2+} and Ni^{2+}, failed to find high-affinity interactions. In order to extract functional information, many efforts have been devoted to the structural characterization of Cu^{2+} binding sites [79, 82–85].

The emerging consensus is that most copper binds in the octarepeat domain, comprising the HGGGW segment as the fundamental unit involved in Cu^{2+} coordination [86, 87]. The crystal structure of the complex reveals equatorial coordination of Cu^{2+} by the histidine imidazole, two deprotonated glycine amides, and a glycine carbonyl, along with an axial water bridging to the Trp indole, consistently with companion experiments in solution [87]. This somewhat unusual copper binding site is by no means unprecedented. In most copper binding proteins, side chain moieties such as histidine imidazole or cysteine thiol enter into contact with the metal ion [88]. Previous studies showed that unstructured peptides containing histidines coordinate in a fashion similar to that now identified for PrP [89]. The pK_a of amide protons is typically 13–15, and consequently the amide nitrogen is not ionised at pH 7. However, nitrogen and Cu^{2+} are well matched on the hard-soft scale of Lewis acid–base interactions. Thus, with the histidine imidazole anchoring the metal ion close to the polypeptide backbone, Cu^{2+} may be uniquely able to displace a nearby amide proton at physiological pH [90]. Modelling, electron paramagnetic resonance spectroscopy (EPR) and companion spectroscopic studies on peptides as well as full length protein provide, however, evidence that additional copper sites are located in the region connecting the unstructured N-terminal segment to the C-terminal globular portion of PrP [83]. Accordingly, it has been proposed that an additional copper-binding site compared with the four of the octarepeat domain binds around His 96 and/or His 111, a region of the PrP molecule known to be crucial for prion propagation [91–93]. In fact, proteolysis of PrP^{Sc} at approximately residue 90 does not result in loss of infectivity. This suggests that the octarepeat domain, and hence copper, do not play a role in TSEs and may not be necessary to PrP conversion and disease, but a modulating role in kinetics and pathology cannot be excluded. Indeed, the octarepeat domain and copper have been directly implicated in neurological disease. Finally, recent studies show that binding of a single copper rapidly and reversibly induces PrP^C to become protease-resistant and detergent-insoluble [94]. There is experimental evidence that binding takes place at His96 in full-length PrP, that is outside both the octarepeat and the C-globular domains [92, 93]. The amyloidogenicity and neurotoxicity of PrP[106–126] are common to the Alzheimer's disease amyloid β peptide. Given that the biophysical behaviour and activity of amyloid β peptide are governed by transition metals, the effect of metals has been also studied on PrP[106–126]. The fibrillization of this peptide is completely inhibited in a metal-depleted buffer, and Cu^{2+} and to a lesser extent Zn^{2+} have been found to restore its aggregation [85]. The metal binding site was found to comprise the N-terminal amino group, His111 and Met112. This supports the view of a common structure-function mechanism of amyloid generation in spongiform encephalopathies and Alzheimer's disease [73–75].

Most recently, the stimulatory potential of Cu^{2+}, Mn^{2+}, Zn^{2+}, and Al^{3+} in inducing defective conformational rearrangements that trigger aggregation and fibrillogenesis has been investigated in the recombinant hPrP fragment spanning residues 82–146 [95]. This region has been identified as a major component of the amyloid deposits in the brain of patients affected by GSS disease. Amino acid substitution in the neurotoxic 106–126 core sequence, reduced its amyloidogenic potential. However, alteration of the 127–146 sequence, which comprises a segment of the C-terminal globular domain, also caused strong inhibition of the fibrillogenesis, thus suggesting that integrity of this region was essential both to confer amyloidogenic properties on GSS peptides and to activate the stimulatory potential of the metal ions. Notably, only a few studies have been carried out on metal interaction with peptides derived from the C-*terminus* of PrP, which contains the histidine residues (H140, H177 and H187) and the helical region. Recent spectroscopic experiments [77, 96] exclude the involvement of H140 in Cu^{2+} binding, but the aggregation of model peptides hampered characterization of the metal interaction with H177 and caused uncertainty about Cu^{2+} binding to His 187 at physiological pH. Incidentally, it has been found [97, 98] that the only known histidine variant associated with familial encephalopathy could be associated with the H187R mutation in the PrP gene.

6.9 The α-2 Helix Domain: What Role?

A computational analysis illustrates that native PrP exhibits large regions of conformational ambivalence and suggest that it is only a marginally stable protein [76]. Other simulations also indicate that the conformational variability of the entire prion protein sequence is unusually high compared with other proteins of similar length [99]. Moreover, the tendency to increase the β-structure content is very likely an intrinsic characteristic of the prion protein fold, irrespective of thermodynamic or structural conditions [100]. In the C-globular domain, unusually low α-helical and β-sheet propensities feature the segment 173–195, corresponding to α-helix 2, in spite of the fact that this segment retains a helical conformation in the whole protein. In addition, the unusually high density of disease-promoting mutations in α-helix 2 also points to the particular importance of this helix for conformational transition of PrP. More specifically, it seems reasonable that a single amino acid replacement in the vicinity of the α-helix 2 may significantly affect the organization of the entire $\alpha2$-$\alpha3$ helical part, enhancing the propension of this region for the β-conformation and facilitating structural rearrangements.

Further support to this hypothesis comes from the finding that the $hPrP^C$ mutants T183A and F198S, which are associated to inherited prion diseases, severely affect folding and maturation of PrP^C in the secretory pathway of neuronal cells *in vitro*, adopting misfolded and partially protease-resistant conformations [101]. These pathogenic mutations interfere with folding and attachment of the GPI anchor [101]. Indeed, based on a refined NMR structure, it was predicted that they would specifically destabilize the PrP C-terminal globular domain, because they involve key interactions in the hydrophobic core [80]. The resulting three-dimensional arrangement could account for the defect in maturation and the efficiency of the GPI anchor attachment. The hypothesis that the segment comprising the C-*terminus* of α-helix 2 and the adjacent loop may be partially unfolded and represent a potential oligomerization site is also supported by crystallographic data [52]. Furthermore, the α-helix

2 fragment, also depending on the glycosylation state and the presence of metals [48, 77, 102] can be toxic to neuronal cells and strongly fibrillogenic, adding a further clue to the working hypothesis that it is involved in the protein aggregation process and in the toxicity associated to the scrapie variant. On this basis, it is apparent that the human prion protein helix 2 domain may influence a conformational change of the whole protein. Indeed, most recent literature data point to a crucial role for the α-2 helix in the misfolding mechanism of the whole PrP [103–106].

6.10 Solution Structure of Prp[173–195] and Its Analogues

Following the thread of these arguments, large efforts have been devoted to the structural characterization of α-helix 2-derived peptides. CD spectra of peptides derived from the N-terminal and the C-terminal part of the full length α-helix 2, PrP[173–179] and PrP[180–195], respectively, are those typical of random and β-type organization, respectively. Thus, it seems reasonable to infer that the occurrence of disordered structure in the full length α-helix 2 peptides, which both include the 173–179 segment, is associated with the N-terminal segment. As a matter of fact, CD spectra of PrP[173–195] and PrP[173–195]D178N, do not show the β-type peculiarities exhibited by PrP[180–195] [103]. SDS titration of the D178N mutant goes to completion in a range of detergent concentration much narrower than that observed for the wild type peptide. Probably, the absence of the negative charge carried by the Asp178 side-chain permits stronger electrostatic interaction between SDS and the protonated His177 side-chain. Moreover, the higher β-inducing propensity of Asn may also contribute to favour reorganization of PrP[173–195]D178N into a β-type conformation. Overall, this suggests that the Asn side-chain renders the mutant peptide more prone to form β-structure. In agreement with that, the TFE-induced recovery of α-helical conformation is larger for the wild type peptide as compared to the D178N mutant [107].

In addition, the fact that PrP[180–195] is able to assume β-arrangement at neutral pH even without SDS and that conformation doesn't dramatically change even in presence of TFE, suggests that the 180–195 parental region in PrP^C strongly contributes to the chameleon conformational behaviour of the segment corresponding to the full-length α-helix 2 [78] and plays a role in determining the structural rearrangements of the entire PrP^C-globular domain.

CD spectra show that these α-helix 2 derived peptides do not possess the same α-helical conformation as that observed in the cellular prion protein. It is known that the lack of mutual interactions has dramatic effects on the integrity of the whole helical domain of the prion protein, and the stability of one single helical region strongly suffers from ablation of the other helical segments as well as of the disulphide bridge. However, native-like conditions can be to some extent restored choosing a medium that may help extract useful information using the peptide fragment approach. Thus, TFE has been used as the most suitable environment to investigate structural similarities between these synthetic peptides by NMR spectroscopy. As

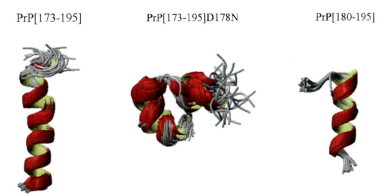

Fig. 6.4 The bundles of PrP helix-2 regions of the best 30 DYANA structures obtained by best fitting of the backbone [107] [pdb codes: 2iv4; 2iv5; 2iv6]

a matter of fact, both PrP[173–195] and PrP[180–195] were found to be helices, whereas the PrP[173–195]D178N mutant was not even able to retain a fully helical arrangement in an α-inducing environment (Fig. 6.4) [107].

The major result from these NMR studies is that the conformation of the wild type peptide is significantly affected by replacing the negatively charged Asp178 with a neutral Asn residue. In the mutant peptide, increased conformational freedom characterizes all residues downstream Gln188, which ultimately causes unwinding and bending of the wild type fully helical structure. As a consequence, structural rearrangement leads to the formation of two short helices separated by a kink centred on Lys185 and Gln186. In this bent structure, His177 and His187 approach to each other as compared to the parent helical peptide, forming two major conformational families, characterized by proximal and distal imidazole rings, respectively. The negative charge of Asp178 plays a key role in forcing the entire 173–195 fragment to assume a full helical conformation.

Notably, the shorter 180–195 fragment still retains an almost fully helical structure, whether or not it is embedded in the 173–195 sequence, suggesting that helix unwinding in the region 180–187 is provoked by the D178N substitution. Indeed, though the sequence of the shorter peptide includes residues involved in the bending of the D178N analogue (Lys185 and Gln186), it does not show any kink in its structure, nor does the wild type peptide. This confirms that the D178N mutation destabilizes the region in which it is located, thus causing unwinding and bending of the wild type fully helical structure.

These observations are best compared with the recent finding that helix-2 unwinding influences the stability of the whole PrP [105]. Analysis of H/D exchange data permitted to map the H-bonded β-sheet core of PrP amyloid to the C-terminal region that in the native structure of PrP monomer corresponds to α-helix 2, a major part of α-helix 3, and the loop between these two helices, whereas no extensive hydrogen bonding was detected in the N-terminal part of PrP90–231 fibrils, arguing against the involvement of residues within this region in stable β-structure [106].

6.11 Metal Ion Titration

Titration with Cu^{2+} of PrP[173–195] and PrP[173–195]D178N dissolved in the same medium as that used for the above NMR investigations showed that increasing metal cation aliquots did not perturb NMR spectra in any specific way. The chemical shifts of all resonances did not vary, as it could be expected in the case of metal-peptide complex formation, and the overall effect was a progressive broadening of all relevant resonances. Indeed, increasingly higher metal aliquots caused irreversible concentration-dependent aggregation, possibly owing to ionic strength increase and/or to water addition on metal cation titration. However, CD spectra, where aggregation did not occur due to the lower peptide concentration, were unaltered on Zn^{2+} or Cu^{2+} addition, thus apparently confirming that the interaction of the metal with the peptide backbone is non-specific. Further experiments in mixed water/TFE solvent suggested that water-induced effects largely dominate structural rearrangements, rendering metal-induced modifications, if any, hard to discriminate.

It is worth emphasizing here that, in studies performed on metal interaction with peptides derived from the hPrP C-*terminus*, the use of the α-helix-inducer TFE may lead to conclusions at odds with those obtained in buffer solution. For example, characterizing the formation of metal complexes in blocked and free C- and N-*termini* analogues of the hPrP^C α-helix 2-derived peptide 180–193 (VNITKQHTVTTTT), Brown et al. [75] suggested that the His187 residue in the structured region of the protein is a binding site for Cu^{2+} and drives the metal coordination environment towards a common binding motif in different regions of the prion protein. Other studies [4] showed that the PrP[178–193] peptide has both structural and bioactive properties in common with the amyloidogenic Alzheimer's disease βA[25–35] peptide and that the second helical region of PrP could be involved in modulation of Cu(II)-mediated toxicity in neurons during prion disease.

We would suggest that, in the peptide fragment approach, it is unlikely that aqueous buffer is the most suitable environment to analyse metal interaction with peptide fragments, because the parent segments in the native protein may experience different environmental conditions. Thus, it can be argued that the conclusion that metal cations bind to peptide fragments derived from the C-terminal globular domain is ambiguous, owing to the fact that the structural organization of these peptides may be rather different from that assumed in PrP^C. We believe that it is crucial to force peptides into a conformation close to the one that has been found in PrP^C when designing experiments aimed at investigating peptide-metal cation interaction. Although it is currently believed that the major structural modifications involved in PrP protein misfolding involve the unstructured N-terminal region, this evidence also suggests that the N- and C-terminal prion domains play a different role in the protein conversion, stressing that the former region is most likely the natural target of metal binding [83].

6.12 Anion-Induced Effects

Environmental conditions, like pH, salts, and presence of nucleic acids or glycosaminoglycans, seem to affect the structural stability of prion proteins to a much larger extent than other proteins [108, 109]. It was suggested that this unusual behaviour could be ascribed to the ability of PrP^C N-terminal region to interact with anions, which leads to destabilization of the prion core structure [110]. Also, in an extensive analysis on the interaction of anti-prion compounds and amyloid-binding dyes with a carboxy-terminal domain of prion protein, it has been found that sulphonates, like Congo red and phthalocyanine tetrasulphonate, bind with high affinity [111]. It has also been reported that anion identity affects far UV CD spectra of two PrP^C α-helix 2-derived wild type and mutant analogues, PrP[180–195] and PrP[180–195]H187A, which strongly hydrated anions cause to fold to β-structure and α-helix, respectively [112]. Such large conformational modifications can be rationalized via the ion charge density dependence that is typical of Hofmeister effects [113, 114] (Fig. 6.5).

Anions like Cl^-, ClO_4^-, and $H_2PO_4^-$ are weakly hydrated because of their low charge density, and their interaction with water molecules is weaker than that of water with itself. This causes them to behave as water structure breakers (chaotropes), which make the bulk solution a better solvent. As a consequence, peptides maximize their solvent accessible surface area, favouring the formation of the unstructured conformation. On the other hand, multiply charged ions, like SO_4^{2-} and HPO_4^{2-}, exhibit stronger interactions with water molecules than water with itself because of their high charge density. These ions are water structure makers (kosmotropes) and make the bulk solution a poorer solvent. Thus, they encourage helix-2 related prion peptides to minimize their solvent accessible surface area and assume

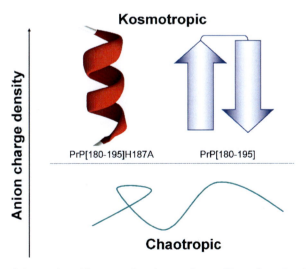

Fig. 6.5 Effect of chaotropic and kosmotropic anions on the peptide conformation [112]

β-sheet-like or α-helix-like conformations. It is likely that the compactness of these structures predisposes peptides to self-association, which may be a preliminary step toward fibril formation.

The different conformational behaviour of the PrP[180–195] wild type peptide as compared to the PrP[180–195]H187A mutant peptide in the presence of kosmotropic anions at neutral pH is likely caused by the His side chain, which displays an α-inducing ability lower than the Ala side chain. Nevertheless, in the prion protein, the segment containing the His187 residue is still able to retain an α-helical conformation owing to tertiary interactions [78]. Furthermore, in acidic solution, His protonation and increased proton exchange [109] could play a role in the destabilization of the helical architecture. These data also confirm that the helix 2 domain possesses chameleon-like character [78], suggesting that preferential binding with naturally occurring anions, rather than non-specific interactions, such as ionic strength-dependent ones, plays an important role in prion protein misfolding and amyloid fibril growth.

Understanding ion-specific effects is a central theme of biology. Unfortunately, the complication of 'ion confounding' is extremely common in all disciplines concerned with ion-based research because ions are generally manipulated through the use of salts. It occurs because changing the concentration of a single cation or anion using a single salt results in a simultaneous change of the associated co-ion, which causes the main effect associated with that ion to be confounded with the effects caused by changing the concentration of the co-ion [115]. It is therefore worth stressing that, even in studies on prion and derived peptides, anion-bound effects may overlap the largely explored cation-bound effects. Unfortunately, experiments are not always designed so that the species that cause the modification can be unequivocally perceived. In conclusion, we highlight that the sensitivity of prion derived peptides, as well as the whole protein [77, 82, 96] or other amyloidogenic systems [116], to environmental modifications suggests that the complication of anion involvement cannot be neglected anymore, either in investigating metal effects on peptide conformation [77, 82, 83, 96, 116] or in checking the inhibition of amyloid formation by unusual agents [117, 118].

6.13 Conclusions

As the general features of prion diseases are common to other amyloid disorders [2], the prion protein can be used as a model to provide the bases for a comprehensive evaluation of the general protein misfolding mechanism. Conformation based approaches to the study of PrP can give useful hints both on the region/residues potentially important for the $PrP^C \rightarrow PrP^{Sc}$ conversion and on the identification or development of anti-prion compounds.

Interest in the study of α-helix 2 comes from evidence that this segment possesses chameleon conformational behaviour [78], gathering several disease-promoting point mutations, and therefore it can be strongly fibrillogenic and toxic to neuronal cells [4], suggesting its involvement in the protein aggregation process and in the toxicity associated to the scrapie variant. Overall, all peptide-based data on the conformational landscape of α-helix 2 strongly suggest a crucial role played by this domain in the misfolding mechanism of PrP^C to the scrapie isoform.

A single amino acid replacement is sufficient to affect the organization of the 173–195 α-helix 2-derived peptide, enhancing the propension of this region for the β-conformation and facilitating structural rearrangements. Indeed, in the CJD-associated D178N mutant peptide, the substitution of a neutral Asn for an Asp residue weakens the helical arrangement of the 173–195 segment in TFE. Also, the PrP[173–195]D178N peptide shows a higher β-type propensity in SDS compared to that exhibited by the wild type peptide in the same condition. Furthermore, neurotoxicity assays have shown that the PrP[180–195] and PrP[173–195]D178N peptides display a toxicity higher than that of the wild type peptide, suggesting a linkage between β-conformation propensity and toxicity [103]. Finally, environmental conditions, such as pH and salts, can affect the conformational behaviour of the α-helix 2 domain. This confirms the chameleon-like character of this PrP fragment [78], suggesting that preferential binding with naturally occurring anions and pH changes may play an important role in prion protein misfolding and amyloid fibril growth.

NMR and CD titrations have shown that no specific interaction of Zn^{2+} or Cu^{2+} with helical α-helix 2-derived peptides occurs, providing further support to evidence accumulated in the literature that N- and C-*termini* domains play a different role in the PrP conversion, and stressing that the N-terminal domain is likely the natural target of metal binding [83]. The intriguing structural properties of the α-helix 2 domain make it a primary target for therapeutic strategies and a suitable model to investigate rational structure-based drug design of compounds able to block or prevent prion diseases.

Abbreviations

BSE, Bovine Spongiform Encephalopathy; CJD, Creutzfeldt-Jakob Disease; DRM, detergent-resistant microdomains; GPI, glycosylphosphatidylinositol; GSS, Gerstmann-Sträussler-Scheinker; hPrP, human Prion Protein; PrP^C, Cellular form of Prion Protein; PrP^{Sc}, Scrapie form of Prion Protein; sPrP, sheep Prion Protein; SDS, Sodium Dodecyl Sulphate; TC, Tetracycline; TSE, Transmissible Spongiform Encephalopathies

References

1. Carrell RW, Lomas DA (1997) Conformational disease. Lancet 350:134–138
2. Temussi PA, Masino L, Pastore A (2003) From Alzheimer to Huntington: why is a structural understanding so difficult? EMBO J 22:355–361
3. Blake C, Serpell L (1996) Synchrotron X-ray studies suggest that the core of the transthyretin amyloid fibril is a continuous beta-sheet helix. Structure 4:989–998

4. Thompson A, White AR, McLean C, Masters CL, Cappai R, Barrow CJ (2000) Amyloidogenicity and neurotoxicity of peptides corresponding to the helical regions of PrP(C). J Neurosci Res 62:293–301
5. Tan SY, Pepys MB (1994) Amyloidosis. Histopathology 25:403–414
6. Sunde M, Serpell LC, Bartlam M, Fraser PE, Pepys MB, Blake CC (1997) Common core structure of amyloid fibrils by synchrotron X-ray diffraction. J Mol Biol 273: 729–739
7. Booth DR, Sunde M, Bellotti V, Robinson CV, Hutchinson WL, Fraser PE, Hawkins PN, Dobson CM, Radford SE, Blake CC et al. (1997) Instability, unfolding and aggregation of human lysozyme variants underlying amyloid fibrillogenesis. Nature 385:787–793
8. Prusiner SB (1998) Prions. Proc Natl Acad Sci USA 95:13363–13383
9. Harris DA (2003) Trafficking, turnover and membrane topology of PrP. Br Med Bull 66: 71–85
10. Campana V, Sarnataro D, Zurzolo C (2005) The highways and byways of prion protein trafficking. Trends Cell Biol 15:102–111
11. Peters PJ, Mironov A Jr, Peretz D, van Donselaar E, Leclerc E, Erpel S, DeArmond SJ, Burton DR, Williamson RA, Vey M et al. (2003) Trafficking of prion proteins through a caveolae-mediated endosomal pathway. J Cell Biol 162:703–717
12. Harris DA (1998) Clathrin-coated vesicles and detergent-resistant rafts in prion biology. Bull Inst Pasteur 96: 207–212
13. Caughey B, Raymond GJ, Ernst D, Race RR (1991) N-terminal truncation of the scrapie-associated form of PrP by lysosomal protease(s): implications regarding the site of conversion of PrP to the protease-resistant state. J Virol 65:6597–6603
14. Ivanova L, Barmada S, Kummer T, Harris DA (2001) Mutant prion proteins are partially retained in the endoplasmic reticulum. J Biol Chem 276:42409–42421
15. Baron GS, Wehrly K, Dorward DW, Chesebro B, Caughey B (2002) Conversion of raft associated prion protein to the protease-resistant state requires insertion of PrP-res (PrP(Sc)) into contiguous membranes. EMBO J 21:1031–1040
16. Ma J, Wollmann R, Lindquist S (2002) Neurotoxicity and Neurodegeneration When PrP Accumulates in the Cytosol. Science 298:1781–1785
17. Ma J, Lindquist S (2002) Conversion of PrP to a self-perpetuating PrPSc-like conformation in the cytosol. Science 298:1785–1788
18. Kocisko DA, Come JH, Priola SA, Chesebro B, Raymond GJ, Lansbury PT, Caughey B (1994) Cell-free formation of protease- resistant prion protein. Nature 370:471–474
19. Weissmann C, Aguzzi A (2005) Approaches to therapy of prion diseases. Annu Rev Med 56:321–344
20. Farquhar CF, Dickinson AG (1986) Prolongation of scrapie incubation period by an injection of dextran sulphate 500 within the month before or after infection. J Gen Virol 67: 463–473
21. Kimberlin RH, Walker CA (1986) Suppression of scrapie infection in mice by heteropolyanion 23, dextran sulfate, and some other polyanions. Antimicrob Agents Chemother 30:409–413
22. Caughey B, Raymond GJ (1993) Sulfated polyanion inhibition of scrapie-associated PrP accumulation in cultured cells. J Virol 67:643–650
23. May BC, Fafarman AT, Hong SB, Rogers M, Deady LW, Prusiner SB, Cohen FE (2003) Potent inhibition of scrapie prion replication in cultured cells by bis-acridines. Proc Natl Acad Sci USA 100:3416–3421
24. Barret A, Tagliavini F, Forloni G, Bate C, Salmona M, Colombo L, De Luigi A, Limido L, Suardi S, Rossi G et al. (2003) Evaluation of quinacrine treatment for prion diseases. J Virol 77:8462–8469
25. Caughey WS, Raymond LD, Horiuchi M., Caughey B (1998) Inhibition of protease-resistant prion protein formation by porphyrins and phthalocyanines. Proc Natl Acad Sci USA 95:12117–12122

26. Lorenzo A, Yankner BA (1994) Beta-amyloid neurotoxicity requires fibril formation and is inhibited by congo red. Proc Natl Acad Sci USA 91:12243–12247
27. Poli G, Ponti W, Carcassola G, Ceciliani F, Colombo L, Dall'Ara P, Gervasoni M, Giannino ML, Martino PA, Pollera C et al. (2003) In vitro evaluation of the anti-prionic activity of newly synthesized congo red derivatives. Arzneimittelforschung 53:875–888
28. Adjou KT, Privat N, Demart S, Deslys JP, Seman M, Hauw JJ, Dormont D (2003) J Comp Pathol 122:3–8
29. Supattapone S, Nguyen HO, Cohen FE, Prusiner SB, Scott MR (1999) Elimination of prions by branched polyamines and implications for therapeutics. Proc Natl Acad Sci USA 96:14529–14534
30. Supattapone S, Wille H, Uyechi L, Safar J, Tremblay P, Szoka FC, Cohen FE, Prusiner SB, Scott MR (2001) Branched polyamines cure prion-infected neuroblastoma cells. J Virol 75:3453–3461
31. Soto C, Saborio GP, Permanne B (2000) Inhibiting the conversion of soluble amyloid-beta peptide into abnormally folded amyloidogenic intermediates: relevance for Alzheimer's disease therapy. Acta Neurol Scand Suppl 176:90–95
32. Forloni G, Angeretti N, Chiesa R, Monzani E, Salmona M, Bugiani O, Tagliavini F, (1993) Neurotoxicity of a prion protein fragment. Nature 362:543–546
33. Tagliavini F, Forloni G, Colombo L, Rossi G, Girala L, Canciani B, Angeretti N, Giampaolo L, Peressini E, Awan T et al. (2000) Tetracycline affects abnormal properties of synthetic PrP peptides and PrP(Sc) in vitro. J Mol Biol 300:1309–1322
34. Ronga L, Langella E, Palladino P, Marasco D, Tizzano B, Saviano M, Pedone C, Improta R, Ruvo M (2007) Does tetracycline bind helix 2 of prion? An integrated spectroscopical and computational study of the interaction between the antibiotic and alpha helix 2 human prion protein fragments. Proteins 66:707–715
35. Forloni G, Varì MR, Colombo L, Bugiani O, Tagliavini F, Salmona M (2003) Prion disease: time for a therapy? Curr Med Chem Imun, Endoc Metab Agents 3:185–197
36. Polymenidou M, Heppner FL, Pellicioli EC, Urich E, Miele G, Braun N, Wopfner F, Schatzl HM, Becher B, Aguzzi A (2004) Humoral immune response to native eukaryotic prion protein correlates with anti-prion protection. Proc Natl Acad Sci USA 101:14670–14676
37. Gilch S, Wopfner F, Renner-Muller I, Kremmer E, Bauer C, Wolf E, Brem G, Groschup MH, Schatzl HM (2003) Polyclonal anti-PrP auto-antibodies induced with dimeric PrP interfere efficiently with PrPSc propagation in prion-infected cells. J Biol Chem 278:18524–18531
38. Sigurdsson EM, Brown DR, Daniels M, Kascsak RJ, Kascsak R, Carp R, Meeker HC, Frangione B, Wisniewski T (2002) Immunization delays the onset of prion disease in mice. Am J Pathol 161:13–17
39. Bade S, Frey A (2007) Potential of active and passive immunizations for the prevention and therapy of transmissible spongiform encephalopathies. Expert Rev Vaccines 6:153–168
40. Horiuchi M, Caughey B (1999) Specific binding of normal prion protein to the scrapie form via a localized domain initiates its conversion to the protease-resistant state. EMBO J 18:3193–3203
41. Enari M, Flechsig E, Weissmann C (2001) Scrapie prion protein accumulation by scrapie-infected neuroblastoma cells abrogated by exposure to a prion protein antibody. Proc Natl Acad Sci USA 98:9295–9299
42. Peretz D, Williamson RA, Kaneko K, Vergara J, Leclerc E, Schmitt-Ulms G, Mehlhorn IR, Legname G, Wormald MR, Rudd PM et al. (2001) Antibodies inhibit prion propagation and clear cell cultures of prion infectivity. Nature 412:739–743
43. Perrier V, Solassol J, Crozet C, Frobert Y, Mourton-Gilles C, Grassi J, Lehmann S (2004) Anti-PrP antibodies block PrPSc replication in prion-infected cell cultures by accelerating PrPC degradation J Neurochem 89:454–463
44. Kim CL, Karino A, Ishiguro N, Shinagawa M, Sato M, Horiuchi M (2004) Cell-surface retention of PrPC by anti-PrP antibody prevents protease-resistant PrP formation. J Gen Virol 85:3473–3482

45. Eghiaian F, Grosclaude J, Lesceu S, Debey P, Doublet B, Tréguer E, Rezaei H, Knossow M (2004) Insight into the PrPC → PrPSc conversion from the structures of antibody-bound ovine prion scrapie-susceptibility variants. Proc Natl Acad Sci USA 101:10254–10259
46. Stahl N, Borchelt DR, Hsiao K, Prusiner SB (1987) Scrapie prion protein contains a phosphatidylinositol glycolipid. Cell 51: 229–240
47. Chesebro B, Trifilo M, Race R, Meade-White K, Teng C, LaCasse R, Raymond L, Favara C, Baron G, Priola S et al. (2005) Anchorless prion protein results in infectious amyloid disease without clinical scrapie. Science 308: 1435–1439
48. Bosques CJ, Imperiali B (2003) The interplay of glycosylation and disulfide formation influences fibrillization in a prion protein fragment. Proc Natl Acad Sci USA 100:7593–7598.
49. Calzolai L, Lysek DA, Perez DR, Güntert P, Wüthrich K (2005) Prion protein NMR structures of chickens, turtles, and frogs. Proc Natl Acad Sci USA 102:651–655
50. Lysek DA, Schorn C, Nivon LG, Esteve-Moya V, Christen B, Calzolai L, von Schroetter C, Fiorito F, Herrmann T, Güntert P et al. (2005) Prion protein NMR structures of cats, dogs, pigs, and sheep. Proc Natl Acad Sci USA 102:640–645
51. Gossert AD, Bonjour S, Lysek DA, Fiorito F, Wüthrich K (2005) Prion protein NMR structures of elk and of mouse/elk hybrids. Proc Natl Acad Sci USA 102:646–650
52. Knaus KJ, Morillas M, Swietnicki W, Malone M, Surewicz WK, Yee VC (2001) Crystal structure of the human prion protein reveals a mechanism for oligomerization. Nat Struct Biol 8:770–774
53. Jackson GS, Hosszu LLP, Power A, Hill AF, Kenney J, Saibil H, Craven CJ, Waltho JP, Clarke AR, Collinge J (1999) Reversible inter-conversion of monomeric human prion protein between native and fibrilogenic conformations. Science 283:1935–1937
54. Shaked GM, Shaked Y, Kariv-Inbal Z, Halimi M, Avraham I, Gabizon R (2001) A protease-resistant prion protein isoform is present in urine of animals and humans affected with prion diseases. J Biol Chem 276:31479–31482
55. Pan K-H, Baldwin M, Nguyen J, Gasset M, Serban A, Groth D, Mehlhorn I, Huang Z, Fletterick RJ, Cohen FE, Prusiner SB (1993) Conversion of alpha-helices into beta-sheets features in the formation of the scrapie prion proteins. Proc Natl Acad Sci USA 90:10962–10966
56. Huang Z, Prusiner SB, Cohen FE (1996) Scrapie prions: a three-dimensional model of an infectious fragment. Fold Des 1:13–19
57. Frankenfield KN, Powers ET, Kelly JW (2005) Influence of the N-terminal domain on the aggregation properties of the prion protein. Protein Sci 14:2154–2166
58. Cordeiro Y, Kraineva J, Gomes MP, Lopes MH, Martins VR, Lima LM, Foguel D, Winter R, Silva JL (2005) The amino-terminal PrP domain is crucial to modulate prion misfolding and aggregation. Biophys J 89:2667–2676
59. Fioriti L, Quaglio E, Massignan T, Colombo L, Stewart RS, Salmona M, Harris DA, Forloni G, Chiesa R (2005) The neurotoxicity of prion protein (PrP) peptide 106–126 is independent of the expression level of PrP and is not mediated by abnormal PrP species. Mol Cell Neurosci 28:165–176
60. Jobling MF, Stewart LR, White AR, McLean C, Friedhuber A, Maher F, Beyreuther K, Master CL, Barrow J, Collins SJ et al. (1999) The hydrophobic core sequence modulates the neurotoxic and secondary structure properties of the prion peptide 106–126. J Neurochem 73:1557–1565
61. Gu Y, Fujioka H, Mishra RS, Li R, Singh N (2002) Prion peptide 106–126 modulates the aggregation of cellular prion protein and induces the synthesis of potentially neurotoxic transmembrane PrP. J Biol Chem 277:2275–2286
62. Haire LF, Whyte SM, Vasisht N, Gill AC, Verma C, Dodson EJ, Dodson GG, Bayley PM (2004) The crystal structure of the globular domain of sheep prion protein. J Mol Biol 336:1175–1183
63. Swietnicki W, Petersen R, Gambetti P, Surewicz WK (1997) pH-dependent stability and conformation of the recombinant human prion protein PrP(90–231). J Biol Chem 272: 27517–27520.

64. Hornemann S, Glockshuber R (1998) A scrapie-like unfolding intermediate of the prion protein domain PrP(121–231) induced by acidic pH. Proc Natl Acad Sci USA 95:6010–6014
65. Swietnicki W, Morillas M, Chen SG, Gambetti P, Surewicz WK (2000) Aggregation and fibrillization of the recombinant human prion protein huPrP90–231. Biochemistry 39: 424–431
66. Zou W-Q, Cashman NR (2002) Acidic pH and detergents enhance in vitro conversion of human brain PrPC to a PrPSc-like form. J Biol Chem 277:43942–43947
67. Langella E, Improta R, Barone V (2004) Checking the pH-induced conformational transition of prion protein by molecular dynamics simulations: effect of protonation of histidine residues. Biophys J 87:3623–3632
68. Borchelt DR, Taraboulos A, Prusiner SB (1992) Evidence for synthesis of scrapie prion proteins in the endocytic pathway. J Biol Chem 267:16188–16199
69. Wopfner F, Weidenhöfer G, Schneider R, von Brunn A, Gilch S, Schwarzl TF, Werner T, Schätz HM (1999) Analysis of 27 mammalian and 9 avian PrPs reveals high conservation of flexible regions of the prion protein. J Mol Biol 289:1163–1178
70. Minor DL Jr, Kim PS (1994) Measurement of the beta-sheet forming propensities of amino acids. Nature 367:660–663
71. Welker E, Raymond LD, Scheraga HA, Caughey B (2002) Intramolecular versus intermolecular disulfide bonds in prion proteins. J Biol Chem 277:33477–33481
72. Liu A, Riek R, Zahn R, Hornemann S, Glockshuber R, Wuthrich K (1999) Peptides and proteins in neurodegenerative disease: helix propensitiy of a polypeptide containing helix 1 of the mouse prion protein studied by NMR and CD spectroscopy. Biopolymers 51: 145–152
73. Ziegler J, Sticht H, Marx UC, Muller W, Rosch P, Schwarzinger S (2003) CD and NMR studies of prion protein (PrP) helix 1. Novel implications for its role in the PrPC → PrPSc conversion process. J Biol Chem 278:50175–50181
74. Gallo M, Paludi D, Cicero DO, Chiovitti K, Millo E, Salis A, Damonte G, Corsaro A, Thellung S, Schettini G et al. (2005) Identification of a conserved N-capping box important for the structural autonomy of the prion alpha 3-helix: the disease associated D202N mutation destabilizes the helical conformation. Int J Immunopathol Pharmacol 18:95–112
75. Winklhofer KF, Heske J, Heller U, Reintjes A, Muranyi IM, Tatzelt J (2003) Determinants of the in vivo folding of the prion protein. A bipartite function of helix 1 in folding and aggregation. J Biol Chem 278:14961–14970
76. Hirschberger T, Stork M, Schropp B, Winklhofer KF, Tatzelt J, Tavan P (2006) Structural instability of the prion protein upon M205S/R mutations revealed by molecular dynamics simulations. Biophys J 90:3908–3918
77. Brown DR, Guantieri V, Grasso G, Impellizzeri G, Pappalardo G, Rizzarelli E (2004) Copper(II) complexes of peptide fragments of the prion protein. Conformation changes induced by copper(II) and the binding motif in C-terminal protein region. J Inorg Biochem 98:133–143
78. Tizzano B, Palladino P, De Capua A, Marasco D, Rossi F, Benedetti E, Pedone C, Ragone R, Ruvo M (2005) The human prion protein alpha2 helix: a thermodynamic study of its conformational preferences. Proteins 59:72–79
79. Ikeda K, Higo J (2003) Free-energy landscape of a chameleon sequence in explicit water and its inherent alpha/beta bifacial property.Protein Sci 12:2542–2548
80. Riek R, Wider G, Billeter M, Hornemann S, Glockshuber R, Wütrich K (1998) Free-energy landscape of a chameleon sequence in explicit water and its inherent alpha/beta bifacial property. Proc Natl Acad Sci USA 95:11667–11672
81. Salmona M, Morbin M, Massignan T, Colombo L, Mazzoleni G, Capobianco R, Diomede L, Thaler F, Mollica L., Musco G et al. (2003) Structural properties of Gerstmann-Straussler-Scheinker disease amyloid protein. J Biol Chem 278:48146–48153
82. Brown DR, Qin K, Herms JW, Madlung A, Manson J, Strome R, Fraser PE, Kruck T, von Bohlen A, Schulz- Schaeffer W et al. (1997) The cellular prion protein binds copper in vivo. Nature 390:684–687

83. Millhauser GL (2007) Copper and the prion protein: methods, structures, function, and disease. Annu Rev Phys Chem 58:299–320
84. Luczkowski M, Kozlowski H, Stawikowski M, Rolka K, Gaggelli E, Valensin D, Valensin G (2002) Is a monomeric prion octapeptide repeat PHGGGWGQ specific ligand for Ca^{2+} Ions? J Chem Soc Dalton Trans 2269–2274
85. Jobling MF, Huang X, Stewart LR, Barnham KJ, Curtain C, Volitakis I, Perugini M, White AR, Cherny RA, Masters CL et al. (2001) Copper and zinc binding modulates the aggregation and neurotoxic properties of the prion peptide PrP106–126. Biochemistry 40:8073–8084
86. Aronoff-Spencer E, Burns CS, Avdievich NI, Gerfen GJ, Peisach J, Antholine WE, Ball HL, Cohen FE, Prusiner SB, Millhauser GL (2000) Identification of the Cu2+ binding sites in the N-terminal domain of the prion protein by EPR and CD spectroscopy. Biochemistry 39:13760–13771
87. Burns CS, Aronoff-Spencer E, Dunham CM, Lario P, Avdievich NI, Antholine WE, Olmstead MM, Vrielink A, Gerfen GJ, Peisach J et al. (2002) Molecular features of the copper binding sites in the octarepeat domain of the prion protein. Biochemistry 41:3991–4001
88. Lippard SJ, Berg JM (1994) Principles of Bioinorganic Chemistry, University Science Books, Mill Valley, CA
89. Bryce GF, Gurd FRN (1966) Visible spectra and optical rotatory properties of cupric ion complexes of L-histidine-containing peptides. J Biol Chem 241:122–129
90. Sundberg RJ, Martin RB (1974) Interactions of histidine and other imidazole derivatives with transition metal ions in chemical and biological systems. Chem Rev 74:471–517
91. Kramer ML, Kratzin HD, Schmidt B, Römer A, Windl O, Liemann S., Hornemann S, Kretzschmar H (2001) Prion protein binds copper within the physiological concentration range. J Biol Chem 276:16711–16719
92. Qin K, Yang Y, Mastrangelo P, Westaway D (2002) Mapping Cu(II) binding sites in prion proteins by diethyl pyrocarbonate modification and matrix-assisted laser desorption ionization-time of flight (MALDI-TOF) mass spectrometric footprinting. J Biol Chem 277:1981–1990
93. Burns CS, Aronoff-Spencer E, Legname G, Prusiner SB, Antholine WE, Gerfen GJ, Peisach J, Millhauser GL (2003) Copper coordination in the full-length, recombinant prion protein. Biochemistry 42:6794–6803
94. Quaglio E, Chiesa R, Harris DA (2001) Copper converts the cellular prion protein into a protease-resistant species that is distinct from the scrapie isoform. J Biol Chem 276: 11432–11438
95. Ricchelli F, Buggio R, Drago D, Salmona M, Forloni G, Negro A, Tognon G, Zatta P (2006) Aggregation/fibrillogenesis of recombinant human prion protein and Gerstmann-Straussler-Scheinker disease peptides in the presence of metal ions. Biochemistry 45: 6724–6732
96. Cereghetti GM, Schweiger A, Glockshuber R, van Doorslaer S (2003) Stability and Cu(II) binding of prion protein variants related to inherited human prion diseases. Biophys J 84:1985–1997
97. Cervenakova L, Buetefisch C, Lee HS, Taller I, Stone G, Gibbs CJ Jr, Brown P, Hallett M, Goldfarb LG (1999) Novel PRNP sequence variant associated with familial encephalopathy. Am J Med Genet 88:653–656
98. Bütefisch CM, Gambetti P, Cervenakova L, Park K-Y, Hallett M, Goldfarb LG (2000) Inherited prion encephalopathy associated with the novel PRNP H187R mutation: a clinical study. Neurology 55:517–522
99. Kuznetsov IB, Rackovsky S (2004) Comparative computational analysis of prion proteins reveals two fragments with unusual structural properties and a pattern of increase in hydrophobicity associated with disease-promoting mutations. Protein Sci 13:3230–3244
100. Barducci A, Chelli R, Procacci P, Schettino V (2005) Misfolding pathways of the prion protein probed by molecular dynamics simulations. Biophys J 88:1334–1343

101. Kiachopoulos S, Bracher A, Winklhofer KF, Tatzelt J (2005) Pathogenic mutations located in the hydrophobic core of the prion protein interfere with folding and attachment of the glycosylphosphatidylinositol anchor. J Biol Chem 280:9320–9329
102. Gasset M, Baldwin MA, Lloyd DH, Gabriel J-M, Holtzman DM, Cohen F, Fletterick R, Prusiner SB (1992) Predicted alpha-helical regions of the prion protein when synthesized as peptides form amyloid. Proc Natl Acad Sci USA 89:10940–10944
103. Ronga L, Tizzano B, Palladino P, Ragone R, Urso E, Maffia M, Ruvo M, Benedetti E, Rossi F (2006) The prion protein: Structural features and related toxic peptides. Chem Biol Drug Des 68:139–147
104. Cobb NJ, Sönnichsen FD, Mchaourab H, Surewicz WK (2007) Molecular architecture of human prion protein amyloid: a parallel, in-register beta-structure. Proc Natl Acad Sci USA 104:18946–18951
105. Fitzmaurice TJ, Burke DF, Hopkins L, Yang S, Yu S, Sy M-S, Thackray AM, Bujdoso R (2008) The stability and aggregation of ovine prion protein associated with classical and atypical scrapie correlates with the ease of unwinding of helix-2. Biochem J 409:367–375
106. Lu X, Wintrode PL, Surewicz WK (2007) Beta-sheet core of human prion protein amyloid fibrils as determined by hydrogen/deuterium exchange. Proc Natl Acad Sci USA 104: 1510–1515
107. Ronga L, Palladino P, Saviano G, Tancredi T, Benedetti E, Ragone R, Rossi F (2007) NMR Structure and CD titration with metal cations of human prion alpha2-helix-related peptides. Bioinorg Chem Appl 2007:10720
108. Nandi PK, Nicole JC (2004) Nucleic acid and prion protein interaction produces spherical amyloids which can function in vivo as coats of spongiform encephalopathy agent. J Mol Biol 344:827–837
109. Calzolai L, Zahn R (2003) Influence of pH on NMR structure and stability of the human prion protein globular domain. J Biol Chem 278:35592–35596.
110. Wong C, Xiong LW, Horiuchi M, Raymond L, Wehrly K, Chesebro B, Caughey B (2001) Sulfated glycans and elevated temperature stimulate PrP(Sc)-dependent cell-free formation of protease-resistant prion protein. EMBO J 20:377–386
111. Kawatake S, Nishimura Y, Sakaguchi S, Iwaki T, Doh-ura K (2006) Surface plasmon resonance analysis for the screening of anti-prion compounds. Biol Pharm Bull 29:927–932
112. Ronga L, Palladino P, Tizzano B, Marasco D, Benedetti E, Ragone R, Rossi F (2006) Effect of salts on the structural behavior of hPrP alpha2-helix-derived analogues: the counterion perspective. J Pept Sci 12:790–795
113. Collins KD (1997) Charge density-dependent strength of hydration and biological structure. Biophys J 72:65–76
114. Collins KD (2004) Ions from the Hofmeister series and osmolytes: effects on proteins in solution and in the crystallization process. Methods 34:300–311
115. Niedz RP, Evens TJ (2006) A solution to the problem of ion confounding in experimental biology. Nat Methods 3:417
116. Chen YR, Huang HB, Chyan CL, Shiao MS, Lin TH, Chen YC (2006) The Effect of A{beta} Conformation on the metal affinity and aggregation mechanism studied by circular dichroism spectroscopy. J Biochem (Tokyo) 139:733–740
117. Roberts MF (2005) Organic compatible solutes of halotolerant and halophilic microorganisms. Saline Syst 1:5
118. Arora A, Ha C, Park CB (2004) Inhibition of insulin amyloid formation by small stress molecules. FEBS Lett 564:121–125

Chapter 7
Synuclein Structure and Function in Parkinson's Disease

David Eliezer

Abstract The protein alpha-synuclein is implicated in the etiology of both sporadic and hereditary Parkinson's disease. Structural studies of both the intrinsically disordered free state of the protein and of more ordered states, adopted when alpha-synuclein self assembles into fibrils or binds to lipid membranes or detergent micelles, have begun to provide insights into factors that likely influence both the pathological aggregation of the protein and its normal functions. Residual secondary structure and transient long-range interactions within the free state can be detected and may influence alpha-synuclein aggregation pathways. Structure within the amyloid fibril form of alpha-synuclein can also provide clues regarding the assembly pathways of the protein. Alpha-synuclein folds upon binding to lipid membranes and the experimentally determined topology of the bound protein likely mediates its physiological functions. The influence of disease-linked mutations on the structural properties of the free, fibrillar, and bound states has also been evaluated in order to examine the basis for altered aggregation kinetics and possible functional impairments of the mutant proteins. Comparative structural studies of the other human synuclein family members, β-synuclein and γ-synuclein have also been performed to clarify features that differentiate them from alpha-synuclein in both pathological and functional contexts. This chapter provides an up to date review of structural studies of the human synuclein family and of the implications of these studies for understanding synuclein pathways in biology and disease.

7.1 Background

7.1.1 The Discovery of Synucleins

Alpha-synuclein (αS) was discovered as a protein highly enriched in synaptosome preparations from the electric organ of the electric ray *T. californica* [1]. The protein was determined to localize to presynaptic nerve terminals, as well as to the

D. Eliezer (✉)
Weill Cornell Medical College, 1300 York Avenue, New York, NY 10065, USA
e-mail: eliezer@med.cornell.edu

interior of the nuclear membrane, giving rise to the name synuclein. The partial segregation of the protein with synaptic vesicles and its localization to the nuclear membrane suggested a membrane-associated function for the protein. Subsequently, the protein and two close homologues, β-synuclein (βS) and γ-synuclein (γS) were discovered in other vertebrates and humans [2–6]. All synuclein proteins contain a series of imperfect 11-residue repeats, each of which contains a more highly conserved 6-residue motif. These repeats bear similarity to those found in the apolipoprotein family [7], further suggesting that synucleins are involved in lipid interactions. The primary structure of all synucleins consists of two regions, an N-terminal membrane-binding domain and a highly acidic C-terminal tail. Within the membrane-binding domain of αS is a region commonly referred to as the NAC (non-amyloid-β component) region, based on its identification as a proteolytic fragment in Alzheimer's disease plaques [2].

7.1.2 Synuclein Mutations in Parkinson's Disease

The gene encoding αS (*SNCA/PARK1*) was the first gene to be associated with familial Parkinson's disease (PD) [8]. Mutations in the gene, leading to a point mutation (A53T) in the primary sequence of the protein, were shown to be responsible for an autosomal dominant form of PD in the Italian-American Contursi kindred [9]. Subsequent to the discovery of the A53T mutation, two further mutations in αS, A30P [10] and E46K [11] were linked to familial forms of Parkinsonism. Furthermore, either triplication [12] or duplication [13] of the *SNCA* gene also lead to familial PD In contrast, to date no mutations in either βS or γS have been clearly associated with PD or other human disease [14–16].

7.1.3 Synuclein in Lewy Bodies

Synuclein mutations are a rare cause of PD, with most cases of the disease considered to be sporadic or idiopathic. Nevertheless, αS was implicated in the latter forms of PD as well by the discovery that it is the primary protein component of the amyloid fibril aggregates found in Lewy body deposits [17]. Lewy bodies are spheroidal proteinaceous deposits found in surviving neurons of the substantia nigra region of PD brains. Indeed, the presence of Lewy bodies is considered to be a requirement for the post-mortem diagnosis of PD [18]. The identification of αS as the amyloid component of Lewy bodies therefore suggests a role for αS-related pathways in both familial and sporadic forms of PD.

7.1.4 Synuclein Toxicity

PD results from the death of dopamine-producing neurons in the substania nigra region of the brain. The cause of death of these cells remains unclear, but a number of hypotheses regarding the involvement of αS in cellular toxicity have been proposed. Perhaps the most obvious proposal is that the amyloid fibril filled Lewy body deposits themselves may mechanically perturb cellular homeostasis and cause cell death. This idea, however, is countermanded by recent evidence that deposits formed of mature amyloid fibrils may in fact be protective [19]. The fact that Lewy bodies are found primarily in surviving neurons of the substantia nigra may also be interpreted as supporting this idea. Numerous other proposals for how αS may be toxic exist, including its potential involvement in oxidative reactions, its inhibition of proteasomal protein degradation, and its possible involvement in forming membrane-permeabilizing channels or pores, or otherwise influencing lipid bilayer integrity, but all of these mechanisms require further confirmation.

7.1.5 Synuclein Function

Just at as the precise mechanism of αS toxicity remains to be determined, the precise physiological role or function of the synucleins remains unclear. The association of αS with synaptic vesicles suggests a role in the regulation of neurotransmission, and indeed knockout or knockdown of αS has been shown to influence neurotransmitter release [20–23]. Such studies have also suggested that the absence of αS affects the size of synaptic vesicle pools in neurons [21, 24], although this result was not obtained in another study [25]. Furthermore, αS is capable of rescuing a severe neurodegenerative phenotype associated with the absence of a protein (cysteine string protein α) that functions as part of the chaperone system that resets the synaptic vesicle fusion machinery [26]. The absence of αS in mice also leads to increased levels of the protein complexin, which is thought to regulate SNARE-mediated membrane fusion [25]. Yeast models of synuclein function have suggested a role in vesicular trafficking as well [27, 28]. αS has also been shown to be an inhibitor of the enzyme phospholipase D [29], which itself may play a role in facilitating synaptic vesicle formation or fusion [30]. Most recently, attention has returned to a possible function of αS in the nucleus, with a number of reports documenting the presence of the protein in this organelle [31–34], but a specific nuclear function for synuclein remains to be determined. An early study in songbirds also identified synucleins as playing a role in learning processes [6].

7.2 Free State Structure

7.2.1 Residual Secondary Structure

Early studies of synuclein structure based on optical methods such as circular dichorism and Fourier-transform infrared spectroscopy revealed that αS was "natively unfolded", or intrinsically disordered [35], a relatively unusual observation at the time. Subsequent solution NMR studies confirmed this conclusion and provided a highly detailed picture of the local structural preferences of αS, including a weak preference for helical structure throughout the N-terminal lipid-binding domain, and for more extended conformations in the C-terminal tail [36]. The strongest indications for helical structure were found near the position of the A30P PD-linked mutation, which was shown to locally disrupt the helical propensity of the protein [37]. The A53T mutation also led to a local decrease in helical character. Similar NMR studies of the free states of βS [38, 39] and γS [38, 40] showed that both proteins retain a weak preference for helical structure in their lipid-binding domains, and more extended structure in their C-terminal tails.

7.2.2 Role of Residual Structure in Aggregation

Based on the observed effects of PD-linked mutations on residual secondary structure, two models were proposed for the role of local structure preferences in modulating αS aggregation [37]. The first model suggested simply that by decreasing the local preference of the protein for helical conformations, mutations might facilitate the conversion of the protein into the highly β-sheet rich amyloid fibril structure. It was pointed out that the PD-linked mutations (A30P in particular) may concurrently decrease the propensity of the protein to adopt the highly helical membrane-bound conformation (see below) and thereby could increase the population of aggregation-competent free protein. Alternately, a second model for the effects of the A30P mutation postulated that transient amphipathic helical structure formed around position 30 could interact in an intramolecular fashion with the hydrophobic NAC region of αS and that such interactions could decrease the probability of NAC-mediated intermolecular interactions leading to aggregation and fibril formation. Perturbation of the transient amphipathic helical structure by the A30P mutation would disrupt such intramolecular contacts and facilitate intermolecular interactions and oligomerization. Finally, the incompatibility of proline with β structure could account for the previously observed [41] reduction in the rate of conversion of A30P oligomers to mature fibrils. A weakness of this model is that position 30 does not appear to be integrated into the structured core of αS amyloid fibrils [42–44], but recent studies [44] suggest that the region around position 30 may still play a role in modulating fibril formation and/or structure.

7.2.3 Transient Long-Range Interactions

Subsequent to the suggestion that long-range intramolecular contacts could influence αS aggregation, the first detailed evidence for such contacts was obtained using paramagnetic relaxation enhancement (PRE) studies [45, 46], which suggested the presence of transient contacts between the C-terminal tail of αS and either the N-terminal or NAC regions of the protein. Following the logic of the previous model described above, it was proposed that release of such long-range interactions would increase exposure of the hydrophobic NAC region and thereby favor aggregation. Such a release could be induced by the direct binding of polycations [47–49] or metals such as copper [50, 51] to the C-terminal tail of αS, or possibly by PD-linked mutations. Indeed, a subsequent study [52] argued, based on changes in PRE and residual dipolar coupling (RDC) data, that the PD-linked A30P and A53T mutations led to such a release of contacts between the C-terminal tail and the rest of the protein. Thus, the idea that transient long-range intramolecular interactions could protect αS from aggregation has gained widespread attention.

Despite the emergence of the long-range contact model, evidence also exists that contraindicates this hypothesis. Earlier studies using X-ray scattering and size exclusion chromatography [53] and NMR diffusion measurements [54], as well as more recent experiments employing fluorescence or electron transfer methods [55–57] showed that αS was more compact than would be expected if the protein populated a fully unfolded conformational ensemble, consistent with the detection of long-range intramolecular contacts by NMR. However, compaction was shown to correlate with an increased propensity to aggregate, rather than with protection from aggregation [53]. Also, PRE measurements from βS and γS show that both of these proteins exhibit few or no long-range contacts [38], despite the fact that both proteins aggregate less readily than αS [53, 58]. Furthermore, PRE data from PD-linked mutants of αS continue to show robust indications of long-range contacts between the C-terminal and lipid-binding domains [52], and suggest at best a small decrease in such contacts. The βS and γS studies [38] also strongly indicate that large RDCs in the C-terminal tails of the synucleins are not necessarily related to long-range contacts, calling into question whether the RDC changes observed in the PD-linked αS mutants [52] reflect the release of long-range contacts. Finally, recent studies of the effects αS phosphorylation at serine 129 show that such phosphorylation profoundly interferes with long-range contacts between the C-terminal tail and the N-terminal lipid-binding domain, but rather than enhancing the self-assembly of the protein, Ser 129 phosphorylation dramatically decreases the rate at which αS aggregates [59].

At present, it appears clear that transient long-range intramolecular contacts exist in the free state of αS. However, the role of such contacts in either limiting aggregation by protecting hydrophobic regions from inter-molecular contacts or enhancing oligomerization by favoring compact aggregation-prone states, remain to be clarified. Other properties of αS that are likely to profoundly influence the aggregation rate of the protein include local properties of the polypeptide such as secondary structure propensity, hydrophobicity and charge, and the electrostatic self-repulsion that is likely to result from the large negative charge carried by the C-terminal tail of the protein [60].

7.3 Fibril Structure

The structure of fibrillar αS was shown early on to belong the class of amyloid aggregates [61, 62], characterized by the so-called cross-beta structure consisting of β-strands running perpendicular to the long axis of the fibrils, with inter-strand hydrogen bonds running in the direction of the fibril axis, forming β-sheets that span the length of the fibrils [63]. Nevertheless, higher-resolution information regarding the detailed structures of amyloid fibrils has only recently become available, primarily as a result of advances in solid state NMR (SSNMR) methods [64], but also complemented by the application of electron spin resonance (ESR) methods [65].

For αS, a number of recent reports have shed light on which regions of the protein are involved in the core cross-β structure, where precisely β-strand structure exists in this core, and where it is interrupted. It was shown that the N-terminal ~35 residues of αS are not a part of the rigid fibril core [42, 43] although this region of the protein does exhibit somewhat restricted mobility [44, 66]. The C-terminal ~40 residues of the protein are also not part of the core region, but unlike the N-terminal region, they remain highly mobile. Within the core region of approximately 65 residues, the combined SSNMR [44, 67, 68] and ESR [42, 66] data suggest extensive and highly immobile β-strand structure, with, however, a number of interruptions. Most prominent appears to be a turn or loop region around positions 63–65 (ESR data) or 66–68 (SSNMR data), and a second such break likely exists around positions 83–86 (both SSNMR and ESR data). The most robust indications for a continuous β-strand are found between these two break regions, from positions 69–83. Other probable β-strand locations, based on the SSNMR data in combination with hydrogen exchange measurements [68], include positions 37–43, 52–59, 62–66 and 86–92, although the thirrd of these strands exhibits significant mobility according to ESR data [66]. It is also clear from the ESR data that the strands in αS fibrils are arranged in a parallel fashion. Although the effects of PD-linked mutations on the structure of αS amyloid fibrils have not yet been characterized in complete detail, one study of A53T fibrils suggests the possible presence of a sixth β-strand, extending the C-terminal end of the core region of the fibrils [69].

The most recent model to emerge from SSNMR, hydrogen exchange, and electron microscopy studies of αS fibrils [68] posits that the five individual β-strands identified are stacked sequentially on top of one another in a single layer of the protofibril, which is therefore proposed to consist of five stacked β-sheets. This geometry is consistent with that observed in earlier studies of the structure of fibrils formed by the smaller amyloid-beta (Aβ) peptide [70, 71], but contrasts with a recent structure of the comparably sized prion domain of the fungal protein HET-s, in which different β-strands pair with each other in the plane of the fibril-long β-sheets, such that each individual polypeptide chain forms two sheet layers rather than one [72]. Further investigation will be required to verify and/or refine the current model of fibrillar αS.

7.4 Lipid-Bound Structure

7.4.1 A Role for Synuclein Function in Disease?

Because the function of αS appears to revolve around neurotransmitter-carrying synaptic vesicles, and because PD is fundamentally characterized by a deficit in the brain of the neurotransmitter dopamine, it remains possible that perturbation of some aspect of normal synuclein function is responsible for the involvement of the protein in PD. This idea is not inconsistent with a role for αS aggregation in PD, as aggregation would almost certainly interfere with αS function, and could easily do so in a genetically dominant fashion. It is also not necessarily inconsistent with the observations that αS knockout animals are essentially normal and do not exhibit a PD-like phenotype because PD, despite being the second most common neurodegenerative disorder, occurs relatively rarely and only at an advanced age, and it is therefore unclear under what circumstances knockout animals might be expected to exhibit a PD phenotype. Furthermore, while the fact that transgenic overexpression of αS has led to a number of PD models in mice, flies and worms [73–75], these models do not entirely recapitulate salient aspects of the human disease. Therefore, a clear understanding of the normal physiological function of αS may ultimately prove crucial for understanding the role of the protein in PD.

Because αS binds to synaptic vesicles and appears to function in the regulation of the formation, trafficking and fusion of such vesicles, the lipid-bound form of the protein is thought to be predominantly responsible for mediating its physiological function(s). This reasoning is supported by the observation that the membrane-bound structure of αS is required for its interactions with its best characterized interaction partner, the lipid-modifying enzyme phospholipase D [76], as well as with more recently identified partners [77]. Thus, insights into the conformation adopted by αS in the presence of lipid membranes should improve our understanding of the basis for αS function.

7.4.2 Secondary Structure

Early studies using circular dichroism spectroscopy showed that in the presence of phospholipid vesicles, αS undergoes a disorder-to-order transition involving the adoption of helical structure by a significant fraction of the polypeptide chain [78–80]. Although the precise disposition of this helical structure could not be delineated by this method, a model was presented, based on the previously noted similarity [6] between the repeats found in the sequence of αS and those found in apolipoproteins [7], in which lipid-bound αS was predicted to form 5 independent helices, all of which were surface associated rather than adopting a trans-membrane position [81].

Higher resolution studies of the structure of lipid-bound αS have largely relied on the use of small spheroidal detergent micelles to mimic larger phospholipid vesicles while retaining a particle size that is accessible to solution state NMR methods. The first such study [36] demonstrated that the micelle-binding domain of αS did indeed adopt a helical structure and extended to approximately residue 102, while the remaining C-terminal tail of the protein remained largely unstructured and did not interact strongly with detergent micelles. This same study also confirmed that the same domain boundaries were observed when the protein was bound to phospholipid vesicles, supporting some degree of similarity between the conformations of micelle-bound and vesicle-bound αS. Follow up studies were able to delineate more precisely the location of helical structure in the micelle-bound N-terminal domain and showed that this domain formed two long helices interrupted by a single break [82, 83], rather than five shorter helical segments, but that the helices did indeed remain on the micelle surface, as predicted, rather than adopting a trans-micelle orientation [84, 85]. It was subsequently shown that the two helices did not form inter-helical contacts [84], and a detailed structure for micelle-bound αS was ultimately obtained using new NMR methods developed for analyzing structures of non-compact proteins [86].

7.4.3 Topology

The two helices of micelle-bound αS were shown to adopt an anti-parallel orientation despite the absence of any tertiary contacts between them [86]. A study using pulsed ESR to measure distances in αS bound to detergent micelles of differing diameters also demonstrated an anti-parallel orientation of the helices, but further showed that when the protein was bound to larger micelles, the two helices splayed further apart, suggesting that their anti-parallel orientation was at least in part imposed by the geometry of the micelle surface [87]. Earlier studies which noted the unusual periodicity of the helices formed by αS [82, 84, 88] proposed that the protein could adopt a single extended helix encompassing the entire lipid-binding domain when bound to the surface of lipid vesicles. Direct evidence for the extended helical conformation has not been published as yet, but an indication that this conformation likely exists is provided by a recent and elegant circular dichroism study [89].

7.4.4 Effects of Parkinson's Disease-Linked Mutations

The effects of PD-linked mutations on the structure of lipid-bound αS have also been investigated. Earlier work demonstrated that the A30P mutation decreases the affinity of the protein for membrane surfaces [80, 90, 91], while neither the A53T nor the E46K mutations do so [90, 92]. Structurally, the A30P mutation was shown to result in a highly localized perturbation of the helical structure of the micelle-bound protein, while the A53T mutation did not result in any evident structural changes [93, 94]. The micelle-bound structure of the E46K mutant has not been characterized in detail but a preliminary study indicates that both helices are likely intact, suggesting that any structural perturbations are likely to be minor for this mutant as well [95]. From these studies, it appears that the A30P mutation could perturb αS function by interfering with the membrane localization of the protein, both by reducing the helical propensity of the free state and by perturbing the structure of the bound state. In contrast, to the extent that the A53T and E46K mutations disrupt αS function, they likely do so without dramatically altering the presumably functional membrane-bound conformation. Instead, they may exert their effects either by interfering with protein-protein interactions that are necessary for synuclein function or possibly by perturbing the ability of the protein to undergo conformational changes that are required for its proper function.

7.4.5 β- and γ-Synuclein

Because βS and γS bear a very high sequence homology to αS, it seems likely that the three proteins have somewhat similar functions. βS in particular exhibits a similar distribution to that of αS, being found predominantly in the central nervous system [2, 4, 96], while γS appears to be more broadly distributed [5, 97–99]. While structural studies of the free states of βS and γS have been performed to gain insights into the relationship between synuclein primary sequence, residual structure, and aggregation [38–40], the structures of the micelle- and membrane-bound forms of both proteins may be expected to provide insights into differences between the functions of the three proteins.

NMR studies of βS and γS bound to spheroidal detergent micelles [100] revealed that the lipid-binding domains of αS and γS adopt highly similar conformations, with both of the helices that were observed in the case of αS being preserved in the micelle-bound structure of γS. Based on this observation, it was suggested that functional differences between αS and γS may be primarily related to the high degree of sequence divergence in the C-terminal tails of the two proteins. Since the C-terminal tail of αS is thought to mediate protein-protein interactions, such sequence variations would suggest that γS interacts with a different set of binding partners than αS, which could result in the presumed functional differences.

In the case of βS, it was found that the micelle-bound structure of the protein differs significantly from that of αS in the second of the two micelle-bound helices. In αS, helix 2 extends from residue 45 through 92, but exhibits a kink at positions 66–68 [82, 86]. In βS, the first part of helix 2 remains intact, but the second part, past position 65, exhibits clear indications of decreased helicity and increased mobility, suggesting that the C-terminal half of this helix is destabilized and becomes decoupled from the N-terminal half. This effect is most probably caused by the absence of 11 residues in the βS sequence corresponding to αS positions 74–85. Because of this structural difference in the lipid-binding domains of αS and βS, and because their C-terminal tails are more highly similar to each other than in the case of γS, it was postulated [100] that functional differences between αS and βS may relate more strongly to their membrane interactions, affinity, and localization than to their protein binding partners. Supporting this hypothesis is the observation that both αS and βS were identified in a screen for inhibitors of the protein phospholipase D, while γS was not [29]. More detailed investigations of differences in the lipid preferences and interactions of αS and βS are underway.

7.5 New Model for Synuclein Function

Despite extensive efforts, the physiological function of αS remains poorly understood. A consensus exists that the protein is involved in the regulation of the synaptic vesicle life cycle, including the crucial fusion step. However, apparently conflicting observations have suggested that αS can both down- and up-regulate synaptic vesicle fusion and neurotransmitter release. Overexpression of αS leads to a reduction in vesicle fusion in several model systems [28, 101] and synuclein knockouts exhibit enhanced dopamine release under repeated stimulation [20, 102], suggesting that αS acts to inhibit vesicle fusion. At the same time, αS can compensate for deficits in the SNARE vesicle fusion machinery [26] and deletion of αS and βS leads to the up-regulation of the protein complexin, which is thought to be required for proper SNARE function [25], suggesting that αS can play a role in promoting vesicle fusion. In addition to this apparent contradiction, it has also been puzzling that despite its interaction with synaptic vesicles, αS appears to bind to vesicles weakly and reversibly [1, 90, 93] and is highly mobile in synaptic nerve terminals [103].

These observations, combined with the structural studies of micelle- and lipid-bound αS described above, have led to the formulation of a new model for αS function [60]. This model postulates that αS can adopt both an extended helix conformation and the anti-parallel broken helix conformation when bound to membranes, and that the specific conformation is determined by the membrane topology. When presented with a (synaptic) vesicle surface, the extended conformation is adopted. However, when presented with two closely apposed membrane surfaces, the broken helix conformation is favored. *In vitro*, the latter situation occurs when the protein is presented with the small highly curved spheroidal micelle surface. *In vivo*, a similar situation is postulated to occur when the vesicles to which the protein is bound (in the extended conformation) approach the plasma membrane. When the two membranes are sufficiently close, the model posits that αS can adopt the broken helix conformation to bridge the two membranes. The extended helix conformation is proposed to be weakly bound, in part because of the mismatch between the periodicity of the synuclein sequence and that of a canonical α-helix [82, 84, 88] while the broken helix conformation is proposed to be more tightly bound. When the protein is in the bridging configuration, it can act to stabilize docked vesicles, thereby down-regulating the efficiency of vesicle fusion. In the absence of the proper vesicle fusion machinery, however, αS could still stabilize vesicles in close proximity to the plasma membrane, facilitating a low level of spontaneous or assisted vesicle fusion that can support viability. Such a model makes a number of testable predictions, and provides an example of how structural studies of αS can be used to develop new hypotheses regarding the function of this unusual and important PD-associated protein.

Abbreviations

alpha-synuclein (αS)
electron spin resonance (ESR)
NAC (non-amyloid-β component)
paramagnetic relaxation enhancement (PRE)
Parkinson's disease (PD)
residual dipolar coupling (RDC)
solid state NMR (SSNMR)
β-synuclein (βS)
γ-synuclein (γS)

References

1. Maroteaux L, JT Campanelli, RH Scheller (1988) Synuclein: a neuron-specific protein localized to the nucleus and presynaptic nerve terminal. J Neurosci 8:2804–2815
2. Ueda K, H Fukushima, E Masliah, Y Xia, A Iwai, M Yoshimoto, DA Otero, J Kondo, Y Ihara, T Saitoh (1993) Molecular cloning of cDNA encoding an unrecognized component of amyloid in Alzheimer disease. Proc Natl Acad Sci USA 90:11282–11286
3. Tobe T, S Nakajo, A Tanaka, A Mitoya, K Omata, K Nakaya, M Tomita, Y Nakamura (1992) Cloning and characterization of the cDNA encoding a novel brain-specific 14-kDa protein. J Neurochem 59:1624–1629
4. Jakes R, MG Spillantini, M Goedert (1994) Identification of two distinct synucleins from human brain. FEBS Lett 345:27–32
5. Ji H, YE Liu, T Jia, M Wang, J Liu, G Xiao, BK Joseph, C Rosen, YE Shi (1997) Identification of a breast cancer-specific gene, BCSG1, by direct differential cDNA sequencing. Cancer Res. 57:759–764
6. George JM, H Jin, WS Woods, DF Clayton (1995) Characterization of a novel protein regulated during the critical period for song learning in the zebra finch. Neuron 15: 361–372.
7. Segrest JP, MK Jones, H De Loof, CG Brouillette, YV Venkatachalapathi, GM Anantharamaiah (1992) The amphipathic helix in the exchangeable apolipoproteins: a review of secondary structure and function. J Lipid Res 33:141–166
8. Polymeropoulos MH, JJ Higgins, LI Golbe, WG Johnson, SE Ide, G Di Iorio, G Sanges, ES Stenroos, LT Pho, AA Schaffer et al. (1996) Mapping of a gene for Parkinson's disease to chromosome 4q21-q23. Science 274:1197–1199
9. Polymeropoulos MH, C Lavedan, E Leroy, SE Ide, A Dehejia, A Dutra, B Pike, H Root, J Rubenstein, R Boyer et al. (1997) Mutation in the alpha-synuclein gene identified in families with Parkinson's disease. Science 276:2045–2047
10. Kruger R, W Kuhn, T Muller, D Woitalla, M Graeber, S Kosel, H Przuntek, JT Epplen, L Schols, O Riess (1998) Ala30Pro mutation in the gene encoding alpha-synuclein in Parkinson's disease. Nat Genet 18:106–108
11. Zarranz JJ, J Alegre, JC Gomez-Esteban, E Lezcano, R Ros, I Ampuero, L Vidal, J Hoenicka, O Rodriguez, B Atares et al. (2004) The new mutation, E46K, of alpha-synuclein causes Parkinson and Lewy body dementia. Ann Neurol 55:164–173
12. Singleton AB, M Farrer, J Johnson, A Singleton, S Hague, J Kachergus, M Hulihan, T Peuralinna, A Dutra, R Nussbaum et al. (2003) alpha-Synuclein locus triplication causes Parkinson's disease. Science 302:841
13. Chartier-Harlin MC, J Kachergus, C Roumier, V Mouroux, X Douay, S Lincoln, C Levecque, L Larvor, J Andrieux, M Hulihan et al. (2004) Alpha-synuclein locus duplication as a cause of familial Parkinson's disease. Lancet 364:1167–1169
14. Lavedan C, S Buchholtz, G Auburger, RL Albin, A Athanassiadou, J Blancato, JA Burguera, RE Ferrell, V Kostic, E Leroy et al. (1998) Absence of mutation in the beta- and gamma-synuclein genes in familial autosomal dominant Parkinson's disease. DNA Res 5: 401–402
15. Lincoln S, R Crook, MC Chartier-Harlin, K Gwinn-Hardy, M Baker, V Mouroux, F Richard, E Becquet, P Amouyel, A Destee, J Hardy, M Farrer (1999) No pathogenic mutations in the beta-synuclein gene in Parkinson's disease. Neurosci Lett 269:107–109
16. Brighina L, NU Okubadejo, NK Schneider, TG Lesnick, M de Andrade, JM Cunningham, MJ Farrer, SJ Lincoln, WA Rocca, DM Maraganore (2007) Beta-synuclein gene variants and Parkinson's disease: A preliminary case-control study. Neurosci Lett 420:229–234
17. Spillantini MG, ML Schmidt, VM Lee, JQ Trojanowski, R Jakes, M Goedert (1997) Alpha-synuclein in Lewy bodies. Nature 388:839–840
18. Forno LS (1996) Neuropathology of Parkinson's disease. J Neuropathol Exp Neurol 55: 259–272

19. Arrasate M, S Mitra, ES Schweitzer, MR Segal, S Finkbeiner (2004) Inclusion body formation reduces levels of mutant huntingtin and the risk of neuronal death. Nature 431:805–810
20. Abeliovich A, Y Schmitz, I Farinas, D Choi-Lundberg, WH Ho, PE Castillo, N Shinsky, JM Verdugo, M Armanini, A Ryan et al. (2000) Mice lacking alpha-synuclein display functional deficits in the nigrostriatal dopamine system. Neuron 25:239–252
21. Cabin DE, K Shimazu, D Murphy, NB Cole, W Gottschalk, KL McIlwain, B Orrison, A Chen, CE Ellis, R Paylor et al. (2002) Synaptic vesicle depletion correlates with attenuated synaptic responses to prolonged repetitive stimulation in mice lacking alpha-synuclein. J Neurosci 22:8797–8807.
22. Martin, ED, C Gonzalez-Garcia, M Milan, I Farinas, V Cena (2004) Stressor-related impairment of synaptic transmission in hippocampal slices from alpha-synuclein knockout mice. Eur J Neurosci 20:3085–3091
23. Liu S, I Ninan, I Antonova, F Battaglia, F Trinchese, A Narasanna, N Kolodilov, W Dauer, RD Hawkins, O Arancio (2004) alpha-Synuclein produces a long-lasting increase in neurotransmitter release. EMBO J 23:4506–4516
24. Murphy DD, SM Rueter, JQ Trojanowski, VM Lee (2000) Synucleins are developmentally expressed, and alpha-synuclein regulates the size of the presynaptic vesicular pool in primary hippocampal neurons. J Neurosci 20:3214–3220
25. Chandra S, F Fornai, HB Kwon, U Yazdani, D Atasoy, X Liu, RE Hammer, G Battaglia, DC German, PE Castillo et al. (2004) Double-knockout mice for alpha- and beta-synucleins: effect on synaptic functions. Proc Natl Acad Sci USA 101:14966–14971
26. Chandra S, G Gallardo, R Fernandez-Chacon, OM Schluter, TC Sudhof (2005) Alpha-synuclein cooperates with CSPalpha in preventing neurodegeneration. Cell. 123: 383–396
27. Cooper AA, AD Gitler, A Cashikar, CM Haynes, KJ Hill, B Bhullar, K Liu, K Xu, KE Strathearn, F Liu et al. (2006) Alpha-synuclein blocks ER-Golgi traffic and Rab1 rescues neuron loss in Parkinson's models. Science 313:324–328
28. Gitler AD, BJ Bevis, J Shorter, KE Strathearn, S Hamamichi, LJ Su, KA Caldwell, GA Caldwell, JC Rochet, JM McCaffery et al. (2008) The Parkinson's disease protein alpha-synuclein disrupts cellular Rab homeostasis. Proc Natl Acad Sci USA 105:145–150
29. Jenco JM, A Rawlingson, B Daniels, AJ Morris (1998) Regulation of phospholipase D2: selective inhibition of mammalian phospholipase D isoenzymes by alpha- and beta-synucleins. Biochemistry 37:4901–4909
30. McDermott, M, MJ Wakelam, AJ Morris (2004) Phospholipase D Biochem Cell Biol 82:225–253
31. Yu S, X Li, G Liu, J Han, C Zhang, Y Li, S Xu, C Liu, Y Gao, H Yang, K Ueda, P Chan (2007) Extensive nuclear localization of alpha-synuclein in normal rat brain neurons revealed by a novel monoclonal antibody. NeuroScience 145:539–555
32. McLean PJ, S Ribich, BT Hyman (2000) Subcellular localization of alpha-synuclein in primary neuronal cultures: effect of missense mutations. J Neural Transm Suppl 53–63
33. Goers J, AB Manning-Bog, AL McCormack, IS Millett, S Doniach, DA Di Monte, VN Uversky, AL Fink (2003) Nuclear localization of alpha-synuclein and its interaction with histones. Biochemistry 42:8465–8471
34. Kontopoulos E, JD Parvin, MB Feany (2006) Alpha-synuclein acts in the nucleus to inhibit histone acetylation and promote neurotoxicity. Hum Mol Genet 15:3012–3023
35. Weinreb PH, W Zhen, AW Poon, KA Conway, PT Lansbury Jr (1996) NACP, a protein implicated in Alzheimer's disease and learning, is natively unfolded. Biochemistry 35:13709–13715
36. Eliezer D, E Kutluay, R Bussell Jr, G Browne (2001) Conformational properties of alpha-synuclein in its free and lipid- associated states. J Mol Biol. 307:1061–1073
37. Bussell R Jr, D Eliezer (2001) Residual structure and dynamics in Parkinson's disease-associated mutants of alpha-synuclein. J Biol Chem 276:45996–46003.

38. Sung YH, D Eliezer (2007) Residual structure, backbone dynamics, and interactions within the synuclein family. J Mol Biol 372:689–707
39. Bertoncini CW, RM Rasia, GR Lamberto, A Binolfi, M Zweckstetter, C Griesinger, CO Fernandez (2007) Structural characterization of the intrinsically unfolded protein beta-synuclein, a natural negative regulator of alpha-synuclein aggregation. J Mol Biol 372:708–722
40. Marsh JA, VK Singh, Z Jia, and JD Forman-Kay (2006) Sensitivity of secondary structure propensities to sequence differences between alpha- and gamma-synuclein: implications for fibrillation. Protein Sci 15:2795–2804
41. Conway KA, SJ Lee, JC Rochet, TT Ding, RE Williamson, PT Lansbury Jr. (2000) Acceleration of oligomerization, not fibrillization, is a shared property of both alpha-synuclein mutations linked to early-onset Parkinson's disease: implications for pathogenesis and therapy. Proc Natl Acad Sci USA 97:571–576
42. Der-Sarkissian A, CC Jao, J Chen, R Langen (2003) Structural organization of alpha-synuclein fibrils studied by site-directed spin labeling. J Biol Chem 278:37530–37535
43. Del Mar C, EA Greenbaum, L Mayne, SW Englander, VL Woods Jr (2005) Structure and properties of alpha-synuclein and other amyloids determined at the amino acid level. Proc Natl Acad Sci USA 102:15477–15482
44. Heise H, W Hoyer, S Becker, OC Andronesi, D Riedel, M Baldus (2005) Molecular-level secondary structure, polymorphism, and dynamics of full-length alpha-synuclein fibrils studied by solid-state NMR Proc Natl Acad Sci USA 102:15871–15876
45. Dedmon MM, K Lindorff-Larsen, J Christodoulou, M Vendruscolo, CM Dobson (2005) Mapping long-range interactions in alpha-synuclein using spin-label NMR and ensemble molecular dynamics simulations. J Am Chem Soc 127:476–477
46. Bertoncini CW, YS Jung, CO Fernandez, W Hoyer, C Griesinger, TM Jovin, M Zweckstetter (2005) Release of long-range tertiary interactions potentiates aggregation of natively unstructured alpha-synuclein. Proc Natl Acad Sci USA 102:1430–1435
47. Goers J, VN Uversky, AL Fink (2003) Polycation-induced oligomerization and accelerated fibrillation of human alpha-synuclein in vitro. Protein Sci 12:702–707
48. Antony T, W Hoyer, D Cherny, G Heim, TM Jovin, V Subramaniam (2003) Cellular polyamines promote the aggregation of alpha-synuclein. J Biol Chem 278:3235–3240
49. Fernandez CO, W Hoyer, M Zweckstetter, EA Jares-Erijman, V Subramaniam, C Griesinger, TM Jovin (2004) NMR of alpha-synuclein-polyamine complexes elucidates the mechanism and kinetics of induced aggregation. EMBO J 23:2039–2046
50. Rasia RM, CW Bertoncini, D Marsh, W Hoyer, D Cherny, M Zweckstetter, C Griesinger, TM Jovin, CO Fernandez (2005) Structural characterization of copper(II) binding to alpha-synuclein: Insights into the bioinorganic chemistry of Parkinson's disease. Proc Natl Acad Sci USA 102:4294–4299
51. Sung YH, C Rospigliosi, D Eliezer (2006) NMR mapping of copper binding sites in alpha-synuclein. Biochim Biophys Acta 1764:5–12
52. Bertoncini CW, CO Fernandez, C Griesinger, TM Jovin, M Zweckstetter (2005) Familial mutants of alpha-synuclein with increased neurotoxicity have a destabilized conformation. J Biol Chem 280:30649–30652
53. Uversky VN, J Li, P Souillac, IS Millett, S Doniach, R Jakes, M Goedert, AL Fink (2002) Biophysical properties of the synucleins and their propensities to fibrillate: inhibition of alpha-synuclein assembly by beta- and gamma-synucleins. J Biol Chem 277:11970–11978
54. Morar AS, A Olteanu, GB Young, GJ Pielak (2001) Solvent-induced collapse of alpha-synuclein and acid-denatured cytochrome c. Protein Sci 10:2195–2199
55. Lee JC, R Langen, PA Hummel, HB Gray, JR Winkler (2004) Alpha-synuclein structures from fluorescence energy-transfer kinetics: implications for the role of the protein in Parkinson's disease. Proc Natl Acad Sci USA 101:16466–16471
56. Lee JC, HB Gray, JR Winkler (2005) Tertiary contact formation in alpha-synuclein probed by electron transfer. J Am Chem Soc. 127:16388–16389

57. Lee JC, BT Lai, JJ Kozak, HB Gray, JR Winkler (2007) Alpha-synuclein tertiary contact dynamics. J Phys Chem B 111:2107–2112
58. Biere AL, SJ Wood, J Wypych, S Steavenson, Y Jiang, D Anafi, FW Jacobsen, MA Jarosinski, GM Wu, JC Louis et al. (2000) Parkinson's disease-associated alpha-synuclein is more fibrillogenic than beta- and gamma-synuclein and cannot cross-seed its homologs. J Biol Chem 275:34574–34579
59. Paleologou KE, AW Schmid, CC Rospigliosi, HY Kim, GR Lamberto, RA Fredenburg, PT Lansbury Jr, CO Fernandez, D Eliezer, M Zweckstetter et al. (2008) Phosphorylation at Ser-129 but not the phosphomimics S129E/D inhibits the fibrillation of alpha-synuclein. J Biol Chem 283:16895–16905
60. Eliezer D (2008) Protein folding and aggregation in in vitro models of Parkinson's disease: Structure and function of α–synuclein. In: Nass R, Prezedborski S (eds) Parkinson's disease: molecular and therapeutic insights from model systems, Academic Press, New York. pp. 575–595
61. Conway KA, JD Harper, PT Lansbury Jr (2000) Fibrils formed in vitro from alpha-synuclein and two mutant forms linked to Parkinson's disease are typical amyloid. Biochemistry 39:2552–63
62. Serpell LC, J Berriman, R Jakes, M Goedert, RA Crowther (2000) Fiber diffraction of synthetic alpha-synuclein filaments shows amyloid- like cross-beta conformation. Proc Natl Acad Sci USA 97:4897–4902
63. Serpell LC, M Sunde, CC Blake (1997) The molecular basis of amyloidosis. Cell Mol Life Sci 53:871–887.
64. Tycko R (2006) Molecular structure of amyloid fibrils: insights from solid-state NMR Q Rev Biophys. 39:1–55
65. Margittai M, R Langen (2006) Spin labeling analysis of amyloids and other protein aggregates. Methods Enzymol 413:122–139
66. Chen M, M Margittai, J Chen, R Langen (2007) Investigation of alpha-synuclein fibril structure by site-directed spin labeling. J Biol Chem 282:24970–24979
67. Kloepper KD, KL Hartman, DT Ladror, CM Rienstra (2007) Solid-state NMR spectroscopy reveals that water is nonessential to the core structure of alpha-synuclein fibrils. J Phys Chem B 111:13353–13356
68. Vilar M, HT Chou, T Luhrs, SK Maji, D Riek-Loher, R Verel, G Manning, H Stahlberg, R Riek (2008) The fold of alpha-synuclein fibrils. Proc Natl Acad Sci USA 105:8637–8642
69. Heise H, MS Celej, S Becker, D Riedel, A Pelah, A Kumar, TM Jovin, M Baldus (2008) Solid-state NMR reveals structural differences between fibrils of wild-type and disease-related A53T mutant alpha-synuclein. J Mol Biol 380:444–450
70. Petkova AT, Y Ishii, JJ Balbach, ON Antzutkin, RD Leapman, F Delaglio, R Tycko (2002) A structural model for Alzheimer's beta -amyloid fibrils based on experimental constraints from solid state NMR Proc Natl Acad Sci USA 99:16742–16747
71. Luhrs T, C Ritter, M Adrian, D Riek-Loher, B Bohrmann, H Dobeli, D Schubert, R Riek (2005) 3D structure of Alzheimer's amyloid-beta(1-42) fibrils. Proc Natl Acad Sci USA 102:17342–17347
72. Wasmer C, A Lange, H Van Melckebeke, AB Siemer, R Riek, BH Meier (2008) Amyloid fibrils of the HET-s(218-289) prion form a beta solenoid with a triangular hydrophobic core. Science 319:1523–1526
73. Masliah E, E Rockenstein, I Veinbergs, M Mallory, M Hashimoto, A Takeda, Y Sagara, A Sisk, L Mucke (2000) Dopaminergic loss and inclusion body formation in alpha-synuclein mice: implications for neurodegenerative disorders. Science 287:1265–1269
74. Feany MB, WW Bender (2000) A Drosophila model of Parkinson's disease. Nature 404:394–398
75. Lakso M, S Vartiainen, AM Moilanen, J Sirvio, JH Thomas, R Nass, RD Blakely, G Wong (2003) Dopaminergic neuronal loss and motor deficits in Caenorhabditis elegans overexpressing human alpha-synuclein. J Neurochem 86:165–172

76. Payton JE, RJ Perrin, WS Woods, JM George (2004) Structural determinants of PLD2 inhibition by alpha-synuclein. J Mol Biol 337:1001–1009
77. Woods WS, JM Boettcher, DH Zhou, KD Kloepper, KL Hartman, DT Ladror, Z Qi, CM Rienstra, JM George (2007) Conformation-specific binding of alpha-synuclein to novel protein partners detected by phage display and NMR spectroscopy. J Biol Chem 282:34555–34567
78. Davidson WS, A Jonas, DF Clayton, JM George (1998) Stabilization of alpha-synuclein secondary structure upon binding to synthetic membranes. J Biol Chem 273:9443–9449
79. Jo E, J McLaurin, CM Yip, P St George-Hyslop, PE Fraser (2000) alpha-Synuclein membrane interactions and lipid specificity. J Biol Chem 275:34328–34334
80. Perrin RJ, WS Woods, DF Clayton, JM George (2000) Interaction of human alpha-Synuclein and Parkinson's disease variants with phospholipids. Structural analysis using site-directed mutagenesis. J Biol Chem 275:34393–34398.
81. Davidson WS, K Arnvig-McGuire, A Kennedy, J Kosman, TL Hazlett, A Jonas (1999) Structural organization of the N-terminal domain of apolipoprotein A-I: studies of tryptophan mutants. Biochemistry 38:14387–14395
82. Bussell R Jr, D Eliezer (2003) A structural and functional role for 11-mer repeats in alpha-synuclein and other exchangeable lipid binding proteins. J Mol Biol 329:763–778
83. Chandra S, X Chen, J Rizo, R Jahn, TC Sudhof (2003) A broken alpha-helix in folded alpha-synuclein. J Biol Chem 278:15313–15318
84. Bussell R Jr, TF Ramlall, D Eliezer (2005) Helix periodicity, topology, and dynamics of membrane-associated alpha-synuclein. Protein Sci 14:862–872
85. Bisaglia M, I Tessari, L Pinato, M Bellanda, S Giraudo, M Fasano, E Bergantino, L Bubacco, S Mammi (2005) A topological model of the interaction between alpha-synuclein and sodium dodecyl sulfate micelles. Biochemistry 44:329–339
86. Ulmer TS, A Bax, NB Cole, RL Nussbaum (2005) Structure and dynamics of micelle-bound human alpha-synuclein. J Biol Chem 280:9595–9603
87. Borbat P, TF Ramlall, JH Freed, D Eliezer (2006) Inter-helix distances in lysophospholipid micelle-bound alpha-synuclein from pulsed ESR measurements. J Am Chem Soc. 128:10004–10005
88. Jao CC, A Der-Sarkissian, J Chen, R Langen (2004) Structure of membrane-bound alpha-synuclein studied by site-directed spin labeling. Proc Natl Acad Sci USA 101:8331–8336
89. Ferreon AC, AA Deniz (2007) Alpha-synuclein multistate folding thermodynamics: implications for protein misfolding and aggregation. Biochemistry 46:4499–4509
90. Jensen PH, MS Nielsen, R Jakes, CG Dotti, M Goedert (1998) Binding of alpha-synuclein to brain vesicles is abolished by familial Parkinson's disease mutation. J Biol Chem 273:26292–26294
91. Jo E, N Fuller, RP Rand, P St George-Hyslop, PE Fraser (2002) Defective membrane interactions of familial Parkinson's disease mutant A30P alpha-synuclein. J Mol Biol 315:799–807
92. Choi W, S Zibaee, R Jakes, LC Serpell, B Davletov, RA Crowther, M Goedert (2004) Mutation E46K increases phospholipid binding and assembly into filaments of human alpha-synuclein. FEBS Lett 576:363–368
93. Bussell R Jr, D Eliezer (2004) Effects of Parkinson's disease-linked mutations on the structure of lipid-associated alpha-synuclein. Biochemistry 43:4810–4818
94. Ulmer TS, A Bax (2005) Comparison of structure and dynamics of micelle-bound human alpha-synuclein and Parkinson disease variants. J Biol Chem 280:43179–43187
95. Fredenburg RA, C Rospigliosi, RK Meray, JC Kessler, HA Lashuel, D Eliezer, PT Lansbury Jr (2007) The impact of the E46K mutation on the properties of alpha-synuclein in its monomeric and oligomeric states. Biochemistry 46:7107–7118
96. Nakajo S, K Tsukada, K Omata, Y Nakamura, K Nakaya (1993) A new brain-specific 14-kDa protein is a phosphoprotein. Its complete amino acid sequence and evidence for phosphorylation. Eur J Biochem 217:1057–1063

97. Akopian AN, JN Wood (1995) Peripheral nervous system-specific genes identified by subtractive cDNA cloning. J Biol Chem 270:21264–2170
98. Lavedan C, E Leroy, A Dehejia, S Buchholtz, A Dutra, RL Nussbaum, MH Polymeropoulos (1998) Identification, localization and characterization of the human gamma-synuclein gene. Hum Genet. 103:106–112
99. Buchman VL, HJ Hunter, LG Pinon, J Thompson, EM Privalova, NN Ninkina, AM Davies (1998) Persyn, a member of the synuclein family, has a distinct pattern of expression in the developing nervous system. J Neurosci 18:9335–9341
100. Sung YH, D Eliezer (2006) Secondary structure and dynamics of micelle bound beta- and gamma-synuclein. Protein Sci 15:1162–1174
101. Larsen KE, Y Schmitz, MD Troyer, E Mosharov, P Dietrich, AZ Quazi, M Savalle, V Nemani, FA Chaudhry, RH Edwards et al. (2006) Alpha-synuclein overexpression in PC12 and chromaffin cells impairs catecholamine release by interfering with a late step in exocytosis. J Neurosci 26:11915–11922
102. Yavich L, H Tanila, S Vepsalainen, P Jakala (2004) Role of alpha-synuclein in presynaptic dopamine recruitment. J Neurosci 24:11165–11170
103. Fortin DL, VM Nemani, SM Voglmaier, MD Anthony, TA Ryan, RH Edwards (2005) Neural activity controls the synaptic accumulation of alpha-synuclein. J Neurosci 25:10913–10921

Chapter 8
Inhibition of α-Synuclein Aggregation by Antioxidants and Chaperones in Parkinson's Disease

Jean-Christophe Rochet and Fang Liu

Abstract Parkinson's disease (PD) is a neurodegenerative disorder involving a loss of dopaminergic neurons from the substantia nigra. A characteristic feature of the post-mortem brains of PD patients is the presence in surviving neurons of Lewy bodies, cytosolic inclusions enriched with fibrillar forms of the presynaptic protein α-synuclein. Upon prolonged incubation at physiological temperature, α-synuclein converts from a natively unfolded protein to β-sheet-rich fibrils. α-Synuclein fibrillization involves a transient buildup of 'protofibrils', prefibrillar oligomers that may elicit neurotoxicity by permeabilizing phospholipid membranes and/or by interfering with cellular protein clearance mechanisms. The formation of α-synuclein protofibrils is stimulated by post-translational modifications (e.g. tyrosine nitration, dopamine adduct formation, methionine oxidation) that occur readily under conditions of oxidative stress. α-Synuclein self-assembly is inhibited by the antioxidant repair enzyme methionine sulfoxide reductase A, antioxidant compounds, and various proteins with molecular chaperone activity. The upregulation of antioxidant- and chaperone-dependent mechanisms may be a reasonable therapeutic strategy for suppressing α-synuclein aggregation and toxicity in PD.

8.1 Introduction

Parkinson's disease (PD) is a progressive neurodegenerative disorder characterized by muscle rigidity, slow movements, and reduced balance [1]. These symptoms result largely from a loss of dopaminergic neurons in the *substantia nigra*. A hallmark of PD is the presence in surviving neurons of cytosolic inclusions named Lewy bodies [1]. Biochemical studies have revealed a defect of mitochondrial complex I in the postmortem brains of PD patients [2, 3]. The decrease in complex I activity is

---

J.-C. Rochet (✉)
Department of Medicinal Chemistry and Molecular Pharmacology, Purdue University
West Lafayette, IN 47907-2091, USA
e-mail: rochet@pharmacy.purdue.edu

predicted to cause an accumulation of reactive oxygen species (ROS) that damage proteins, lipids, and DNA [4, 5]. In addition, dopaminergic neurons in the *substantia nigra* have high basal levels of ROS due to the metabolism and auto-oxidation of dopamine [5, 6].

Lewy bodies are enriched with fibrillar forms of α-synuclein, a 14 kDa presynaptic protein expressed exclusively in the nervous system [7, 8]. Autosomal-dominant mutations in the α-synuclein gene have been identified in patients with early-onset, familial PD. The mutations include substitutions (A30P, E46K, and A53T [9–11]) and mutations that increase the copy number of α-synuclein [12, 13]. These neuropathological and genetic data have prompted the hypothesis that α-synuclein aggregation is involved in PD pathogenesis.

The goal of this chapter is to provide an overview of current knowledge relating to the role of α-synuclein aggregation in PD and other diseases involving α-synuclein aggregation (collectively referred to as 'synucleinopathy disorders'). In Part I, we focus on the molecular details of α-synuclein aggregation, addressing the following questions: (i) Which species are formed on the α-synuclein self-assembly pathway? (ii) Which of these species are responsible for neurotoxicity? and (iii) How is α-synuclein aggregation modulated by cellular perturbations such as oxidative stress and membrane binding? In Parts II and III, we review evidence that α-synuclein aggregation and toxicity are inhibited by antioxidants and molecular chaperones. From this evidence, we conclude that upregulation of antioxidant and chaperone-dependent mechanisms represents a powerful strategy for the treatment of PD and other synucleinopathy disorders.

8.2 Molecular Details of α-Synuclein Aggregation

8.2.1 α-Synuclein Is a Natively Unfolded, Presynaptic Protein

α-Synuclein was first identified as the precursor of the 'non amyloid-beta component' (NAC), a peptide that is present in addition to the amyloid-beta peptide in amyloid preparations from the brains of Alzheimer's disease patients [14]. α-Synuclein exists as cytosolic and membrane-bound forms and is co-localized with synaptophysin in the presynaptic terminal [7, 15, 16]. This subcellular distribution is consistent with evidence that the protein modulates synaptic vesicle release [17–19], dopamine uptake [20], and SNARE complex assembly [21]. Data from far-ultraviolet circular dichroism (far-UV CD), Fourier transform infrared (FTIR), and nuclear magnetic resonance (NMR) measurements indicate that recombinant α-synuclein exists as a 'natively unfolded' protein with a primarily random-coil structure in dilute aqueous solution [22, 23]. However, the protein has been shown to adopt an α-helical structure upon binding to anionic phospholipids by far-UV CD, NMR, and electron paramagnetic resonance [23–27].

8.2.2 α-Synuclein Forms Fibrils and Protofibrils

A number of groups have shown that purified, recombinant α-synuclein forms fibrils in aqueous solution upon prolonged incubation at elevated temperatures. The fibrils have several properties of classic amyloid fibrils, including elevated β-sheet content [28, 29], the ability to bind dyes with a high affinity for extended β-sheet structure, including thioflavin T and Congo red [29–31], and X-ray and electron diffraction patterns consistent with a cross-β structure [32]. Both the A53T and E46K mutants of α-synuclein have been shown to fibrillize more readily than wild-type α-synuclein via measurements of thioflavin T fluorescence or by monitoring the loss of soluble protein [28, 32–38]. In contrast, A30P fibrillizes less rapidly than the wild-type protein, E46K, or A53T [35–38].

α-Synuclein fibrillization involves the formation of pre-fibrillar, oligomeric intermediates, referred to as 'protofibrils'. Evidence from atomic force microscopy (AFM) and electron microscopy imaging suggests that protofibrils consist of spheres, chains, rings, and tubular species, and far-UV CD data indicate that they are enriched with β-sheet secondary structure [29, 31, 35, 39–43]. Both the A30P and A53T mutants form protofibrils more rapidly than wild type α-synuclein [35, 44].

8.2.3 α-Synuclein Protofibrils Permeabilize Membranes

Because α-synuclein interacts with phospholipids, Lansbury and colleagues [39, 45] characterized α-synuclein protofibrils in terms of their ability to bind and permeabilize lipid membranes. The authors showed via surface plasmon resonance (SPR) that protofibrillar α-synuclein binds to phosphatidylglycerol vesicles with much higher affinity than the monomeric or fibrillar protein, and far-UV CD data revealed that the membrane-bound protofibrils are enriched with β-sheet structure. In addition, α-synuclein protofibrils (but not the monomer or fibrils) triggered the release of molecules encapsulated in phosphatidylglycerol vesicles, with an apparent preference for low-molecular weight species [39, 45]. Protofibrillar α-synuclein also bound brain-derived membrane fractions with higher affinity than the monomeric protein, and AFM analyses revealed the presence of membrane-bound annular protofibrils [40]. Together, these data suggested that α-synuclein protofibrils elicit neurotoxicity by permeabilizing membranes, possibly via a mechanism involving the formation of ion channels similar to β-sheet-rich pores generated by bacterial toxins [39, 40, 42, 45].

Protofibrillar A30P, A53T, and mouse α-synuclein have a greater propensity to permeabilize phosphatidylglycerol vesicles (on a per mole basis) than the human recombinant wild-type protein [45]. In contrast, E46K has a decreased propensity to form protofibrils compared to wild-type α-synuclein, and protofibrillar E46K exhibits a decreased specific membrane permeabilization activity [38]. These findings suggest that E46K elicits neurotoxicity via a mechanism unrelated to protofibril formation or membrane permeabilization. Alternatively, cellular perturbations in addition to the E46K substitution (e.g. post-translational modifications) may increase the propensity of this variant to form protofibrils and/or permeabilize membranes *in vivo* [38].

The results outlined above suggest a model in which α-synuclein protofibrils elicit neurotoxicity by forming pore-like structures that permeabilize membranes and perturb ionic gradients required for cellular homeostasis [39, 42, 45]. Consistent with this hypothesis, other investigators have reported that α-synuclein oligomers elicit an increase in phospholipid bilayer conductance [46, 47]. The data from one study indicated that α-synuclein oligomers caused an increase in conductance across a membrane without forming discrete, channel-like structures [46]. In a second study, treatment of a bilayer with oligomers formed from the NAC peptide led to an increase in conductance and the appearance of channel-like structures detectable by AFM [47]. Because these analyses were carried out using the NAC peptide, it is unclear whether full length α-synuclein would exhibit similar channel-like activity in this bilayer model. Finally, a recent report revealed that monomeric α-synuclein forms well-defined conducting channels enriched with α-helical structure in phosphatidylethanolamine-rich bilayers subjected to a transmembrane potential [48]. In summary, these findings indicate that α-synuclein has a membrane permeabilizing activity which may involve the formation of pore-like structures by β-sheet-rich oligomers, although different mechanisms of membrane disruption may be operative under different conditions.

8.2.4 α-Synuclein Self-Assembly Is Promoted by Membranes

Data from several groups suggest that phospholipid membranes stimulate α-synuclein aggregation. Membrane-bound α-synuclein was shown to form oligomeric species more rapidly than the cytosolic protein in rat brain homogenates [49]. Other studies revealed that long-chain polyunsaturated fatty acids promote α-synuclein self-assembly [50, 51], and electron microscopy data indicate that α-synuclein forms fibrils at the surface of synaptosomal membranes [52]. In addition, the protein undergoes accelerated fibrillization in the presence of anionic detergent micelles or when incubated with phospholipid vesicles at high protein-lipid ratios [53–56]. Finally, synthetic vesicles enriched with the ganglioside GM1 inhibit α-synuclein fibrillization but promote the formation of β-sheet-rich oligomers [57].

8.2.5 α-Synuclein Aggregates Disrupt Protein Clearance Mechanisms

Emerging evidence suggests that α-synuclein elicits toxicity by interfering with cellular mechanisms of protein degradation. Several groups have shown that α-synuclein is degraded by the ubiquitin-proteasome pathway (UPP), chaperone-mediated autophagy (CMA), and macroautophagy [58–62]. However, the UPP is impaired in cells over-expressing α-synuclein [63–65], and the 26S proteasome exhibits reduced proteolytic activity upon incubation with protofibrillar α-synuclein in cell-free systems [66–68]. In addition, the A30P and A53T mutants inhibit CMA, apparently by binding to the lysosomal LAMP2A receptor [60]. Because these mutant forms of α-synuclein are readily converted to protofibrils, the latter species may be responsible for inhibiting CMA. In turn, impairment of the UPP and CMA by α-synuclein may induce neurotoxicity by triggering a build-up of misfolded polypeptides.

8.2.6 α-Synuclein Is Post-Translationally Modified in Parkinson's Disease

Neuropathological data indicate that α-synuclein is post-translationally modified in the brains of patients with PD and other synucleinopathies, including dementia with Lewy bodies (DLB), the Lewy body variant of Alzheimer's disease, and multiple system atrophy (MSA). Lee, Trojanowski, and colleagues [69] showed that α-synuclein is nitrated at various tyrosine residues in synucleinopathy inclusions by staining with site-specific antibodies. The protein was also shown to be phosphorylated at serine 129 in DLB cortical inclusions via matrix-assisted laser desorption ionization mass spectrometry (MALDI-MS) and immunohistochemical analysis [70]. Subsequently, Anderson and colleagues [71] identified α-synuclein phosphorylated at serine 129 ("P-S129 α-synuclein") and a number of truncated α-synuclein species in inclusions from patients with DLB, MSA, and familial PD. In this thorough study, the authors characterized post-translationally modified α-synuclein isoforms via immunoblotting, ELISA, and liquid chromatography coupled with tandem mass spectrometry (LC-MS/MS). The results indicated that P-S129 α-synuclein is the most abundant modified isoform in Lewy bodies [71].

Post-translational modifications of α-synuclein have also been characterized in cellular and animal models of PD. Using an affinity purification method coupled with MS/MS, Mirzaei et al. [72] showed that α-synuclein is modified at various C-terminal residues in dopaminergic PC12 cells exposed to the complex I inhibitor rotenone. Specific modifications identified using this model include: (i) methionine oxidation at positions 116 and 127; (ii) tyrosine phosphorylation, nitration, or amination at positions 125, 133, and 136; and (iii) serine phosphorylation at position 129 [72] (Schieler and Rochet, unpublished observations). Data from other studies suggest that α-synuclein undergoes an increase in serine 129 phosphorylation in SH-SY5Y neuroblastoma cells treated with H_2O_2 [73], an increase in tyrosine nitration in mice exposed to the mitochondrial toxin, 1-methyl-4-phenyl-1,2,3,6-tetrahydropyridine (MPTP) [74], and an increase in both modifications in A30P α-synuclein transgenic mice [75].

8.2.7 α-Synuclein Self-Assembly Is Stimulated by Oxidative Modifications

A number of groups have reported that oxidative stress promotes α-synuclein aggregation and/or toxicity. Data from early works revealed that recombinant α-synuclein undergoes accelerated oligomerization in solutions containing H_2O_2 and Fe^{2+} or

Cu^{2+} [76, 77]. Subsequently it was demonstrated that the rate-limiting step in the nucleation of α-synuclein fibrillization is the oxidative formation of dityrosine cross-linked dimers [78, 79]. Consistent with this finding, Zhou and Freed [80] showed that α-synuclein variants in which tyrosine 39 or tyrosine 125 is replaced with a cysteine residue (Y39C and Y125C, respectively) have an increased propensity to form dimers and inclusions and to elicit neurotoxicity in a dopaminergic cell line compared to the wild-type protein, especially in the presence of an oxidizing insult. The authors' rationale of replacing tyrosine 39 or 125 with cysteine was that disulfide bonds are predicted to form more rapidly than dityrosine under conditions of oxidative stress [80].

Data from other studies indicate that α-synuclein oligomer formation is stimulated, whereas fibrillization is inhibited, as a result of tyrosine nitration [81, 82] or upon exposing the protein to metal ions [83] or dopamine [84, 85]. Several groups have reported that α-synuclein reacts with quinone derivatives of dopamine (e.g. indole-5,6-quinone) to form covalent adducts, which in turn inhibit the conversion of oligomers to fibrils [84–87]. The dopamine-α-synuclein adduct inhibits CMA, resulting in a buildup of misfolded polypeptides [88]. Other data suggest that the dopamine oxidation product, dopaminochrome, promotes α-synuclein oligomerization and suppresses fibril formation via non-covalent interactions with a C-terminal segment of the protein [89, 90]. Dopamine oxidation products have also been shown to induce the dissociation of preformed α-synuclein fibrils [89, 91]. Cyclooxygenase-2, an enzyme that is upregulated in PD brain, induces dopamine oxidation and α-synuclein aggregation in a dopaminergic cell line [92]. Together, these findings suggest that oxidized derivatives of dopamine modulate α-synuclein self-assembly via covalent and non-covalent interactions.

α-Synuclein oligomerization and fibrillization are inhibited following the oxidation of all four methionine residues (M1, M5, M116, and M127) to methionine sulfoxide. This inhibitory effect is proportional to the number of methionine sulfoxides [93] and is reversed when the oxidized protein is incubated with various metal ions, including Ti^{3+}, Zn^{2+}, Al^{3+}, and Pb^{2+} [94]. Interestingly, a buildup of oligomers but not fibrils occurs in mixtures of methionine-oxidized and unoxidized α-synuclein [93, 95, 96], suggesting that interactions between oxidized and unoxidized isoforms result in inhibition of the protofibril-to-fibril conversion.

The aggregation of α-synuclein is also modulated by oxidation products of molecules found in biological membranes. Two products of lipid peroxidation, 4-hydroxy-2-nonenal (HNE) [97] and acrolein [98], stimulate α-synuclein oligomerization but inhibit fibrillization via covalent modification of the protein. In contrast, oxidized cholesterol metabolites promote the formation of α-synuclein protofibrils and fibrils, apparently without a requirement for covalent adduct formation [99]. These effects of HNE, acrolein, and oxidized cholesterol metabolites on α-synuclein aggregation are likely to be relevant *in vivo* given the affinity of the protein for biological membranes.

8.2.8 α-Synuclein Aggregation Is Modulated by Phosphorylation

A number of groups have examined whether α-synuclein aggregation is modulated by the phosphorylation of serine 129, a modification induced by oxidative stress (see above). In one study, P-S129 α-synuclein was found to generate oligomers and fibrils more rapidly than the wild-type protein in a test-tube model [70]. However, Lashuel and colleagues [100] recently showed via far-UV CD and NMR that the phosphorylated protein has increased flexibility and a diminished ability to form fibrils. The reason for this discrepancy is unclear, although even subtle differences in the experimental conditions used for α-synuclein fibrillization may have accounted for the different reported outcomes [101].

Other research has addressed whether α-synuclein aggregation and neurotoxicity are affected by the phosphorylation of serine 129 *in vivo*. α-Synuclein toxicity was shown to be enhanced in transgenic *Drosophila* by replacing serine 129 with aspartate, a phosphoserine mimic, or by co-expressing the wild-type protein with G protein-coupled receptor kinase 2, previously shown to phosphorylate serine 129 [102, 103]. In contrast, the S129A mutant exhibited decreased neurotoxicity and enhanced inclusion formation in the transgenic fly model. These results suggested that the phosphorylation of serine 129 promotes neurodegeneration but inhibits α-synuclein aggregation [103]. Contrary to these results, recombinant adeno-associated virus (rAAV) encoding human S129A produced more rapid dopaminergic cell death upon injection in rat *substantia nigra* than rAAV encoding human wild-type α-synuclein, whereas S129D-encoding rAAV did not exhibit neurotoxicity in this model [104]. It is unclear why opposite results were obtained in rats versus flies, although interactions between the over-expressed human variants and the endogenous protein in rats (but not in flies, which lack α-synuclein) may have contributed to this discrepancy. Importantly, data from studies of S129D and S129E *in vivo* should be interpreted with caution, based on recent evidence that these variants do not faithfully recapitulate the conformational properties and self-assembly behavior of P-S129 α-synuclein in a test-tube model [100].

8.2.9 Summary (Part I)

α-Synuclein self-assembly involves the formation of protofibrils and fibrils (Fig. 8.1). Multiple lines of evidence suggest that protofibrils are more toxic than fibrils, perhaps due to their ability to permeabilize membranes and/or interfere with protein degradation via the UPP or CMA. The formation of α-synuclein protofibrils is stimulated by the familial mutations A30P and A53T and by binding of the protein to phospholipid membranes. In addition, data from studies involving test-tube, cellular, and animal models indicate that post-translational modifications induced by oxidative stress, including tyrosine nitration, dopamine adduct formation, and methionine oxidation, promote the conversion of α-synuclein to potentially toxic oligomers. Accordingly, α-synuclein aggregation and neurotoxicity may be

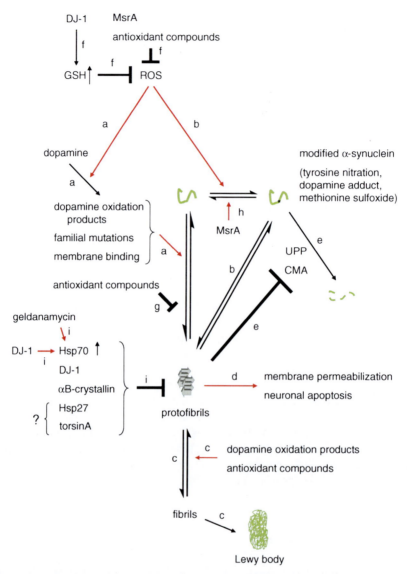

Fig. 8.1 Model illustrating α-synuclein self-assembly and neuroprotective mechanisms. (**a**) Reactive oxygen species (ROS) react with dopamine to form dopamine oxidation products, which stimulate α-synuclein oligomerization via non-covalent interactions. The formation of potentially toxic α-synuclein oligomers is also accelerated by familial mutations (A30P, A53T) and by membrane binding. (**b**) ROS also induce α-synuclein protofibril formation via direct oxidative modification of the protein, yielding variants with nitrated tyrosine residues, covalently attached dopamine, or oxidized methionine residues. (**c**) α-Synuclein protofibrils undergo further self-assembly to form fibrils, which ultimately become incorporated into Lewy bodies. Dopamine oxidation products and antioxidant compounds promote the dissociation of fibrils to oligomers. (**d**) α-Synuclein protofibrils may elicit neurotoxicity by permeabilizing phospholipid membranes.

8 Inhibition of α-Synuclein Aggregation

mitigated by upregulating cellular antioxidant mechanisms. Moreover, molecular chaperones may suppress α-synuclein toxicity by inhibiting protofibril formation and/or by improving the efficiency of protein homeostasis.

8.3 Inhibition of α-Synuclein Aggregation by Molecules with Antioxidant Activity

As described above, oxidative stress promotes the conversion of α-synuclein to potentially neurotoxic aggregates. Accordingly, intensive research has been carried out to determine whether α-synuclein aggregation and toxicity are suppressed by proteins or small molecules with antioxidant activity. In this section, we highlight studies of (i) methionine sulfoxide reductase A (MsrA), an enzyme involved in the cellular response to oxidative stress, and (ii) antioxidant compounds that interfere with α-synuclein fibrillization.

8.3.1 Inhibition of α-Synuclein Aggregation by MsrA

The enzyme MsrA converts the *S*-stereoisomer of methionine sulfoxide (including protein-bound methionine sulfoxide) to reduced methionine. MsrA over-expression in *Saccharomyces cerevisiae* and mammalian cells leads to enhanced resistance to oxidative stress, whereas the downregulation of MsrA increases sensitivity to oxidative insults [105–108]. In addition, MsrA knock-out mice have a reduced lifespan and are sensitized to oxidative stress [109, 110]. MsrA activity decreases with age in rat [111], whereas increased expression of the enzyme in *Drosophila* results in increased longevity [112]. Two mechanisms have been proposed for the protective effect of MsrA [113]. First, the enzyme may preserve the functions of essential cellular proteins via the repair of oxidized methionine residues. Second, MsrA engages in cycles of methionine oxidation and reduction, with the net effect of depleting (or 'scavenging') intracellular ROS. Exposed methionine residues on

---

Fig. 8.1 (**e**) Oxidatively damaged forms of α-synuclein may be eliminated by the ubiquitin proteasome pathway (UPP) or by chaperone-mediated autophagy (CMA). However, these clearance pathways are disrupted by α-synuclein protofibrils, resulting in a vicious cycle that triggers further α-synuclein oligomerization. (**f**) MsrA and antioxidant compounds deplete cellular ROS, thereby preventing a buildup of dopamine oxidation products and oxidatively damaged α-synuclein. DJ-1 also suppresses ROS by upregulating glutathione (GSH). (**g**) Some antioxidant compounds (e.g. EGCG) inhibit the formation of toxic, β-sheet-rich protofibrils by instead promoting the formation of unstructured (non-toxic) oligomers. (**h**) MsrA suppresses α-synuclein self-assembly by repairing oxidized forms of the protein. (**i**) α-Synuclein self-assembly is abrogated by molecular chaperones. Hsp70, αB-crystallin, and DJ-1 may inhibit α-synuclein fibrillization by binding protofibrils and promoting their conversion to non-toxic species. DJ-1 also suppresses α-synuclein aggregation and toxicity by upregulating Hsp70. Geldanamycin mitigates α-synuclein toxicity by inhibiting Hsp90, resulting in increased Hsp70 expression

the surfaces of proteins are thought to play an important role in this cycling mechanism, and in so doing they may protect other essential residues from oxidative damage [113].

MsrA is abundant throughout the brain including the *substantia nigra* [114], the major region of neuronal loss in the brains of PD patients. This observation suggests that the enzyme may inhibit dopaminergic cell death elicited by PD-related insults. To address this hypothesis, Liu et al. [115] tested whether MsrA over-expression protects dopaminergic neurons against α-synuclein aggregation and toxicity in primary cultures isolated from embryonic rat mesencephalon. These cultures consist of a mixed population of dopaminergic and non-dopaminergic neurons and glial cells, similar to the native environment of the midbrain. Therefore, the use of this cellular model is a powerful approach to investigate neurotoxic and neuroprotective mechanisms in PD. MsrA and the A53T mutant of α-synuclein were expressed in the midbrain cultures from lentiviruses previously shown to transduce neurons with an efficiency of 80–90% [116]. The relative viability of dopaminergic neurons (stained by an antibody specific for tyrosine hydroxylase) was dramatically reduced in cultures expressing A53T α-synuclein, and this toxic effect correlated with the appearance of SDS-resistant oligomers in the soluble fraction of the cell lysate (detectable via Western blotting). A53T aggregation and neurotoxicity were both partially rescued by co-expressing MsrA, whereas the small-molecule antioxidant N-acetyl-cysteine inhibited α-synuclein aggregation and neurotoxicity to a much lesser degree. Because N-acetyl-cysteine protects against oxidative stress by scavenging free radicals [117–119], this finding suggests that MsrA interferes with α-synuclein toxicity and oligomerization by directly repairing oxidized α-synuclein rather than by depleting ROS. In support of this idea, Liu et al. [115] showed that recombinant bovine MsrA repaired a variant of human wild-type α-synuclein that had been oxidized at all four of its methionine residues via pretreatment with H_2O_2.

These results suggest an apparent inconsistency: N-acetyl-cysteine was not protective against α-synuclein, yet this antioxidant molecule should suppress oxidative stress upstream of MsrA-catalyzed reduction. One possibility is that the antioxidant activity of N-acetyl-cysteine may not be sufficiently high to fully suppress α-synuclein methionine oxidation. In support of this idea, α-synuclein was found to readily undergo methionine oxidation in PC12 cells in the absence of an oxidizing treatment, most likely because these dopaminergic cells have high basal levels of ROS and because the natively unfolded structure of α-synuclein renders the protein susceptible to oxidative damage [72, 115]. Therefore, MsrA may play a critical role in preventing methionine oxidation of α-synuclein in dopaminergic neurons, even under basal conditions [115].

Preferential expression of human α-synuclein in the dopaminergic neurons of transgenic *Drosophila* results in a loss of tyrosine hydroxylase-immunoreactive cells and reduced spontaneous locomotor activity, and these deficits are overcome by co-expression of bovine MsrA [120]. The protective effect of MsrA was reproduced by supplementing the fly diet with *S*-methyl-L-cysteine (SMLC), a methionine analog that is readily oxidized to a species which is then a substrate for MsrA period. The authors inferred that SMLC causes a net depletion of ROS in the fly brain by activating a redox cycle catalyzed by endogenous MsrA [120]. These findings suggest that MsrA inhibits α-synuclein-induced cell death by scavenging ROS, a result that conflicts with evidence that MsrA protects primary dopaminergic neurons against α-synuclein toxicity primarily via its repair activity (see above). The discrepancy between the results obtained from studies of *Drosophila* and primary midbrain cultures may indicate that different MsrA-dependent protective mechanisms are operative in the two models. Evidence suggests that α-synuclein over-expression does not trigger a buildup of ROS, determined by measuring glutathione depletion and an increase in 8-oxo-deoxyguanosine levels, in primary midbrain cultures (Liu and Rochet, unpublished observations) [116, 121]. In contrast, oxidative stress may play a more central role in dopaminergic cell death in the brains of α-synuclein transgenic flies [122, 123], and, therefore, the consumption of ROS via repeated rounds of SMLC oxidation and reduction may be beneficial. It is also possible that MsrA-mediated repair and ROS scavenging occur to some degree in both models [115]. Ultimately, these issues will be resolved by investigating whether SMLC protects against α-synuclein neurotoxicity in primary midbrain cultures, and whether α-synuclein over-expression causes increased oxidative damage in the fly model.

Previous studies revealed that methionine-oxidized α-synuclein exhibits a decreased propensity to form fibrils and interferes with the fibrillization of the unoxidized protein [95]. In addition, potentially toxic, soluble oligomers have been shown to accumulate in mixtures of unoxidized and methionine-oxidized α-synuclein [84, 96]. Accordingly, MsrA may protect against α-synuclein neurotoxicity by preventing a buildup of oxidized isoforms with a high propensity to generate harmful oligomers.

A goal of future studies will be to identify methionine residues of α-synuclein that are repaired in dopaminergic cells, and to assess whether substitutions that mimic or prevent the oxidation of these residues promote or suppress the formation of toxic α-synuclein oligomers. Additional research is also necessary to assess whether the neuroprotective function of MsrA against α-synuclein toxicity occurs in the cytosol or mitochondria, given that each protein is present in both compartments [115, 124–126]. Finally, it is unknown whether methionine sulfoxide reductase B (MsrB), the enzyme responsible for reducing the *R*-stereoisomer of methionine sulfoxide, also protects dopaminergic neurons from PD-related insults.

8.3.2 Inhibition of α-Synuclein Aggregation by Small-Molecule Antioxidants

Various antioxidant compounds have been tested for their effects on α-synuclein self-assembly. In one set of studies, Fink and colleagues [127, 128] showed that the flavonoid baicalein (5–50 μM) and the antibiotic rifampicin (10–100 μM) inhibit α-synuclein fibrillization and disaggregate preformed fibrils, resulting in a buildup of non-fibrillar α-synuclein oligomers. The inhibition of fibrillization was more pronounced under oxidizing conditions, which favor auto-oxidation of the compounds to a quinone derivative. Moreover, spectrophotometric data revealed that both compounds form a covalent adduct with α-synuclein, and the results of mass spectral analyses implied that the quinone form of baicalein forms a Schiff base with a lysine residue [127, 128].

Other groups have reported that the fibrillization of recombinant α-synuclein is inhibited by a broad spectrum of antioxidant compounds, generally with a half-maximal effective concentration (EC_{50}) in the low micromolar range [129–132]. These compounds include: (i) polyphenolics (e.g. baicalein, curcumin, (–)-epigallocatechin gallate (EGCG), ferulic acid, myricetin, nordihydroguaiaretic acid (NDGA), rosmarinic acid, tannic acid) (Fig. 8.2); and (ii) non-polyphenolics (e.g. amphotericin B, perphenazine, rifampicin) (Fig. 8.3). Curcumin also inhibits the aggregation of A53T α-synuclein fused to red fluorescent protein in a neuroblastoma cell line [132]. The inhibitory effect of some molecules (e.g. curcumin, EGCG, myricetin) on the fibrillization of recombinant α-synuclein correlates with a buildup of soluble oligomers detectable as high molecular weight species by SDS-PAGE or gel filtration [129, 131, 132]. In addition, several polyphenolic compounds (e.g. curcumin, myricetin, NDGA, rosmarinic acid, tannic acid) induce the dissociation of preformed fibrils with EC_{50} values in the low micromolar range [130]. Some of these compounds consist of two 3,4-dihydroxyphenyl rings (NDGA, rosmarinic acid) or 4-hydroxy-3-methoxyphenyl rings (curcumin) joined by a hydrocarbon linker, and these structural motifs may promote binding of the compounds to soluble and/or fibrillar α-synuclein [130].

Collectively, these observations suggest that the antioxidant compounds described above may alleviate α-synuclein toxicity in synucleinopathy disorders, both by suppressing oxidative stress and by interfering with α-synuclein fibrillization. However, many of these compounds promote the accumulation of α-synuclein oligomers, and it is critical to determine whether these species are benign or neurotoxic. Epidemiological evidence suggests that flavonoids reduce the risk of neurodegenerative disease – for example, there is a decreased risk of PD in India, where the typical diet is enriched with curcumin [128, 132]. From these observations, one would infer that α-synuclein oligomers formed in the presence of these compounds are non-toxic. In addition, Wanker and colleagues [131] demonstrated via far-UV CD measurements that α-synuclein oligomers generated in the presence of EGCG lack significant β-sheet structure, and, therefore, they are conformationally distinct from the β-sheet-rich, toxic protofibrils previously shown to permeabilize membranes [39, 45]. α-Synuclein oligomers formed in the presence of EGCG and other compounds were also shown to be non-toxic when added to neuronal cell lines in tissue-culture models [129, 131]. Another study revealed that baicalein protects differentiated PC12 cells from the toxic effects of E46K α-synuclein [133], suggesting that the α-synuclein oligomers stabilized by this compound are non-toxic. A potential limitation of the cell-culture models used in these studies is that they may only provide limited insight about α-synuclein toxicity in the brain. For example, oligomers that are non-toxic when applied to the exterior of cells may in fact elicit toxicity upon

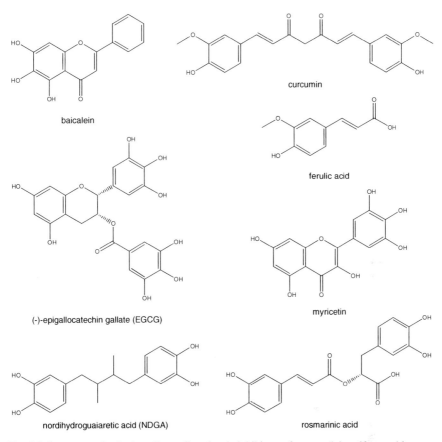

Fig. 8.2 Structures of polyphenolic small-molecule inhibitors of α-synuclein self-assembly

accumulating intracellularly, and oligomers that cause little damage to immortalized neuronal cells may in fact be toxic to post-mitotic neurons. These caveats can be addressed by testing whether the compounds mitigate α-synuclein aggregation and neurotoxicity in primary mesencephalic cultures or in animal models of PD.

8.3.3 Summary (Part II)

α-Synuclein self-assembly is inhibited by the antioxidant enzyme MsrA and by various antioxidant compounds (Fig. 8.1). MsrA suppresses α-synuclein aggregation and neurotoxicity by repairing oxidized methionine residues and by depleting ROS via a redox cycling mechanism. Antioxidant compounds inhibit α-synuclein self-assembly and promote dissociation of preformed fibrils by interfering with

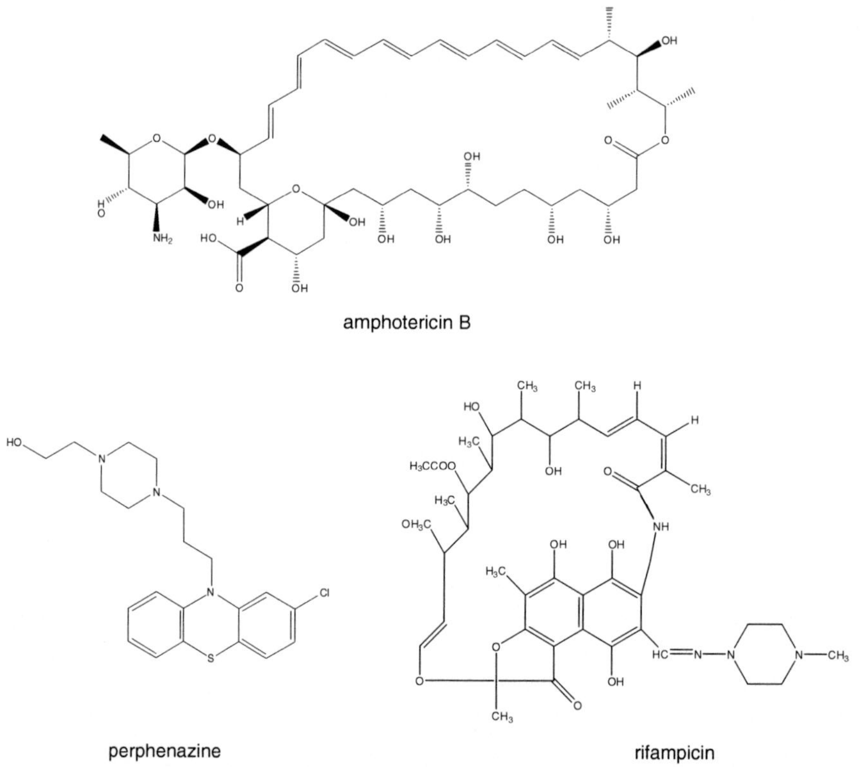

Fig. 8.3 Structures of non-polyphenolic small-molecule inhibitors of α-synuclein self-assembly

intermolecular interactions necessary for fibril formation and/or stability (presumably, these compounds also suppress α-synuclein aggregation in cells by scavenging ROS). A number of antioxidant compounds have been shown to promote a buildup of α-synuclein oligomers, and it is unclear whether these species are benign or neurotoxic. One way to address this problem is to prove that the α-synuclein oligomers formed in the presence of a compound are structurally distinct from toxic, β-sheet-rich protofibrils (as in the case of the high molecular weight α-synuclein species stabilized by EGCG). Ultimately, the best way to validate the therapeutic potential of these small molecules will be to demonstrate that they alleviate α-synuclein neurotoxicity in preclinical models of neurodegenerative disease.

8.4 Inhibition of α-Synuclein Aggregation by Molecular Chaperones

Cells rely on a number of 'surveillance' mechanisms to prevent a buildup of harmful protein aggregates. One such mechanism is provided by molecular chaperones, cellular proteins that interfere with protein misfolding and aggregation [134, 135]. Molecular chaperones play different roles under different conditions. In some cases, they interact with misfolded proteins and promote their correct refolding. Alternatively, chaperones target misfolded or aggregated polypeptides for degradation by the UPP or via autophagy [134]. Many chaperones are heat shock proteins that are upregulated under different stress conditions, including elevated temperatures and other insults that cause a build-up of misfolded proteins. Examples of heat shock proteins in eukaryotic cells include Hsp27, Hsp40, Hsp70, and Hsp90. Because molecular chaperones interfere with protein misfolding and aggregation, substantial research efforts have focused on developing strategies to prevent the formation of toxic α-synuclein assemblies by enhancing chaperone activity. In this section, we highlight studies of Hsp70, Hsp27, αB-crystallin, Hsp90, torsinA, and DJ-1.

8.4.1 Inhibition of α-Synuclein Aggregation by Hsp70

Heat shock protein 70 (Hsp70) has been characterized extensively in terms of inhibition of α-synuclein aggregation and toxicity. Hsp70 was shown to interfere with α-synuclein fibrillization by interacting with prefibrillar oligomers on the α-synuclein self-assembly pathway [136, 137]. In contrast, no interactions between Hsp70 and monomeric α-synuclein were detected by NMR [136]. Results obtained by comparing the rates of 'seeded' α-synuclein fibrillization in the absence or presence of Hsp70 revealed that the chaperone inhibits fibril elongation by binding to prefibrillar oligomers, and gel-filtration data indicated that the chaperone does not induce dissociation of these species to the monomer [136, 137]. Moreover, Hsp70 was found to decrease membrane permeabilization by oligomeric α-synuclein, suggesting that binding of the chaperone converts α-synuclein protofibrils to non-toxic aggregates [137].

Hsp70 was shown to protect against α-synuclein toxicity in the brains of transgenic *Drosophila* [138]. In this model, targeted expression of human α-synuclein to dopaminergic neurons in the fly brain results in progressive dopaminergic cell death and a buildup of Lewy-like inclusions. Co-expression of human Hsp70 rescued the loss of dopaminergic neurons, whereas the number of inclusions was unchanged. In addition, the inclusions stained positive for human Hsp70, suggesting that the chaperone interacted with the α-synuclein aggregates. These findings implied that Hsp70 converts toxic α-synuclein aggregates to non-toxic forms (alternatively, the aggregates detected by immunohistochemistry may not be involved in neurotoxicity). Inactivation of the major constitutively-expressed fly Hsp70 protein, Hsc4, resulted in accelerated dopaminergic cell death in α-synuclein-transgenic flies. This result further supported the idea that Hsp70 plays a critical role in suppressing α-synuclein-mediated neurotoxicity in *Drosophila*.

Subsequently it was discovered that geldanamycin (GA), a small-molecule inhibitor of Hsp90, suppresses α-synuclein-induced neurodegeneration in transgenic *Drosophila* [139]. GA activates the endogenous stress response by releasing the transcription factor heat shock factor 1 (HSF1) from Hsp90, thereby upregulating the expression of HSF1-dependent proteins including Hsp27, Hsp40, and Hsp70 [135]. Interestingly, the protective effect of GA in the transgenic flies was not associated with a decrease in the levels of Lewy-like aggregates. Rather, Western blot analyses revealed that GA treatment increased the levels of detergent-insoluble α-synuclein in the brains of the flies, suggesting that (i) α-synuclein aggregation and toxicity are not necessarily coupled phenomena, and (ii) Lewy body formation may, in fact, be neuroprotective.

Hsp70 also protects against α-synuclein aggregation and toxicity in mammalian cell-culture and animal models. In one study, the chaperone was shown to inhibit α-synuclein oligomer formation in a mesencephalic cell line exposed to rotenone [140]. Another group showed that the accumulation of high molecular weight and detergent-insoluble forms of α-synuclein is abrogated by Hsp70 upregulation in H4 human neuroglioma cells transfected with an Hsp70-encoding construct [141] or treated with GA [142]. The same group showed that Hsp70 interferes with α-synuclein oligomerization monitored in living H4 cells using bimolecular fluorescence complementation (BiFC) [143]. The inhibitory effect of Hsp70 on α-synuclein aggregation in H4 cells correlates with a decrease in α-synuclein toxicity [141, 143]. Hsp70 was also found to suppress α-synuclein aggregation in the brains of α-synuclein transgenic mice [141], although it was not determined whether the chaperone alleviates the motor deficits and loss of dopaminergic nerve terminals characteristic of this transgenic mouse model [144]. Together, these data suggest that Hsp70 plays a major role in mitigating α-synuclein aggregation and toxicity in synucleinopathy disorders.

8.4.2 Inhibition of α-Synuclein Aggregation by αB-Crystallin and Hsp27

The small heat-shock proteins αB-crystallin and Hsp27 are molecular chaperones that consist of low-molecular weight subunits assembled into large multimeric complexes. Data reported by Carver and colleagues [145] indicate that αB-crystallin suppresses α-synuclein fibril formation in a test-tube model, apparently by favoring conversion of the protein to nonfibrillar aggregates. Evidence from gel-filtration, mass spectrometry, and NMR analyses suggests that αB-crystallin and α-synuclein form a complex [145], and kinetic data obtained via dual polarization interferometry indicate that the chaperone interacts with prefibrillar species on the α-synuclein self-assembly pathway [146]. The authors also showed that the inhibitory effect of αB-crystallin on α-synuclein fibrillization is potentiated by lysine, arginine, and guanidine, presumably because these positively charged molecules induce subtle conformational changes in the chaperone and/or the α-synuclein target [147]. Results reported by other groups indicate that the chaperone activity of αB-crystallin against α-synuclein aggregation is: (i) enhanced by phosphorylation of the chaperone at three serine residues (S19, S45, and S59) [148]; and (ii) largely reproduced by two peptides from the conserved α-crystallin core domain, spanning residues 73–85 and 101–110 [149].

Both αB-crystallin and Hsp27 have been shown to inhibit α-synuclein aggregation and toxicity in various cell-culture models. In one study, Hsp27 was found to suppress apoptosis triggered by wild-type or mutant α-synuclein in a neuronal cell line [150]. Subsequently, McLean, Hyman, and colleagues [151] reported that αB-crystallin and Hsp27 interfere with α-synuclein toxicity in H4 cells, and Hsp27 inhibits dopaminergic cell death elicited by A53T α-synuclein in primary midbrain cultures. The authors also reported that Hsp27 suppresses α-synuclein inclusion formation in neuroglioma cells [151]. These findings, together with evidence that αB-crystallin and Hsp27 are present in synucleinopathy inclusions [151, 152], suggest that both chaperones interfere with α-synuclein aggregation and neurotoxicity *in vivo*.

8.4.3 Interaction Between α-Synuclein and Hsp90

Hsp90 was recently shown to be the major heat-shock protein co-localized with α-synuclein in filamentous lesions associated with PD, DLB, MSA, and other synucleinopathy disorders [153]. Western blot analyses of brain homogenates from patients with these diseases revealed a redistribution of α-synuclein and Hsp90, but not Hsp70 or Hsp40, to the detergent-insoluble fraction. A similar co-localization of Hsp90 and α-synuclein was observed in the brains of A53T α-synuclein transgenic mice. Moreover, the authors showed that a complex of Hsp90 and α-synuclein can be immunoprecipitated from an oligodendroglial cell line, and the chaperone was found to be co-localized with α-synuclein fibrils in nigral Lewy bodies from PD patients. From these results, it was inferred that Hsp90 may play a role in suppressing α-synuclein aggregation and/or promoting α-synuclein degradation.

The observations outlined above imply that upregulation of Hsp90 function may protect against α-synuclein neurotoxicity. However, this hypothesis has yet to be tested. A recent study showed that Hsp90 forms a complex with leucine-rich repeat kinase 2 (LRRK2), mutant forms of which are thought to contribute to the pathogenesis of familial and sporadic PD via a gain-of-toxic-function mechanism [154]. Disruption of the Hsp90-LRRK2 interaction with the Hsp90 inhibitor PU-H71 was found to stimulate proteasomal degradation of LRRK2 and rescue axonal growth defects in transgenic mice expressing the LRRK2 mutant G2019S. These findings, together with evidence that the Hsp90 inhibitor GA suppresses α-synuclein aggregation and toxicity by upregulating Hsp70, suggest that inhibition rather than activation of Hsp90 may be a more promising strategy for the treatment of synucleinopathy diseases.

8.4.4 Inhibition of α-Synuclein Aggregation by TorsinA

Another molecular chaperone with the ability to inhibit α-synuclein self-assembly is torsinA, a member of the AAA^+ family of cellular ATPases [155]. Autosomal dominant mutations in the gene encoding torsinA have been identified in patients with early-onset torsion dystonia. Immunohistochemical data indicate that torsinA is closely associated with α-synuclein in Lewy bodies from patients with synucleinopathy disorders [156, 157]. In addition, wild-type torsinA, but not a familial mutant form of the protein, inhibits α-synuclein aggregation in H4 neuroglioma cells [157]. Another study revealed that torsinA suppresses dopaminergic cell death elicited by α-synuclein in a *C. elegans* model [155]. Collectively, these findings suggest that torsinA inhibits α-synuclein aggregation and neurotoxicity.

8.4.5 Inhibition of α-Synuclein Aggregation by DJ-1

Autosomal-recessive mutations in the gene encoding DJ-1 have been identified in patients with early-onset PD [158–163]. The mutations include a deletion of the first six exons and missense mutations encoding the DJ-1 variants M26I, A39S, E64D, E163K, and L166P. DJ-1 is a homodimer of ~20 kDa subunits, each of which has a classic alpha/beta fold [164–168]. The crystal structure reveals the presence of a readily oxidized cysteine residue (cysteine 106) located at the subunit interface, and the protein exhibits a decrease in pI under conditions of oxidative stress due to the conversion of cysteine 106 to the sulfinic acid [168–170]. Studies of the effects of DJ-1 over-expression or silencing in cellular and animal models have demonstrated that the protein protects neurons against various oxidative insults [169–176]. DJ-1 is partially present in mitochondria, implying that it carries out its antioxidant function by quenching ROS which accumulate during electron transport [177]. Data reported by Cookson and colleagues [169] suggest that DJ-1 associates with mitochondria under oxidizing conditions via a mechanism requiring the oxidation of cysteine 106 to the sulfinic acid. DJ-1 may also act as an antioxidant by upregulating glutathione, potentially via stabilization of Nrf2 (nuclear factor erythroid 2-related factor 2), a master regulator of the antioxidant response [116, 121, 178].

DJ-1 was initially hypothesized to have a chaperone function due to its structural similarity with the molecular chaperone Hsp31 [166]. In support of this hypothesis, DJ-1 was shown to inhibit the aggregation of citrate synthase and luciferase in a test-tube model [166]. A subsequent study revealed that wild-type DJ-1, but not the familial mutant L166P, suppresses the heat-induced aggregation of citrate synthase, glutathione-S-transferase, and α-synuclein [179]. DJ-1 interfered with α-synuclein self-assembly by inhibiting the formation of oligomers and fibrils, monitored by Western blotting and Congo red staining, respectively. The DJ-1 chaperone function was ablated upon exposing the protein to the reducing agent dithiothreitol (DTT), but it was restored by treating the reduced protein with H_2O_2. In addition, the C53A mutant of DJ-1 exhibited no chaperone activity (in contrast to C106A, which behaved similarly to the wild-type protein) [179]. These data implied that the formation of a disulfide bond involving cysteine 53 is necessary for the chaperone function of DJ-1 under oxidizing conditions.

Data reported by Fink and colleagues [180] indicated that human wild-type DJ-1 inhibits α-synuclein fibrillization monitored by thioflavin T fluorescence and electron microscopy. In this study, a variant of DJ-1 in which cysteine 106 was converted to the sulfinic acid (termed the '2O' form of DJ-1) exhibited maximal chaperone activity. In contrast, unoxidized DJ-1 or isoforms oxidized more exhaustively than the 2O form had a dramatically reduced propensity to inhibit α-synuclein fibrillization. Despite this evidence that 2O DJ-1 carries out a chaperone function against α-synuclein self-assembly, a complex of the two proteins was not detected by gel filtration, nondenaturing PAGE, far-UV CD, or fluorescence spectrophotometry [180]. One explanation for this result may be that the authors tested the binding of DJ-1 to monomeric α-synuclein, whereas DJ-1 may in fact carry out its chaperone function by sequestering α-synuclein oligomers, similar to Hsp70. In support of this idea, data from our laboratory indicate that DJ-1 interacts with purified α-synuclein protofibrils, and this interaction is diminished by replacing C106 of DJ-1 with alanine (Hulleman and Rochet, unpublished observations). Collectively, these data suggest that DJ-1 suppresses α-synuclein self-assembly via a mechanism involving the oxidation of cysteine 106 to the sulfinic acid.

Results from several groups indicate that DJ-1 alleviates α-synuclein neurotoxicity in cell-culture models [116, 121, 181]. In two of these studies, DJ-1 was also shown to protect dopaminergic neurons against oxidative insults by upregulating glutathione [116, 121]. However, DJ-1 had no impact on glutathione levels in α-synuclein-expressing cells, and the protective effect of DJ-1 against α-synuclein was not reproduced by the small-molecule antioxidant, N-acetyl-cysteine [116, 121]. These observations imply that the suppression of α-synuclein-mediated neurodegeneration by DJ-1 does not involve an antioxidant mechanism. Instead, evidence suggests that DJ-1 inhibits α-synuclein aggregation and toxicity by inducing the expression of Hsp70 [116, 121, 181]. Data from RT-PCR analyses indicate that DJ-1-dependent upregulation of Hsp70 occurs at the transcriptional level [121, 181]. By expressing Hsp70 at the same level as in DJ-1-expressing cells, Liu et al. [116] demonstrated that the DJ-1-mediated upregulation of Hsp70 is sufficient to inhibit α-synuclein aggregation and toxicity. Surprisingly, however, the results of RNA silencing experiments indicated that Hsp70 upregulation is necessary for DJ-1-mediated suppression of α-synuclein aggregation, but not α-synuclein toxicity [116]. This finding suggests that DJ-1 protects against α-synuclein neurotoxicity via multiple pathways – for example, by activating anti-apoptotic signaling mechanisms in addition to preventing the formation of toxic α-synuclein oligomers via Hsp70 upregulation [175, 176, 182–184]. Moreover, DJ-1 may bind α-synuclein oligomers and alter their conformation, thereby reducing their toxicity [179, 180] (Hulleman and Rochet, unpublished observations). Presumably, redundant mechanisms of DJ-1 protection play a critical role in mitigating α-synuclein-induced neurodegeneration *in vivo* [116].

In summary, evidence suggests that DJ-1 suppresses α-synuclein aggregation and neurotoxicity by acting directly as a chaperone and/or by upregulating Hsp70. The extent to which DJ-1 carries out a direct chaperone function in neurons is unclear. Data reported by Liu et al. [116] indicate that DJ-1 induces an increase in Hsp70 solubility, perhaps by relieving the heat shock protein of its load of misfolded substrates. This finding implies that the direct chaperone function of DJ-1 may be operative under some conditions *in vivo*.

8.4.6 Summary (Part III)

A number of molecular chaperones protect against α-synuclein-induced cell death, presumably by preventing a buildup of toxic α-synuclein assemblies (Fig. 8.1). Data from test-tube analyses imply that Hsp70, αB-crystallin, and DJ-1 inhibit α-synuclein fibrillization by binding and sequestering protofibrils. DJ-1 also interferes with α-synuclein aggregation in neurons via a mechanism involving Hsp70 upregulation. Importantly, small molecules that upregulate molecular chaperones (e.g. GA) hold promise as therapies for PD and other synucleinopathy disorders.

8.5 Concluding Remarks

The PD field was transformed a little over ten years ago by the discovery of α-synuclein gene mutations in patients with familial forms of the disease. Since this discovery, intensive research efforts have yielded remarkable insights into the role of α-synuclein in neurodegeneration. A large body of evidence suggests that the protein elicits dopaminergic cell death in the *substantia nigra* by forming neurotoxic aggregates. The question of whether prefibrillar oligomers (protofibrils) or amyloid-like fibrils are responsible for α-synuclein toxicity remains a matter of debate. However, two familial mutations accelerate the formation of protofibrils, suggesting that the latter may play a central role in PD pathogenesis. Protofibrillar α-synuclein may elicit toxicity by permeabilizing membranes and/or interfering with protein clearance mechanisms. Other factors (in addition to familial mutations) that stimulate the conversion of monomeric α-synuclein to protofibrils include phospholipid membranes and oxidative modifications. Conversely, α-synuclein self-assembly is inhibited by molecules with antioxidant or chaperone activity.

Advances in our understanding of α-synuclein pathophysiology have revealed various strategies for the treatment of PD. Notably, small-molecule antioxidants that inhibit α-synuclein aggregation (in addition to quenching ROS) have the potential to alleviate nigral neurodegeneration and, therefore, should be characterized extensively in preclinical models. Moreover, α-synuclein neurotoxicity may be suppressed by upregulating endogenous protective mechanisms, including cellular responses activated by DJ-1 or MsrA. In support of this idea, compounds that prevent the over-oxidation of DJ-1 were recently shown to ameliorate DJ-1-mediated neuroprotection against oxidizing insults [185]. Ultimately, strategies aimed at activating various neuroprotective responses (including antioxidant defenses and chaperone-dependent mechanisms) may represent the best approach for treating PD and other synucleinopathy disorders.

Acknowledgments We are grateful to the National Institute of Neurological Disorders and Stroke, Michael J. Fox Foundation, and Parkinson's Disease Foundation for funding. We also thank the members of the Rochet laboratory for stimulating discussions.

Abbreviations

atomic force microscopy (AFM)
chaperone-mediated autophagy (CMA)
circular dichroism (CD)
dementia with Lewy bodies (DLB)
(−)-epigallocatechin gallate (EGCG)
geldanamycin (GA)
heat shock protein (Hsp)
leucine-rich repeat kinase 2 (LRRK2)
mass spectrometry (MS)
methionine sulfoxide reductase A (MsrA)
multiple system atrophy (MSA)
non amyloid-beta component (NAC)
nordihydroguaiaretic acid (NDGA)
Parkinson's disease (PD)
reactive oxygen species (ROS)
recombinant adeno-associated virus (rAAV)
S-methyl-L-cysteine (SMLC)
ubiquitin-proteasome pathway (UPP)

References

1. Dawson TM, Dawson VL (2003) Molecular pathways of neurodegeneration in Parkinson's disease. Science 302:819–822.
2. Greenamyre JT, Sherer TB, Betarbet R, Panov AV (2001) Complex I and Parkinson's disease. IUBMB Life 52:135–141.
3. Orth M, Schapira AH (2002) Mitochondrial involvement in Parkinson's disease. Neurochem Int 40:533–541.
4. Beal MF (2003) Mitochondria, oxidative damage, and inflammation in Parkinson's disease. Ann N Y Acad Sci 991:120–131.
5. Jenner P (2003) Oxidative stress in Parkinson's disease. Ann Neurol 53 Suppl 3:S26–S36; discussion S36–S28.
6. Graham DG, Tiffany SM, Bell WR Jr, Gutknecht WF (1978) Autoxidation versus covalent binding of quinones as the mechanism of toxicity of dopamine, 6-hydroxydopamine, and related compounds toward C1300 neuroblastoma cells in vitro. Mol Pharmacol 14:644–653.
7. Iwai A, Masliah E, Yoshimoto M, Ge N, Flanagan L, de Silva HA, Kittel A, Saitoh T (1995) The precursor protein of non-A beta component of Alzheimer's disease amyloid is a presynaptic protein of the central nervous system. Neuron 14:467–475.
8. Spillantini MG, Schmidt ML, Lee VM-Y, Trojanowski JQ, Jakes R, Goedert M (1997) α-Synuclein in Lewy bodies. Nature 388:839–840.
9. Polymeropoulos MH, Lavedan C, Leroy E, Ide SE, Dehejia A, Dutra A, Pike B, Root H, Rubenstein J, Boyer R et al (1997) Mutation in the alpha-synuclein gene identified in families with Parkinson's disease. Science 276:2045–2047.
10. Kruger R, Kuhn W, Muller T, Woitalla D, Graeber M, Kosel S, Przuntek H, Epplen JT, Schols L, Riess O (1998) Ala30Pro mutation in the gene encoding alpha-synuclein in Parkinson's disease. Nat Genet 18:106–108.

11. Zarranz JJ, Alegre J, Gomez-Esteban JC, Lezcano E, Ros R, Ampuero I, Vidal L, Hoenicka J, Rodriguez O, Atares B et al (2004) The new mutation, E46K, of alpha-synuclein causes Parkinson and Lewy body dementia. Ann Neurol 55:164–173.
12. Singleton AB, Farrer M, Johnson J, Singleton A, Hague S, Kachergus J, Hulihan M, Peuralinna T, Dutra A, Nussbaum R et al (2003) Alpha-synuclein locus triplication causes Parkinson's disease. Science 302:841.
13. Chartier-Harlin MC, Kachergus J, Roumier C, Mouroux V, Douay X, Lincoln S, Levecque C, Larvor L, Andrieux J, Hulihan M et al (2004) Alpha-synuclein locus duplication as a cause of familial Parkinson's disease. Lancet 364:1167–1169.
14. Ueda K, Fukushima H, Masliah E, Xia Y, Iwai A, Yoshimoto M, Otero DAC, Kondo J, Ihara Y, Saitoh T (1993) Molecular cloning of cDNA encoding an unrecognized component of amyloid in Alzheimer disease. Proc Natl Acad Sci USA 90:11282–11286.
15. Maroteaux L, Campanelli JT, Scheller RH (1988) Synuclein: a neuron-specific protein localized to the nucleus and presynaptic nerve terminal. J Neurosci 8:2804–2815.
16. Fortin DL, Troyer MD, Nakamura K, Kubo S, Anthony MD, Edwards RH (2004) Lipid rafts mediate the synaptic localization of alpha-synuclein. J Neurosci 24:6715–6723.
17. Abeliovich A, Schmitz Y, Farinas I, Choi-Lundberg D, Ho WH, Castillo PE, Shinsky N, Verdugo JM, Armanini M, Ryan A et al (2000) Mice lacking alpha-synuclein display functional deficits in the nigrostriatal dopamine system. Neuron 25:239–252.
18. Murphy DD, Rueter SM, Trojanowski JQ, Lee VM (2000) Synucleins are developmentally expressed, and alpha-synuclein regulates the size of the presynaptic vesicular pool in primary hippocampal neurons. J Neurosci 20:3214–3220.
19. Cabin DE, Shimazu K, Murphy D, Cole NB, Gottschalk W, McIlwain KL, Orrison B, Chen A, Ellis CE, Paylor R et al (2002) Synaptic vesicle depletion correlates with attenuated synaptic responses to prolonged repetitive stimulation in mice lacking α-synuclein. J Neurosci 22:8797–8807.
20. Sidhu A, Wersinger C, Moussa CE, Vernier P (2004) The role of alpha-synuclein in both neuroprotection and neurodegeneration. Ann N Y Acad Sci 1035:250–270.
21. Chandra S, Gallardo G, Fernandez-Chacon R, Schluter OM, Sudhof TC (2005) Alpha-synuclein cooperates with CSPalpha in preventing neurodegeneration. Cell 123:383–396.
22. Weinreb PH, Zhen W, Poon AW, Conway KA, Lansbury PT Jr (1996) NACP, a protein implicated in Alzheimer's disease and learning, is natively unfolded. Biochemistry 35:13709–13715.
23. Eliezer D, Kutluay E, Bussell R Jr, Browne G (2001) Conformational properties of α-synuclein in its free and lipid-associated states. J Mol Biol 307:1061–1073.
24. Davidson WS, Jonas A, Clayton DF, George JM (1998) Stabilization of alpha-synuclein secondary structure upon binding to synthetic membranes. J Biol Chem 273:9443–9449.
25. Bussell R Jr, Eliezer D (2003) A structural and functional role for 11-mer repeats in α-synuclein and other exchangeable lipid binding proteins. J Mol Biol 329:763–778.
26. Chandra S, Chen X, Rizo J, Jahn R, Sudhof TC (2003) A broken α-helix in folded α-synuclein. J Biol Chem 278:15313–15318.
27. Jao CC, Der-Sarkissian A, Chen J, Langen R (2004) Structure of membrane-bound alpha-synuclein studied by site-directed spin labeling. Proc Natl Acad Sci USA 101:8331–8336.
28. Narhi L, Wood SJ, Steavenson S, Jiang Y, Wu GM, Anafi D, Kaufman SA, Martin F, Sitney K, Denis P et al (1999) Both familial Parkinson's disease mutations accelerate alpha-synuclein aggregation. J Biol Chem 274:9843–9846.
29. Conway KA, Harper JD, Lansbury PT Jr (2000) Fibrils formed *in vitro* from α-synuclein and two mutant forms linked to Parkinson's disease are typical amyloid. Biochemistry 39:2552–2563.
30. El-Agnaf OM, Jakes R, Curran MD, Wallace A (1998) Effects of the mutations Ala30 to Pro and Ala53 to Thr on the physical and morphological properties of alpha-synuclein protein implicated in Parkinson's disease. FEBS Lett 440:67–70.

31. Rochet JC, Conway KA, Lansbury PT Jr (2000) Inhibition of fibrillization and accumulation of prefibrillar oligomers in mixtures of human and mouse alpha-synuclein. Biochemistry 39:10619–10626.
32. Serpell LC, Berriman J, Jakes R, Goedert M, Crowther RA (2000) Fiber diffraction of synthetic α-synuclein filaments shows amyloid-like cross-β conformation. Proc Natl Acad Sci USA 97:4897–4902.
33. Conway KA, Harper JD, Lansbury PT (1998) Accelerated *in vitro* fibril formation by a mutant α–synuclein linked to early-onset Parkinson disease. Nat Med 4:1318–1320.
34. Giasson BI, Uryu K, Trojanowski JQ Lee VM-Y (1999) Mutant and wild type human α-synucleins assemble into elongated filaments with distinct morphologies *in vitro*. J Biol Chem 274:7619–7622.
35. Conway KA, Lee S-J, Rochet J-C, Ding TT, Williamson RE, Lansbury PT Jr (2000) Acceleration of oligomerization, not fibrillization, is a shared property of both α-synuclein mutations linked to early-onset Parkinson's disease: Implications for pathogenesis and therapy. Proc Natl Acad Sci USA 97:571–576.
36. Choi W, Zibaee S, Jakes R, Serpell LC, Davletov B, Crowther RA, Goedert M (2004) Mutation E46K increases phospholipid binding and assembly into filaments of human alpha-synuclein. FEBS Lett 576:363–368.
37. Greenbaum EA, Graves CL, Mishizen-Eberz AJ, Lupoli MA, Lynch DR, Englander SW, Axelsen PH, Giasson BI (2005) The E46K mutation in alpha-synuclein increases amyloid fibril formation. J Biol Chem 280:7800–7807.
38. Fredenburg RA, Rospigliosi C, Meray RK, Kessler JC, Lashuel HA, Eliezer D, Lansbury PT Jr (2007) The impact of the E46K mutation on the properties of alpha-synuclein in its monomeric and oligomeric states. Biochemistry 46:7107–7118.
39. Volles MJ, Lee S-J, Rochet J-C, Shtilerman MD, Ding TT, Kessler JC, Lansbury PT Jr (2001) Vesicle permeabilization by protofibrillar α-synuclein: implications for the pathogenesis and treatment of Parkinson's disease. Biochemistry 40:7812–7819.
40. Ding TT, Lee S-J, Rochet J-C, Lansbury PT Jr (2002) Annular α-synuclein protofibrils are produced when spherical protofibrils are incubated in solution or bound to brain-derived membranes. Biochemistry 41:10209–10217.
41. Lashuel HA, Petre BM, Wall J, Simon M, Nowak RJ, Walz T, Lansbury PT Jr (2002) α-Synuclein, especially the Parkinson's disease-associated mutants, forms pore-like annular and tubular protofibrils. J Mol Biol 322:1089–1102.
42. Volles MJ, Lansbury PT Jr (2003) Zeroing in on the pathogenic form of α-synuclein and its mechanism of neurotoxicity in Parkinson's disease. Biochemistry 42:7871–7878.
43. Rochet JC, Outeiro TF, Conway KA, Ding TT, Volles MJ, Lashuel HA, Bieganski RM, Lindquist SL, Lansbury PT (2004) Interactions among alpha-synuclein, dopamine, and biomembranes: some clues for understanding neurodegeneration in Parkinson's disease. J Mol Neurosci 23:23–34.
44. Li J, Uversky VN, Fink AL (2001) Effect of familial Parkinson's disease point mutations A30P and A53T on the structural properties, aggregation, and fibrillation of human α-synuclein. Biochemistry 40:11604–11613.
45. Volles MJ, Lansbury PT Jr (2002) Vesicle permeabilization by protofibrillar alpha-synuclein is sensitive to Parkinson's disease-linked mutations and occurs by a pore-like mechanism. Biochemistry 41:4595–4602.
46. Kayed R, Sokolov Y, Edmonds B, McIntire TM, Milton SC, Hall JE, Glabe CG (2004) Permeabilization of lipid bilayers is a common conformation-dependent activity of soluble amyloid oligomers in protein misfolding diseases. J Biol Chem 279:46363–46366.
47. Quist A, Doudevski I, Lin H, Azimova R, Ng D, Frangione B, Kagan B, Ghiso J, Lal R (2005) Amyloid ion channels: a common structural link for protein-misfolding disease. Proc Natl Acad Sci USA 102:10427–10432.
48. Zakharov SD, Hulleman JD, Dutseva EA, Antonenko YN, Rochet JC, Cramer WA (2007) Helical alpha-synuclein forms highly conductive ion channels. Biochemistry 46: 14369–14379.

49. Lee H-J, Choi C, Lee S-J (2002) Membrane-bound α-synuclein has a high aggregation propensity and the ability to seed the aggregation of the cytosolic form. J Biol Chem 277:671–678.
50. Perrin RJ, Woods WS, Clayton DF George JM (2001) Exposure to long chain polyunsaturated fatty acids triggers rapid multimerization of synucleins. J Biol Chem 276:41958–41962.
51. Sharon R, Bar-Joseph I, Frosch MP, Walsh DM, Hamilton JA Selkoe DJ (2003) The formation of highly soluble oligomers of α-synuclein is regulated by fatty acids and enhanced in Parkinson's disease. Neuron 37:583–595.
52. Jo E, Darabie AA, Han K, Tandon A, Fraser PE, McLaurin J (2004) Alpha-synuclein-synaptosomal membrane interactions: implications for fibrillogenesis. Eur J Biochem 271:3180–3189.
53. Narayanan V, Scarlata S (2001) Membrane binding and self-association of alpha-synucleins. Biochemistry 40:9927–9934.
54. Necula M, Chirita CN, Kuret J (2003) Rapid anionic micelle-mediated alpha-synuclein fibrillization in vitro. J Biol Chem 278:46674–46680.
55. Zhu M, Fink AL (2003) Lipid binding inhibits α-synuclein fibril formation. J Biol Chem 278:16873–16877.
56. Zhu M, Li J, Fink AL (2003) The association of α-synuclein with membranes affects bilayer structure, stability and fibril formation. J Biol Chem 278:40186–40197.
57. Martinez Z, Zhu M, Han S Fink AL (2007) GM1 specifically interacts with alpha-synuclein and inhibits fibrillation. Biochemistry 46:1868–1877.
58. Bennett MC, Bishop JF, Leng Y, Chock PB, Chase TN, Mouradian MM (1999) Degradation of alpha-synuclein by proteasome. J Biol Chem 274:33855–33858.
59. Webb JL, Ravikumar B, Atkins J, Skepper JN, Rubinsztein DC (2003) Alpha-synuclein is degraded by both autophagy and the proteasome. J Biol Chem 278:25009–25013.
60. Cuervo AM, Stefanis L, Fredenburg R, Lansbury PT, Sulzer D (2004) Impaired degradation of mutant alpha-synuclein by chaperone-mediated autophagy. Science 305:1292–1295.
61. Shin Y, Klucken J, Patterson C, Hyman BT, McLean PJ (2005) The co-chaperone carboxyl terminus of Hsp70-interacting protein (CHIP) mediates alpha-synuclein degradation decisions between proteasomal and lysosomal pathways. J Biol Chem 280: 23727–23734.
62. Vogiatzi T, Xilouri M, Vekrellis K, Stefanis L (2008) Wild type α-synuclein is degraded by chaperone mediated autophagy and macroautophagy in neuronal cells. J Biol Chem 283:23542–23556.
63. Stefanis L, Larsen KE, Rideout HJ, Sulzer D, Greene LA (2001) Expression of A53T mutant but not wild-type alpha-synuclein in PC12 cells induces alterations of the ubiquitin-dependent degradation system, loss of dopamine release, and autophagic cell death. J Neurosci 21:9549–9560.
64. Tanaka Y, Engelender S, Igarashi S, Rao RK, Wanner T, Tanzi RE, Sawa A, Dawson VL, Dawson TM, Ross CA (2001) Inducible expression of mutant alpha-synuclein decreases proteasome activity and increases sensitivity to mitochondria-dependent apoptosis. Hum Mol Genet 10:919–926.
65. Petrucelli L, O'Farrell C, Lockhart PJ, Baptista M, Kehoe K, Vink L, Choi P, Wolozin B, Farrer M, Hardy J et al (2002) Parkin protects against the toxicity associated with mutant α-synuclein: proteasome dysfunction selectively affects catecholaminergic neurons. Neuron 36:1007–1019.
66. Snyder H, Mensah K, Theisler C, Lee J, Matouschek A, Wolozin B (2003) Aggregated and monomeric alpha-synuclein bind to the S6' proteasomal protein and inhibit proteasomal function. J Biol Chem 278:11753–11759.
67. Lindersson E, Beedholm R, Hojrup P, Moos T, Gai W, Hendil KB Jensen PH (2004) Proteasomal inhibition by alpha-synuclein filaments and oligomers. J Biol Chem 279: 12924–12934.

68. Zhang NY, Tang Z, Liu CW (2008) Alpha-synuclein protofibrils inhibit 26S proteasome-mediated protein degradation. Understanding the cytotoxicity of protein protofibrils in neurodegenerative diseases pathogenesis. J Biol Chem. 283:20288–20298.
69. Giasson BI, Duda JE, Murray IV, Chen Q, Souza JM, Hurtig HI, Ischiropoulos H, Trojanowski JQ, Lee VM (2000) Oxidative damage linked to neurodegeneration by selective alpha-synuclein nitration in synucleinopathy lesions. Science 290:985–989.
70. Fujiwara H, Hasegawa M, Dohmae N, Kawashima A, Masliah E, Goldberg MS, Shen J, Takio K Iwatsubo T (2002) Alpha-synuclein is phosphorylated in synucleinopathy lesions. Nat Cell Biol 4:160–164.
71. Anderson JP, Walker DE, Goldstein JM, de Laat R, Banducci K, Caccavello RJ, Barbour R, Huang J, Kling K, Lee M et al (2006) Phosphorylation of Ser-129 is the dominant pathological modification of alpha-synuclein in familial and sporadic Lewy body disease. J Biol Chem 281:29739–29752.
72. Mirzaei H, Schieler JL, Rochet JC, Regnier F (2006) Identification of rotenone-induced modifications in alpha-synuclein using affinity pull-down and tandem mass spectrometry. Anal Chem 78:2422–2431.
73. Smith WW, Margolis RL, Li X, Troncoso JC, Lee MK, Dawson VL, Dawson TM, Iwatsubo T, Ross CA (2005) Alpha-synuclein phosphorylation enhances eosinophilic cytoplasmic inclusion formation in SH-SY5Y cells. J Neurosci 25:5544–5552.
74. Przedborski S, Chen Q, Vila M, Giasson BI, Djaldatti R, Vukosavic S, Souza JM, Jackson-Lewis V, Lee VM, Ischiropoulos H (2001) Oxidative post-translational modifications of alpha-synuclein in the 1-methyl-4-phenyl-1,2,3,6-tetrahydropyridine (MPTP) mouse model of Parkinson's disease. J Neurochem 76:637–640.
75. Neumann M, Kahle PJ, Giasson BI, Ozmen L, Borroni E, Spooren W, Muller V, Odoy S, Fujiwara H, Hasegawa M et al (2002) Misfolded proteinase K-resistant hyperphosphorylated alpha-synuclein in aged transgenic mice with locomotor deterioration and in human alpha-synucleinopathies. J Clin Invest 110:1429–1439.
76. Hashimoto M, Hsu LJ, Xia Y, Takeda A, Sisk A, Sundsmo M, Masliah E (1999) Oxidative stress induces amyloid-like aggregate formation of NACP/alpha-synuclein in vitro. Neuroreport 10:717–721.
77. Paik SR, Shin HJ, Lee JH (2000) Metal-catalyzed oxidation of alpha-synuclein in the presence of Copper(II) and hydrogen peroxide. Arch Biochem Biophys 378:269–277.
78. Souza JM, Giasson BI, Chen Q, Lee VM, Ischiropoulos H (2000) Dityrosine cross-linking promotes formation of stable alpha -synuclein polymers. Implication of nitrative and oxidative stress in the pathogenesis of neurodegenerative synucleinopathies. J Biol Chem 275:18344–18349.
79. Krishnan S, Chi EY, Wood SJ, Kendrick BS, Li C, Garzon-Rodriguez W, Wypych J, Randolph TW, Narhi LO, Biere AL et al (2003) Oxidative dimer formation is the critical rate-limiting step for Parkinson's disease alpha-synuclein fibrillogenesis. Biochemistry 42:829–837.
80. Zhou W, Freed CR (2004) Tyrosine-to-cysteine modification of human alpha-synuclein enhances protein aggregation and cellular toxicity. J Biol Chem 279:10128–10135.
81. Norris EH, Giasson BI, Ischiropoulos H, Lee VM (2003) Effects of oxidative and nitrative challenges on α-synuclein fibrillogenesis involve distinct mechanisms of protein modifications. J Biol Chem 278:27230–27240.
82. Yamin G, Uversky VN, Fink AL (2003) Nitration inhibits fibrillation of human alpha-synuclein in vitro by formation of soluble oligomers. FEBS Lett 542:147–152.
83. Cole NB, Murphy DD, Lebowitz J, Di Noto L, Levine RL, Nussbaum RL (2005) Metal-catalyzed oxidation of alpha-synuclein: helping to define the relationship between oligomers, protofibrils, and filaments. J Biol Chem 280:9678–9690.
84. Conway KA, Rochet JC, Bieganski RM, Lansbury PT Jr (2001) Kinetic stabilization of the alpha-synuclein protofibril by a dopamine-alpha-synuclein adduct. Science 294:1346–1349.

85. Cappai R, Leck SL, Tew DJ, Williamson NA, Smith DP, Galatis D, Sharples RA, Curtain CC, Ali FE, Cherny RA et al (2005) Dopamine promotes alpha-synuclein aggregation into SDS-resistant soluble oligomers via a distinct folding pathway. Faseb J 19: 1377–1379.
86. Li HT, Lin DH, Luo XY, Zhang F, Ji LN, Du HN, Song GQ, Hu J, Zhou JW, Hu HY (2005) Inhibition of alpha-synuclein fibrillization by dopamine analogs via reaction with the amino groups of alpha-synuclein. Implication for dopaminergic neurodegeneration. FEBS J 272:3661–3672.
87. Bisaglia M, Mammi S, Bubacco L (2007) Kinetic and structural analysis of the early oxidation products of dopamine. Analysis of the interactions with alpha-synuclein. J Biol Chem 282:15597–15605.
88. Martinez-Vicente M, Talloczy Z, Kaushik S, Massey AC, Mazzulli J, Mosharov EV, Hodara R, Fredenburg R, Wu DC, Follenzi A et al (2008) Dopamine-modified alpha-synuclein blocks chaperone-mediated autophagy. J Clin Invest 118:777–788.
89. Li J, Zhu M, Manning-Bog AB, Di Monte DA, Fink AL (2004) Dopamine and L-dopa disaggregate amyloid fibrils: implications for Parkinson's and Alzheimer's disease. Faseb J 18:962–964.
90. Norris EH, Giasson BI, Hodara R, Xu S, Trojanowski JQ, Ischiropoulos H, Lee VM (2005) Reversible inhibition of alpha-synuclein fibrillization by dopaminochrome-mediated conformational alterations. J Biol Chem 280:21212–21219.
91. Follmer C, Romao L, Einsiedler CM, Porto TC, Lara FA, Moncores M, Weissmuller G, Lashuel HA, Lansbury P, Neto VM et al (2007) Dopamine affects the stability, hydration, and packing of protofibrils and fibrils of the wild type and variants of alpha-synuclein. Biochemistry 46:472–482.
92. Chae SW, Kang BY, Hwang O, Choi HJ (2008) Cyclooxygenase-2 is involved in oxidative damage and alpha-synuclein accumulation in dopaminergic cells. Neurosci Lett 436: 205–209.
93. Hokenson MJ, Uversky VN, Goers J, Yamin G, Munishkina LA, Fink AL (2004) Role of individual methionines in the fibrillation of methionine-oxidized alpha-synuclein. Biochemistry 43:4621–4633.
94. Yamin G, Glaser CB, Uversky VN, Fink AL (2003) Certain metals trigger fibrillation of methionine-oxidized α-synuclein. J Biol Chem 278:27630–27635.
95. Uversky VN, Yamin G, Souillac PO, Goers J, Glaser CB Fink AL (2002) Methionine oxidation inhibits fibrillation of human α-synuclein in vitro. FEBS Lett 517:239–244.
96. Glaser CB, Yamin G, Uversky VN, Fink AL (2005) Methionine oxidation, alpha-synuclein and Parkinson's disease. Biochim Biophys Acta 1703:157–169.
97. Qin Z, Hu D, Han S, Reaney SH, Di Monte DA, Fink AL (2007) Effect of 4-hydroxy-2-nonenal modification on alpha-synuclein aggregation. J Biol Chem 282:5862–5870.
98. Shamoto-Nagai M, Maruyama W, Hashizume Y, Yoshida M, Osawa T, Riederer P, Naoi M (2007) In parkinsonian substantia nigra, alpha-synuclein is modified by acrolein, a lipid-peroxidation product, and accumulates in the dopamine neurons with inhibition of proteasome activity. J Neural Transm 114:1559–1567.
99. Bosco DA, Fowler DM, Zhang Q, Nieva J, Powers ET, Wentworth P Jr, Lerner RA, Kelly JW (2006) Elevated levels of oxidized cholesterol metabolites in Lewy body disease brains accelerate alpha-synuclein fibrilization. Nat Chem Biol 2:249–253.
100. Paleologou KE, Schmid AW, Rospigliosi CC, Kim HY, Lamberto GR, Fredenburg RA, Lansbury PT Jr, Fernandez CO, Eliezer D, Zweckstetter M et al (2008) Phosphorylation at Ser-129 but not the phosphomimics S129E/D inhibits the fibrillation of alpha-synuclein. J Biol Chem 283:16895–16905.
101. Hoyer W, Antony T, Cherny D, Heim G, Jovin TM, Subramaniam V (2002) Dependence of alpha-synuclein aggregate morphology on solution conditions. J Mol Biol 322: 383–393.
102. Pronin AN, Morris AJ, Surguchov A, Benovic JL (2000) Synucleins are a novel class of substrates for G protein-coupled receptor kinases. J Biol Chem 275:26515–26522.

103. Chen L Feany MB (2005) Alpha-synuclein phosphorylation controls neurotoxicity and inclusion formation in a Drosophila model of Parkinson disease. Nat Neurosci 8:657–663.
104. Gorbatyuk OS, Li S, Sullivan LF, Chen W, Kondrikova G, Manfredsson FP, Mandel RJ, Muzyczka N (2008) The phosphorylation state of Ser-129 in human alpha-synuclein determines neurodegeneration in a rat model of Parkinson disease. Proc Natl Acad Sci USA 105:763–768.
105. Moskovitz J, Berlett BS, Poston JM, Stadtman ER (1997) The yeast peptide-methionine sulfoxide reductase functions as an antioxidant in vivo. Proc Natl Acad Sci USA 94:9585–9589.
106. Moskovitz J, Flescher E, Berlett BS, Azare J, Poston JM, Stadtman ER (1998) Overexpression of peptide-methionine sulfoxide reductase in Saccharomyces cerevisiae and human T cells provides them with high resistance to oxidative stress. Proc Natl Acad Sci USA 95:14071–14075.
107. Yermolaieva O, Xu R, Schinstock C, Brot N, Weissbach H, Heinemann SH, Hoshi T (2004) Methionine sulfoxide reductase A protects neuronal cells against brief hypoxia/reoxygenation. Proc Natl Acad Sci USA 101:1159–1164.
108. Marchetti MA, Lee W, Cowell TL, Wells TM, Weissbach H, Kantorow M (2006) Silencing of the methionine sulfoxide reductase A gene results in loss of mitochondrial membrane potential and increased ROS production in human lens cells. Exp Eye Res 83:1281–1286.
109. Moskovitz J, Bar-Noy S, Williams WM, Requena J, Berlett BS, Stadtman ER (2001) Methionine sulfoxide reductase (MsrA) is a regulator of antioxidant defense and lifespan in mammals. Proc Natl Acad Sci USA 98:12920–12925.
110. Stadtman ER, Moskovitz J, Berlett BS, Levine RL (2002) Cyclic oxidation and reduction of protein methionine residues is an important antioxidant mechanism. Mol Cell Biochem 234–235:3–9.
111. Petropoulos I, Mary J, Perichon M, Friguet B (2001) Rat peptide methionine sulphoxide reductase: cloning of the cDNA, and down-regulation of gene expression and enzyme activity during aging. Biochem J 355:819–825.
112. Ruan H, Tang XD, Chen ML, Joiner ML, Sun G, Brot N, Weissbach H, Heinemann SH, Iverson L, Wu CF et al (2002) High-quality life extension by the enzyme peptide methionine sulfoxide reductase. Proc Natl Acad Sci USA 99:2748–2753.
113. Levine RL, Moskovitz J, Stadtman ER (2000) Oxidation of methionine in proteins: roles in antioxidant defense and cellular regulation. IUBMB Life 50:301–307.
114. Moskovitz J, Jenkins NA, Gilbert DJ, Copeland NG, Jursky F, Weissbach H, Brot N (1996) Chromosomal localization of the mammalian peptide-methionine sulfoxide reductase gene and its differential expression in various tissues. Proc Natl Acad Sci USA 93:3205–3208.
115. Liu F, Hindupur J, Nguyen JL, Ruf KJ, Zhu J, Schieler JL, Bonham CC, Wood KV, Davisson VJ, Rochet JC (2008) Methionine sulfoxide reductase A protects dopaminergic cells from Parkinson's disease-related insults. Free Radic Biol Med 45:242–255.
116. Liu F, Nguyen JL, Hulleman JD, Li L, Rochet JC (2008) Mechanisms of DJ-1 neuroprotection in a cellular model of Parkinson's disease. J Neurochem 105:2435–2453.
117. Jenner P, Olanow CW (1998) Understanding cell death in Parkinson's disease. Ann Neurol 44:S72–84.
118. Bharath S, Hsu M, Kaur D, Rajagopalan S, Andersen JK (2002) Glutathione, iron and Parkinson's disease. Biochem Pharmacol 64:1037–1048.
119. Maher P (2005) The effects of stress and aging on glutathione metabolism. Ageing Res Rev 4:288–314.
120. Wassef R, Haenold R, Hansel A, Brot N, Heinemann SH, Hoshi T (2007) Methionine sulfoxide reductase A and a dietary supplement S-methyl-L-cysteine prevent Parkinson's-like symptoms. J Neurosci 27:12808–12816.
121. Zhou W, Freed CR (2005) DJ-1 up-regulates glutathione synthesis during oxidative stress and inhibits A53T alpha-synuclein toxicity. J Biol Chem 280:43150–43158.

122. Botella JA, Bayersdorfer F Schneuwly S (2008) Superoxide dismutase overexpression protects dopaminergic neurons in a Drosophila model of Parkinson's disease. Neurobiol Dis 30:65–73.
123. Trinh K, Moore K, Wes PD, Muchowski PJ, Dey J, Andrews L, Pallanck LJ (2008) Induction of the phase II detoxification pathway suppresses neuron loss in Drosophila models of Parkinson's disease. J Neurosci 28:465–472.
124. Kim HY, Gladyshev VN (2005) Role of structural and functional elements of mouse methionine-S-sulfoxide reductase in its subcellular distribution. Biochemistry 44: 8059–8067.
125. Cole NB, Dieuliis D, Leo P, Mitchell DC Nussbaum RL (2008) Mitochondrial translocation of alpha-synuclein is promoted by intracellular acidification. Exp Cell Res 314:2076–2089.
126. Devi L, Raghavendran V, Prabhu BM, Avadhani NG, Anandatheerthavarada HK (2008) Mitochondrial import and accumulation of alpha-synuclein impair complex I in human dopaminergic neuronal cultures and Parkinson disease brain. J Biol Chem 283:9089–9100.
127. Li J, Zhu M, Rajamani S, Uversky VN, Fink AL (2004) Rifampicin inhibits alpha-synuclein fibrillation and disaggregates fibrils. Chem Biol 11:1513–1521.
128. Zhu M, Rajamani S, Kaylor J, Han S, Zhou F Fink AL (2004) The flavonoid baicalein inhibits fibrillation of alpha-synuclein and disaggregates existing fibrils. J Biol Chem 279:26846–26857.
129. Masuda M, Suzuki N, Taniguchi S, Oikawa T, Nonaka T, Iwatsubo T, Hisanaga S, Goedert M, Hasegawa M (2006) Small molecule inhibitors of alpha-synuclein filament assembly. Biochemistry 45:6085–6094.
130. Ono K, Yamada M (2006) Antioxidant compounds have potent anti-fibrillogenic and fibril-destabilizing effects for alpha-synuclein fibrils in vitro. J Neurochem 97: 105–115.
131. Ehrnhoefer DE, Bieschke J, Boeddrich A, Herbst M, Masino L, Lurz R, Engemann S, Pastore A Wanker EE (2008) EGCG redirects amyloidogenic polypeptides into unstructured, off-pathway oligomers. Nat Struct Mol Biol 15:558–566.
132. Pandey N, Strider J, Nolan WC, Yan SX, Galvin JE (2008) Curcumin inhibits aggregation of alpha-synuclein. Acta Neuropathol 115:479–489.
133. Kostka M, Hogen T, Danzer KM, Levin J, Habeck M, Wirth A, Wagner R, Glabe CG, Finger S, Heinzelmann U et al (2008) Single particle characterization of iron-induced pore-forming alpha-synuclein oligomers. J Biol Chem 283:10992–11003.
134. Muchowski PJ, Wacker JL (2005) Modulation of neurodegeneration by molecular chaperones. Nat Rev Neurosci 6:11–22.
135. Rochet JC (2007) Novel therapeutic strategies for the treatment of protein-misfolding diseases. Expert Rev Mol Med 9:1–34.
136. Dedmon MM, Christodoulou J, Wilson MR, Dobson CM (2005) Heat shock protein 70 inhibits alpha-synuclein fibril formation via preferential binding to prefibrillar species. J Biol Chem 280:14733–14740.
137. Huang C, Cheng H, Hao S, Zhou H, Zhang X, Gao J, Sun QH, Hu H, Wang CC (2006) Heat shock protein 70 inhibits alpha-synuclein fibril formation via interactions with diverse intermediates. J Mol Biol 364:323–336.
138. Auluck PK, Chan HY, Trojanowski JQ, Lee VM, Bonini NM (2002) Chaperone suppression of alpha-synuclein toxicity in a Drosophila model for Parkinson's disease. Science 295: 865–868.
139. Auluck PK, Meulener MC, Bonini NM (2005) Mechanisms of suppression of alpha-synuclein neurotoxicity by geldanamycin in Drosophila. J Biol Chem 280:2873–2878.
140. Zhou Y, Gu G, Goodlett DR, Zhang T, Pan C, Montine TJ, Montine KS, Aebersold RH, Zhang J (2004) Analysis of alpha-synuclein-associated proteins by quantitative proteomics. J Biol Chem 279:39155–39164.
141. Klucken J, Shin Y, Masliah E, Hyman BT, McLean PJ (2004) Hsp70 Reduces alpha-synuclein Aggregation and Toxicity. J Biol Chem 279:25497–25502.

142. McLean PJ, Klucken J, Shin Y, Hyman BT (2004) Geldanamycin induces Hsp70 and prevents alpha-synuclein aggregation and toxicity in vitro. Biochem Biophys Res Commun 321:665–669.
143. Outeiro TF, Putcha P, Tetzlaff JE, Spoelgen R, Koker M, Carvalho F, Hyman BT, McLean PJ (2008) Formation of toxic oligomeric alpha-synuclein species in living cells. PLoS ONE 3:e1867.
144. Masliah E, Rockenstein E, Veinbergs I, Mallory M, Hashimoto M, Takeda A, Sagara Y, Sisk A, Mucke L (2000) Dopaminergic loss and inclusion body formation in alpha-synuclein mice: implications for neurodegenerative disorders. Science 287: 1265–1269.
145. Rekas A, Adda CG, Andrew Aquilina J, Barnham KJ, Sunde M, Galatis D, Williamson NA, Masters CL, Anders RF, Robinson CV et al (2004) Interaction of the molecular chaperone alphaB-crystallin with alpha-synuclein: effects on amyloid fibril formation and chaperone activity. J Mol Biol 340:1167–1183.
146. Rekas A, Jankova L, Thorn DC, Cappai R, Carver JA (2007) Monitoring the prevention of amyloid fibril formation by alpha-crystallin. Temperature dependence and the nature of the aggregating species. FEBS J 274:6290–6304.
147. Ecroyd H, Carver JA (2008) The effect of small molecules in modulating the chaperone activity of alphaB-crystallin against ordered and disordered protein aggregation. FEBS J 275:935–947.
148. Ahmad MF, Raman B, Ramakrishna T, Rao Ch M (2008) Effect of phosphorylation on alpha B-crystallin: differences in stability, subunit exchange and chaperone activity of homo and mixed oligomers of alpha B-crystallin and its phosphorylation-mimicking mutant. J Mol Biol 375:1040–1051.
149. Ghosh JG, Houck SA, Clark JI (2008) Interactive sequences in the molecular chaperone, human alphaB crystallin modulate the fibrillation of amyloidogenic proteins. Int J Biochem Cell Biol 40:954–967.
150. Zourlidou A, Payne Smith MD, Latchman DS (2004) HSP27 but not HSP70 has a potent protective effect against alpha-synuclein-induced cell death in mammalian neuronal cells. J Neurochem 88:1439–1448.
151. Outeiro TF, Klucken J, Strathearn KE, Liu F, Nguyen P, Rochet JC, Hyman BT, McLean PJ (2006) Small heat shock proteins protect against alpha-synuclein-induced toxicity and aggregation. Biochem Biophys Res Commun 351:631–638.
152. Pountney DL, Treweek TM, Chataway T, Huang Y, Chegini F, Blumbergs PC, Raftery MJ, Gai WP (2005) Alpha B-crystallin is a major component of glial cytoplasmic inclusions in multiple system atrophy. Neurotox Res 7:77–85.
153. Uryu K, Richter-Landsberg C, Welch W, Sun E, Goldbaum O, Norris EH, Pham CT, Yazawa I, Hilburger K, Micsenyi M et al (2006) Convergence of heat shock protein 90 with ubiquitin in filamentous alpha-synuclein inclusions of alpha-synucleinopathies. Am J Pathol 168: 947–961.
154. Wang L, Xie C, Greggio E, Parisiadou L, Shim H, Sun L, Chandran J, Lin X, Lai C, Yang WJ et al (2008) The chaperone activity of heat shock protein 90 is critical for maintaining the stability of leucine-rich repeat kinase 2. J Neurosci 28:3384–3391.
155. Cao S, Gelwix CC, Caldwell KA, Caldwell GA (2005) Torsin-mediated protection from cellular stress in the dopaminergic neurons of Caenorhabditis elegans. J Neurosci 25: 3801–3812.
156. Sharma N, Hewett J, Ozelius LJ, Ramesh V, McLean PJ, Breakefield XO, Hyman BT (2001) A close association of torsinA and alpha-synuclein in Lewy bodies: a fluorescence resonance energy transfer study. Am J Pathol 159:339–344.
157. McLean PJ, Kawamata H, Shariff S, Hewett J, Sharma N, Ueda K, Breakefield XO, Hyman BT (2002) TorsinA and heat shock proteins act as molecular chaperones: suppression of alpha-synuclein aggregation. J Neurochem 83:846–854.
158. Abou-Sleiman PM, Healy DG, Quinn N, Lees AJ, Wood NW (2003) The role of pathogenic DJ-1 mutations in Parkinson's disease. Ann Neurol 54:283–286.

159. Bonifati V, Rizzu P, van Baren MJ, Schaap O, Breedveld GJ, Krieger E, Dekker MC, Squitieri F, Ibanez P, Joosse M et al (2003) Mutations in the DJ-1 gene associated with autosomal recessive early-onset parkinsonism. Science 299:256–259.
160. Hering R, Strauss KM, Tao X, Bauer A, Woitalla D, Mietz EM, Petrovic S, Bauer P, Schaible W, Muller T et al (2004) Novel homozygous p.E64D mutation in DJ1 in early onset Parkinson disease (PARK7). Hum Mutat 24:321–329.
161. Annesi G, Savettieri G, Pugliese P, D'Amelio M, Tarantino P, Ragonese P, La Bella V, Piccoli T, Civitelli D, Annesi F et al (2005) DJ-1 mutations and parkinsonism-dementia-amyotrophic lateral sclerosis complex. Ann Neurol 58:803–807.
162. Lev N, Roncevic D, Ickowicz D, Melamed E, Offen D (2006) Role of DJ-1 in Parkinson's disease. J Mol Neurosci 29:215–225.
163. Tang B, Xiong H, Sun P, Zhang Y, Wang D, Hu Z, Zhu Z, Ma H, Pan Q, Xia JH et al (2006) Association of PINK1 and DJ-1 confers digenic inheritance of early-onset Parkinson's disease. Hum Mol Genet 15:1816–1825.
164. Honbou K, Suzuki NN, Horiuchi M, Niki T, Taira T, Ariga H, Inagaki F (2003) The crystal structure of DJ-1, a protein related to male fertility and Parkinson's disease. J Biol Chem 278:31380–31384.
165. Huai Q, Sun Y, Wang H, Chin LS, Li L, Robinson H, Ke H (2003) Crystal structure of DJ-1/RS and implication on familial Parkinson's disease. FEBS Lett 549:171–175.
166. Lee SJ, Kim SJ, Kim IK, Ko J, Jeong CS, Kim GH, Park C, Kang SO, Suh PG, Lee HS et al (2003) Crystal structures of human DJ-1 and Escherichia coli Hsp31, which share an evolutionarily conserved domain. J Biol Chem 278:44552–44559.
167. Tao X, Tong L (2003) Crystal structure of human DJ-1, a protein associated with early onset Parkinson's disease. J Biol Chem 278:31372–31379.
168. Wilson MA, Collins JL, Hod Y, Ringe D, Petsko GA (2003) The 1.1-A resolution crystal structure of DJ-1, the protein mutated in autosomal recessive early onset Parkinson's disease. Proc Natl Acad Sci USA 100:9256–9261.
169. Canet-Aviles RM, Wilson MA, Miller DW, Ahmad R, McLendon C, Bandyopadhyay S, Baptista MJ, Ringe D, Petsko GA, Cookson MR (2004) The Parkinson's disease protein DJ-1 is neuroprotective due to cysteine-sulfinic acid-driven mitochondrial localization. Proc Natl Acad Sci USA 101:9103–9108.
170. Taira T, Saito Y, Niki T, Iguchi-Ariga SM, Takahashi K, Ariga H (2004) DJ-1 has a role in antioxidative stress to prevent cell death. EMBO Rep 5:213–218.
171. Yokota T, Sugawara K, Ito K, Takahashi R, Ariga H, Mizusawa H (2003) Down regulation of DJ-1 enhances cell death by oxidative stress, ER stress, and proteasome inhibition. Biochem Biophys Res Commun 312:1342–1348.
172. Martinat C, Shendelman S, Jonason A, Leete T, Beal MF, Yang L, Floss T, Abeliovich A (2004) Sensitivity to oxidative stress in DJ-1-deficient dopamine neurons: an ES- derived cell model of primary parkinsonism. PLoS Biol 2:e327.
173. Kim RH, Smith PD, Aleyasin H, Hayley S, Mount MP, Pownall S, Wakeham A, You-Ten AJ, Kalia SK, Horne P et al (2005) Hypersensitivity of DJ-1-deficient mice to 1-methyl-4-phenyl-1,2,3,6-tetrahydropyrindine (MPTP) and oxidative stress. Proc Natl Acad Sci USA 102:5215–5220.
174. Menzies FM, Yenisetti SC, Min KT (2005) Roles of Drosophila DJ-1 in survival of dopaminergic neurons and oxidative stress. Curr Biol 15:1578–1582.
175. Xu J, Zhong N, Wang H, Elias JE, Kim CY, Woldman I, Pifl C, Gygi SP, Geula C Yankner BA (2005) The Parkinson's disease-associated DJ-1 protein is a transcriptional co-activator that protects against neuronal apoptosis. Hum Mol Genet 14:1231–1241.
176. Yang Y, Gehrke S, Haque ME, Imai Y, Kosek J, Yang L, Beal MF, Nishimura I, Wakamatsu K, Ito S et al (2005) Inactivation of Drosophila DJ-1 leads to impairments of oxidative stress response and phosphatidylinositol 3-kinase/Akt signaling. Proc Natl Acad Sci USA 102:13670–13675.

177. Andres-Mateos E, Perier C, Zhang L, Blanchard-Fillion B, Greco TM, Thomas B, Ko HS, Sasaki M, Ischiropoulos H, Przedborski S et al (2007) DJ-1 gene deletion reveals that DJ-1 is an atypical peroxiredoxin-like peroxidase. Proc Natl Acad Sci USA 104: 14807–14812.
178. Clements CM, McNally RS, Conti BJ, Mak TW, Ting JP (2006) DJ-1, a cancer- and Parkinson's disease-associated protein, stabilizes the antioxidant transcriptional master regulator Nrf2. Proc Natl Acad Sci USA 103:15091–15096.
179. Shendelman S, Jonason A, Martinat C, Leete T Abeliovich A (2004) DJ-1 is a redox-dependent molecular chaperone that inhibits alpha-synuclein aggregate formation. PLoS Biol 2:e362.
180. Zhou W, Zhu M, Wilson MA, Petsko GA, Fink AL (2006) The oxidation state of DJ-1 regulates its chaperone activity toward alpha-synuclein. J Mol Biol 356:1036–1048.
181. Batelli S, Albani D, Rametta R, Polito L, Prato F, Pesaresi M, Negro A, Forloni G (2008) DJ-1 modulates alpha-synuclein aggregation state in a cellular model of oxidative stress: relevance for Parkinson's disease and involvement of HSP70. PLoS ONE 3:e1884.
182. Junn E, Taniguchi H, Jeong BS, Zhao X, Ichijo H, Mouradian MM (2005) Interaction of DJ-1 with Daxx inhibits apoptosis signal-regulating kinase 1 activity and cell death. Proc Natl Acad Sci USA 102:9691–9696.
183. Gorner K, Holtorf E, Waak J, Pham TT, Vogt-Weisenhorn DM, Wurst W, Haass C, Kahle PJ (2007) Structural determinants of the C-terminal helix-kink-helix motif essential for protein stability and survival promoting activity of DJ-1. J Biol Chem 282:13680–13691.
184. Fan J, Ren H, Jia N, Fei E, Zhou T, Jiang P, Wu M, Wang G (2008) DJ-1 decreases Bax expression through repressing p53 transcriptional activity. J Biol Chem 283:4022–4030.
185. Miyazaki S, Yanagida T, Nunome K, Ishikawa S, Inden M, Kitamura Y, Nakagawa S, Taira T, Hirota K, Niwa M et al (2008) DJ-1-binding compounds prevent oxidative stress-induced cell death and movement defect in Parkinson's disease model rats. J Neurochem 105: 2418–2434.

Chapter 9
Novel Proteins in α-Synucleinopathies

Christine Lund Kragh and Poul Henning Jensen

Abstract α-Synucleinopathies are a group of neurodegenerative disorders characterized by the presence of intracytoplasmic Lewy bodies in Parkinson's disease and dementia with Lewy bodies as well as glial cytoplasmic inclusions in multiple system atrophy. The main component of these inclusions is aggregated α-synuclein which supports a strong link between α-synuclein and disease pathogenesis. The mechanisms responsible for α-synuclein aggregation and subsequent degeneration are largely unknown. However, several factors have been shown to accelerate the aggregation of α-synuclein *in vitro* and suggested to contribute to the pathogenesis of α-synucleinopathies. Several different proteins can stimulate the aggregation process *in vitro* and have been shown to colocalize with aggregated α-synuclein in pathological brain tissue. We review our current knowledge on proteins with a putative involvement in α-synuclein-dependent degeneration based on aggregatory properties, colocalization with aggregated α-synuclein, or genetic evidence.

9.1 Introduction

The α-synucleinopathies comprise Parkinson's disease (PD), Dementia with Lewy bodies (DLB) [1–4], multiple system atrophy (MSA) [1, 5–8] as well as neurodegeneration with brain iron accumulation type I and Lewy body variant of Alzheimer's disease (AD) [9, 10]. The disorders are neuropathologically hallmarked by the presence of intracellular inclusions containing aggregated α-synuclein. These inclusions are deposited in selective populations of neurons and glia varying among the disorders [11–13].

Several lines of evidence demonstrate a key role of α-synuclein in the pathogenesis of α-synucleinopathies; (i) missense mutations in the α-synuclein gene

---

C. Lund Kragh (✉)
Institute of Medical Biochemistry, University of Aarhus, Ole Worms Allé 1.170
DK-8000 Aarhus C, Denmark
e-mail: clp@biokemi.au.dk

(SNCA) cause autosomal dominant PD and DLB [14–16]; (ii) multiplications of the normal SNCA gene resulting in an increased expression of the α-synuclein protein cause autosomal dominant PD and DLB [17, 18]; (iii) the presence of α-synuclein aggregates and intracytoplasmic inclusions in the hereditary cases are indistinguishable from those found in sporadic cases of PD [4, 8, 19, 20]. The mechanism whereby α-synuclein contributes to cellular degeneration is unclear, but aggregation of α-synuclein is thought to play a central role. Firstly, aggregates of α-synuclein are present in all α-synucleinopathies and the disease-causing mutations stimulate the aggregation of the α-synuclein protein *in vitro* [21–23]. The α-synucleinopathies thus resemble other neurodegenerative diseases caused by mutations in specific genes leading to aggregation and neurodegeneration, e.g. tau or amyloid precursor protein (APP) [24]. Secondly, transgenic modelling of the α-synucleinopathies has been achieved in different organisms and inhibition of aggregation has been protective toward the degeneration in a *Drosophila* model of PD [25].

The events responsible for aggregation and subsequent degeneration are largely unknown, but several studies have suggested complex mechanisms encompassing phenomena such as posttranslational modifications e.g. phosphorylation, oxidation, and C-terminal proteolysis [26, 27], and impairment in protein catabolism [28].

The pivotal role of α-synuclein in neurodegeneration raises a fundamental question regarding triggers of α-synuclein aggregation in common sporadic α-synucleinopathies. Other proteins are able to stimulate the aggregation of α-synuclein, e.g. p25α, tau, and histones [29–32], and are colocalized with aggregated α-synuclein in pathological brain tissue. Different scenarios can place α-synuclein

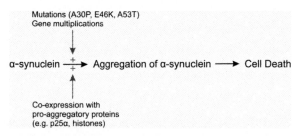

Fig. 9.1 Mechanisms of cytotoxicity caused by α-synuclein aggregation. α-synuclein aggregation and subsequent cell death occur during the development of α-synucleinopathies. Genetic evidence from familial cases harboring mutations in the SNCA gene encoding α-synuclein demonstrates that missense mutations and gene multiplications stimulate the process. The effects can be modeled *in vitro* where the missense mutations stimulate the process of aggregation as does an increased concentration of α-synuclein. Aggregation in the sporadic α-synucleinopathies may be stimulated by additional mechanisms among which the coexpression of α-synuclein with proteins possessing aggregate-promoting activity *in vitro* is a strong candidate. Coexpression with p25α has thus been demonstrated in neurons abnormally expressing p25α in Parkinson's disease and Lewy body dementia and in oligodendroglia in multiple system atrophy where α-synuclein is abnormally expressed with the resident p25α. Coexpression with histones has been shown in neurons where nuclear α-synuclein may play prominent pathogenic roles

in a milieu which triggers its aggregation and thereby initiate a final pathway in the degenerative process. These scenarios could include dysexpression of pro-aggregatory proteins and abnormal expression of α-synuclein within cells, e.g. in the nucleus, or in non-neuronal cells such as oligodendrocytes in MSA (Fig. 9.1).

This review will provide an update on proteins that may stimulate α-synuclein-dependent degeneration based on direct aggregatory properties, colocalization with aggregated α-synuclein, or genetic evidence.

9.2 The α-Synucleinopathies

PD is the most common movement disorder in the elderly and is characterized by tremor, rigidity, and bradykinesia. It is a progressive disorder that involves diverse neurons in the human nervous system [33]. Affected neurons eventually develop inclusions termed Lewy bodies (LBs) in their perikarya and Lewy neurites (LNs) in their processes [34]. A staging procedure for LB and LN pathology has been proposed and states that the pathological process spreads from the lower brain stem and olfactory bulb and progresses in a predictable sequence in six stages. Parts of the limbic system and the motor system have been shown to be particularly vulnerable to damage, although some nuclei in the substantia nigra (SN) also undergo major changes [35].

Genetic screening of families with PD has demonstrated that mutations in specific genes are responsible for a small number of disease cases whereas the vast majority of cases are sporadic [36]. The first gene associated with PD to be identified was the α-synuclein gene in which three pathogenic mutations (A30P, E46K, and A53T) associated with autosomal dominant PD have been discovered [14–16]. Additionally, genomic multiplications of the SNCA gene have been found to cause early-onset PD [17, 18]. As reviewed elsewhere [36, 37], several additional genes and genetic loci have also been implicated in recessive and autosomal PD. The recessive PD genes may trigger special kinds of PD as their relation to α-synuclein cytopathology is dubious. By contrast, mutations in the LRRK2 gene cause autosomal dominant PD and can be associated with α-synuclein pathology [38, 39]. Shortly after it was reported that mutations in α-synuclein were responsible for disease, a series of studies demonstrated that α-synuclein is the major building block in the filaments that form the amyloid inclusions characteristic of the neurodegenerative disorders now known as the α-synucleinopathies [4, 8, 19, 20].

DLB is the second most common cause of dementia in the elderly and is characterized by dementia, parkinsonism, and psychiatric symptoms. Neuropathological findings include widely distributed LBs within the brain [40, 41]. A mutation in α-synuclein (E46K) has been described in a large Spanish pedigree involving patients with both DLB and PD [16].

MSA is characterized clinically by parkinsonism, autonomic failure, and cerebellar ataxia [42]. The neuropathological hallmarks are neuronal loss, gliosis, and myelin pathology particularly in the striatonigral system. In affected white matter, oligodendroglia contain glial cytoplasmic inclusions (GCIs) with deposited, filamentous α-synuclein [43, 44]. Furthermore, neuronal cytoplasmic and nuclear inclusions are also pathological findings in MSA [44, 45].

9.3 α-Synuclein

The common hallmark of hereditary as well as sporadic α-synucleinopathies is the presence of aggregated α-synuclein in LBs, LNs, or GCIs supporting a strong link between α-synuclein and disease pathogenesis [4, 8, 19, 20]. Normally, α-synuclein is localized in presynaptic terminals where it associates with synaptic vesicles [46] and may regulate vesicular release and other synaptic functions in the central nervous system [47, 48].

α-Synuclein is a natively unfolded protein [49] but has an increased propensity to aggregate owing to its hydrophobic domain, which is localized in the middle region of the protein [50, 51]. A direct role for α-synuclein aggregation in the development of α-synucleinopathies is demonstrated by genetic evidence. Autosomal dominant PD can be induced by expression of mutant α-synuclein (A30P, E46K, and A53T) [14–16] demonstrating that a single mutation in the human α-synuclein gene is sufficient to cause the PD phenotype. All three PD-related mutations have been shown to accelerate aggregation of α-synuclein *in vitro* suggesting that accelerated fibrillization of α-synuclein is responsible for the development of early-onset PD in patients harbouring these mutations [21–23]. The link between aggregated α-synuclein and autosomal dominant PD suggests a toxic gain-of-function of α-synuclein due to its structural alteration from monomer to aggregate. The transition from monomeric to filamentous α-synuclein is characterized by a lag phase in which soluble, oligomeric α-synuclein species assemble. These oligomers might then self-assemble into fibrillar structures that are insoluble [52, 53]. Increasing evidence suggests that prefibrillar oligomers and protofibrils, rather than mature fibrils of α-synuclein, are the pathogenic species in α-synucleinopathies (reviewed in [54]). α-synuclein oligomers permeabilize synthetic vesicles and form pore-like assemblies, a putative mechanism of cytotoxicity [55].

The mechanisms underlying the aggregation of α-synuclein are still unknown, but several factors and events have been demonstrated to influence the aggregation process. Recombinant wild-type α-synuclein assembles into amyloid, fibrillar structures *in vitro* [51]. Multiple events have been reported to stimulate the aggregation process *in vitro* including posttranslational modifications such as phosphorylation [26] and C-terminal truncations [3]. α-synuclein is predominantly found in a non-phosphorylated state *in vivo* [26]. However, enhanced levels of Ser129-phosphorylated α-synuclein have been detected in inclusions in all α-synucleinopathies indicating that phosphorylation at this specific residue may be a significant event in aggregation [26, 56, 57]. Other factors known to stimulate the aggregation process include oxidative and nitrative modifications [27, 58, 59] and dopamine conjugation [60]. Interactions with other proteins such as p25α [30] and tau [61] accelerate the aggregation of α-synuclein and hence may be involved in the development of sporadic α-synucleinopathies.

9.4 Proteins Involved in α-Synuclein Aggregation

α-Synuclein is known to interact with a large variety of proteins (reviewed in [62]) perhaps owing to its flexible and dynamic structure. Some of these proteins have been shown to stimulate the aggregation process *in vitro* at substociometric concentrations as well as to colocalize with α-synuclein in inclusions in α-synucleinopathies. The identification of proteins stimulating aggregation of α-synuclein *in vitro* suggests that dysregulation of the expression of such proteins can trigger α-synuclein aggregation *in vivo*. Proteins with a putative involvement in α-synuclein aggregation based on *in vitro* and *in vivo* observations will be presented below.

9.4.1 Brain-Specific Protein p25α/TPPP

A search for proteins interacting with aggregated α-synuclein led to the identification of the brain-specific protein p25α [30]. p25α preferentially binds to aggregated α-synuclein through motifs in its C-terminal and stimulates its aggregation at substoichimetric concentrations *in vitro* [30]. p25α is an oligodendroglial-specific protein present in all parts of the brain [63, 64] although neuronal expression has been demonstrated in nucleus supraopticus in rat [30]. Within the oligodendrocyte, p25α is expressed in the perinuclear cytoplasm [30, 65] as well as in myelin [66]. Investigations of the expression profile of p25α have revealed that it is detectable in the developing rat brain from the time of onset for myelination [67].

The p25α protein was originally isolated from bovine brain co-purified with a tau kinase [63]. It is a heat-stable protein of 219 amino acids, which has been shown by nuclear magnetic resonance (NMR) and circular dichroism (CD) spectroscopy to be a natively folded protein [68]. Two additional members of the p25 protein family have been identified. These proteins are designated p25β and p25γ, and share a very high degree of sequence identity with p25α in the middle and C-terminal regions, including a highly conserved Rossmann fold known to be involved in nucleotide binding [69, 70]. The putative unstructured N-terminal region is missing in p25β and p25γ.

The physiological function of p25α has not yet been clarified, albeit several interaction partners have been identified.

p25α interacts with tubulin in a 1:2 complex [68] and induces aberrant tubulin assemblies and bundling of microtubule and is therefore also known as tubulin polymerization promoting protein (TPPP) [71]. At low expression levels, p25α colocalizes with the microtubule cytoskeleton in transfected HeLa cells in a dynamic process that changes during the phases of mitosis as p25α dissociates from microtubules at the prophase and reassociates at later phases. At higher expression levels, p25α causes the formation of aggresome-like bodies at the centrosome region [72]. Moreover, injection of bovine p25α into dividing *Drosophila* embryos expressing tubulin-GFP fusion protein inhibited mitotic spindle assembly and nuclear envelope breakdown [73].

Glyceraldehyde-3-phosphate dehydrogenase (GAPDH) has been identified as an interacting partner of p25α. GAPDH was long considered exclusively as a housekeeping enzyme, but recent studies have assigned multiple functions to the protein such as membrane function, transcriptional control, DNA repair [74], and involvement in apoptosis [75]. It was demonstrated that GAPDH and p25α interact *in vitro* and colocalize in aggresome-like aggregates formed in HeLa cells expressing p25α. Furthermore, immunohistochemical studies of LBs from PD brain revealed a colocalization between GAPDH and p25α [76].

p25α interacts with complexin [77] which belongs to a family of small brain proteins involved in regulating fast transmitter release at the synapse [78]. The interaction between p25α and complexin was demonstrated in homogenate from rat hippocampus and shown to be decreased in rats exposed to a spatial memory task, suggesting that p25α might be involved in the process of spatial memory [77].

The oligodendroglial protein myelin basic protein (MBP), a major component in myelin, binds directly to p25α and forms a 1:1 complex [66]. p25α colocalizes with MBP in brainstem myelinated fiber tracts and this colocalization is lost in MSA brains, where p25α relocates toward the cell body of the oligodendroglia. Along with the relocalization of p25α from myelin, a degradation and relocalization of MPB occurs. These data suggest that disruption of the normal cellular function of p25α could lead to pathogenic signals by reducing the stability of MBP and accumulating p25α in expanded cell bodies, and thereby favouring subsequent deposition and aggregation of α-synuclein [66].

p25α is subject to phosphorylation by different kinases e.g. ERK2 [79], cyclin-dependent kinase 5 (Cdk5) [63, 80], LIM kinase 1 (LIMK1) [81], glycogen synthase kinase 3β, protein kinase A [80], and protein kinase C isoforms [82]. Phosphorylation by ERK2, Cdk5, and LIMK1 blocks the microtubule-assembling activity of p25α [79, 81].

p25α and α-synuclein are not normally coexpressed in the adult nervous system as p25α is expressed in oligodendrocytes and α-synuclein in neurons [46, 64] However, p25α colocalizes with aggregated α-synuclein in LBs in PD and DLB suggesting an abnormal expression of the protein in affected nerve cells. Additionally, p25α colocalizes with α-synuclein in GCIs in MSA and immunohistochemical analyses of MSA brains show that p25α accumulates in expanded cell bodies of dystrophic oligodendrocytes containing GCIs [30, 65, 66, 83, 84]. Thus, dysregulation of p25α expression could be a contributing factor in cases of sporadic α-synucleinopathies. Based on these findings, it has been suggested that p25α may serve as a new marker of α-synucleinopathies [84]. However, a number of neuronal cytoplasmic inclusions have been shown to contain p25α but not α-synuclein [83, 85] indicating that abnormal p25α expression precedes α-synuclein accumulation or constitutes an independent cellular lesion.

Moreover, p25α induces α-synuclein-dependent degeneration in an oligodendroglial cell culture model which is elicited by aggregation of α-synuclein (Kragh C.L., unpublished data).

9.4.2 Tau

Tau proteins are microtubule-binding proteins that act by stabilizing and promoting microtubule polymerization in neuronal perikarya and processes. Alternative RNA splicing yields six different tau isoforms that differ in part by the number of tandem repeats in the microtubule binding region [86, 87]. Tauopathies are a group of neurodegenerative disorders characterized by the presence of tau inclusions in neurons, oligodendrocytes, or astrocytes [29, 88]. These inclusions are characterized by the presence of hyperphosphorylated tau, which is deposited as neurofibrillary tangles (NFTs) in AD, the most common tauopathy [29]. Other tauopathies include progressive supranuclear palsy (PSP), corticobasal degeneration (CBD), and frontotemporal dementia and parkinsonism linked to chromosome 17 (FTDP-17) in which tau is deposited within neurons and glia [89, 90]. FTDP-17 is caused by mutations in the gene encoding tau [91].

Although rare, the presence of inclusions containing both tau and α-synuclein has been reported in several neurodegenerative diseases including PD, MSA, AD, and Down's syndrome [92]. There is an age-related increase in the occurrence of inclusions composed of both tau and α-synuclein within the brains of normal aged individuals. However, the occurrence of these aggregates in neurodegenerative disorders is greater than that accounted for by normal aging alone [92]. The co-occurrence of inclusions containing α-synuclein and tau indicates that common mechanisms may be involved in their formation. Indeed, both α-synuclein and tau are unfolded proteins in solution, but acquire a β-sheet conformation upon amyloid fibril formation. α-synuclein is capable of self-polymerizing *in vitro* [50, 93], whereas tau requires co-factors to fibrillize [94]. Examination of α-synuclein from LBs and tau from NFTs has shown that both proteins undergo similar post-translational modifications including hyperphosphorylation [26, 29], nitration [58, 95], and ubiquination [29, 96, 97]. It has also been reported that tau and α-synuclein synergistically promote each others aggregation *in vitro* [61]. The ability of α-synuclein to stimulate the formation of tau inclusions has further been demonstrated *in vivo* using transgenic mouse models. Mice harbouring the A53T mutation in transgenic human α-synuclein develop tau inclusions in a subset of cells [61]. These findings clearly demonstrate that α-synuclein can act as a pathological initiator of tau amyloid formation and that fibril formation by α-synuclein and tau may share common mechanisms. However, these mechanisms remain poorly understood.

9.4.3 Synphilin-1

A protein-protein interaction study using yeast two hybrid screening, led to the identification of synphilin-1 as an α-synuclein binding protein [98]. This interaction was later confirmed by fluorescence resonance energy transfer studies using transfected neuroglioma cells [99]. Synphilin-1 is a protein of 919 amino acids, and contains ankyrin-like repeats, a coiled-coil domain, and a putative ATP,GTP-binding domain [100]. Like α-synuclein, synphilin-1 localizes to the presynapse where it binds to synaptic vesicles and may thereby affect dopamine release [101].

Coexpressing α-synuclein and synphilin-1 in human embryonic kidney (HEK) 293 cells, causes the formation of eosinophilic cytoplasmic inclusions resembling LBs [98]. Phosphorylation of Ser129 in α-synuclein has been shown to be critical for inclusion formation as expression of S129A decreased the development of inclusions [102]. Synphilin-1 is a ubiquitination target for various E3 ubiquitin ligases, such as SIAH, parkin, and dorfin [103–105]. Synphilin-1 ubiquitination by SIAH results in proteasomal degradation [105]. Coexpression of SIAH and synphilin-1 in the presence of proteasome inhibitors results in the accumulation of polyubiquinated synphilin-1, and a marked enhancement in the occurrence of inclusion bodies [106]. These inclusions are positive for ubiquitin and can recruit α-synuclein. Due to the presence of SIAH in LBs, it has been suggested that SIAH plays an active role in the recruitment of synphilin-1 and α-synuclein into LBs [106].

Synphilin-1 has been found to colocalize with α-synuclein in LBs from PD and MSA brains [107, 108]. GCIs have also been shown to be synphilin-1-positive, suggesting that this protein is connected with the aggregation of α-synuclein in the different inclusion-bearing cells [108]. Recently, a new isoform of synphilin-1, termed synphilin-1A, has been identified [109]. Synphilin-1A is an alternative splice variant of synphilin-1. It has shown to be an aggregation-prone and neurotoxic protein, which interacts with α-synuclein and synphilin-1, thereby promoting their recruitment into inclusion bodies. The synphilin-1A isoform has also been found in LBs in PD as well as those of other α-synucleinopathies [109].

9.4.4 TAR-DNA-Binding Protein 43 (TDP-43)

TAR-DNA-binding protein 43 (TDP-43) is a main component of the inclusions characterizing TDP-43 proteinopathies, which include frontotemporal lobar degeneration (FTLD) and amyotrophic lateral sclerosis (ALS) [110]. TDP-43 proteinopathies are distinct from most other neurodegenerative disorders because the TDP-43 inclusions do not contain amyloid deposits. TDP-43 is considered a highly specific marker for FTLD and ALS. However, TDP-43-positive inclusions co-existing with LBs and NFTs were recently found within neurons and oligodendroglia in brains from patients with AD and DLB [111, 112]. Thus, TDP-43 pathology may very well be associated with pathogenic pathways in α-synucleinopathies although no direct interaction has been demonstrated between TDP-43 and α-synuclein.

9.4.5 Leucine-Rich Repeat Kinase 2 (LRRK2)

Mutations in the leucine-rich repeat kinase 2 (LRRK2) or dardarin gene cause autosomal dominant PD [38, 39]. LRRK2 encodes a large 2527 amino acid multidomain protein containing a Rho-Ras-like GTPase domain, a protein kinase domain, and leucine-rich, and WD-40 repeat domains. The physiological function of LRRK2 has not been established, but it has been suggested that the protein may regulate neurite maintenance and neuronal survival [39, 113]. LRRK2 mutation carriers display a diverse neuropathology, including α-synuclein- and tau-positive inclusions, indicating that LRRK2 has an upstream role in the aggregation process of pathogenic proteins [39]. Recently, LRRK2 was found to colocalize with α-synuclein in LBs from PD brains [114].

9.4.6 FK506-Binding Proteins

α-Synuclein aggregation is accelerated by the presence of FK506-binding proteins (FKBPs) *in vitro* [115]. FKBPs are members of the immunophilin family of proteins that bind specific immunosuppressant molecules and possess chaperone activities [116]. The immunosuppressant, FK506, has been ascribed neuroregenerative and neuroprotective properties in cell culture and *in vivo* models. It promotes neurite outgrowth in PC12 cells [117] and increases recovery and nerve regeneration following peripheral nerve injury *in vivo* [118]. Oral administration of FK506 to a MPTP mouse model of PD, causes a decreased loss of striatal tyrosine hydroxylase immunoreactivity, which further supports a role for FK506 in neuroprotection [119].

Two members of the FKBP family, FKBP12 and FKBP52, are expressed in the SN and in the grey matter of the human brain [120]. The expression of FKBP12 is increased in the brain of patients with PD, AD, and DLB. Furthermore, FKBP12 has been demonstrated to colocalize with α-synuclein in LBs and LNs suggesting a relation to α-synuclein pathology [120].

9.4.7 Histones

Intranuclear inclusions containing α-synuclein are present in patients with MSA [45] and in dopaminergic cells of transgenic mice overexpressing α-synuclein [121]. Moreover, α-synuclein colocalizes with histones in murine nigral neurons upon paraquat administration [122]. In fact, histones have been reported to form a tight complex with α-synuclein *in vitro*, resulting in enhanced fibrillization of α-synuclein [123]. These observations indicate that the formation of histone-α-synuclein complexes may be relevant in the pathogenesis of α-synucleinopathies. These data have further been corroborated by a recent study showing that α-synuclein targeted to the nucleus promotes neurotoxicity in cell culture and transgenic flies, where α-synuclein associates with histones and inhibit their acetylation resulting in toxicity [124]. Inhibition of histone deacetylase was demonstrated to protect against α-synuclein-dependent neurotoxicity [124, 125]. These data may suggest that α-synuclein acts in the nucleus to promote degeneration of neurons through an interaction with histones.

Another nuclear protein may be implicated in the aggregation of α-synuclein. It has been shown that the transcriptional co-factor, high mobility group protein 1 (HMGB-1), is able to bind to aggregated α-synuclein *in vitro*. HMGB-1 was also shown to be present in LBs isolated from PD and DLB brain [126].

9.4.8 Other Proteins Involved in α-Synuclein Aggregation

Cytoskeletal proteins, including tau, tubulin, microtubule-associated protein 1B (MAP1B), MAP2, and torsinA all interact with α-synuclein *in vitro* and colocalize with α-synuclein aggregates in LBs [127–132]. The relevance of these interactions remains uncertain, but such interactions may contribute to the hypothesized roles for α-synuclein in perturbing axonal transport [48], modulation of synaptic vesicle recycling [133], and modulation of the function of neurotransmitter transporters e.g. the dopamine transporter [134].

Agrin, an extracellular matrix and transmembrane glycoprotein, has been demonstrated to bind to α-synuclein and accelerate the formation of α-synuclein aggregates *in vitro*. Furthermore, agrin and α-synuclein were found to colocalize in LBs in the SN of PD brain [31]. Agrin has also been implicated in AD as it has been found to be associated with lesions characteristic of AD including Aβ senile plaques and NFTs [135, 136].

9.5 Concluding Remarks

Abnormal protein aggregation is a common characteristic of neurodegenerative disorders and α-synuclein aggregation is the hallmark of the α-synucleinopathies. The pivotal role of α-synuclein aggregation in the degenerative process has been underscored by the rare familial cases caused by mutations in the α-synuclein gene. However, the factors initiating the aggregation in the common sporadic cases are unknown but changes in the microenvironment of α-synuclein are likely to be involved. Such changes can be caused by (i) abnormal expression of proteins that trigger α-synuclein aggregation, (ii) expression of α-synuclein in abnormal subcellular compartments such as the nucleus, (iii) ectopic expression of α-synuclein in non-neuronal cells such as oligodendrocytes in MSA.

Our understanding of the nature of these changes in gene expression or in the cellular and subcellular expression of α-synuclein is incomplete. However, further studies will elucidate mechanisms underlying the α-synucleinopathies and facilitate the generation of models amenable for future drug development.

Abbreviations

AD	Alzheimers disease
DLB	Dementia with Lewy bodies
FKBP	FK506-binding protein
GAPDH	Glyceraldehyde-3-phosphate dehydrogenase
GCI	Glial cytoplasmic inclusion
LB	Lewy bodies
LN	Lewy neurite
LRRK2	Leucine-rich repeat kinase 2
MBP	Myelin basic protein
MPTP	1-methyl 4-phenyl 1,2,3,6-tetrahydropyridine
MSA	Multiple system atrophy
NFT	Neurofibrillary tangle
PD	Parkinson's disease
SN	Substantia nigra
TDP-43	TAR-DNA-binding protein 43
TPPP	Tubulin polymerization promoting protein

References

1. Arima K, Ueda K, Sunohara N, Hirai S, Izumiyama Y, Tonozuka-Uehara H, Kawai M (1998) Immunoelectron-microscopic demonstration of NACP/alpha-synuclein- epitopes on the filamentous component of Lewy bodies in Parkinson's disease and in dementia with Lewy bodies. Brain Res 808:93–100
2. Baba M, Nakajo S, Tu PH, Tomita T, Nakaya K, Lee VM, Trojanowski JQ, Iwatsubo T (1998) Aggregation of alpha-synuclein in Lewy bodies of sporadic Parkinson's disease and dementia with Lewy bodies. Am J Pathol 152:879–884
3. Crowther RA, Jakes R, Spillantini MG, Goedert M (1998) Synthetic filaments assembled from C-terminally truncated alpha-synuclein. FEBS Lett 436:309–312
4. Spillantini MG, Schmidt ML, Lee VM, Trojanowski JQ, Jakes R, Goedert M (1997) Alpha-synuclein in Lewy bodies. Nature 388:839–840
5. Gai WP, Power JHT, Blumbergs PC, Blessing WW (1998) Multiple-system atrophy: a new alpha-synuclein disease? Lancet 352:547–548
6. Gai WP, Pountney DL, Power JHT, Li QX, Culvenor JG, Mclean CA, Jensen PH, Blumbergs PC (2003) alpha-Synuclein fibrils constitute the central core of oligodendroglial inclusion filaments in multiple system atrophy. Exp Neurol 181:68–78
7. Spillantini MG, Crowther RA, Jakes R, Cairns NJ, Lantos PL, Goedert M (1998) Filamentous alpha-synuclein inclusions link multiple system atrophy with Parkinson's disease and dementia with Lewy bodies. Neurosci Lett 251:205–208
8. Wakabayashi K, Yoshimoto M, Tsuji S, Takahashi H (1998) alpha-synuclein immunoreactivity in glial cytoplasmic inclusions in multiple system atrophy. Neurosci Lett 249:180–182
9. Arawaka S, Saito Y, Murayama S, Mori H (1998) Lewy body in neurodegeneration with brain iron accumulation type 1 is immunoreactive for alpha-synuclein. Neurology 51:887–889
10. Marti MJ, Tolosa E, Campdelacreu J (2003) Clinical overview of the synucleinopathies. Mov Disorders 18:S21–S27
11. Goedert M, Spillantini MG, Davies SW (1998) Filamentous nerve cell inclusions in neurodegenerative diseases. Curr Opin Neurobiol 8:619–632

12. Spillantini MG, Goedert M (2000) The alpha-synucleinopathies: Parkinson's disease, dementia with Lewy bodies, and multiple system atrophy. Ann N Y Acad Sci 920:16–27
13. Trojanowski JQ, Lee VMY (2002) Parkinson's disease and related alpha-synucleinopathies a new class of nervous system amyloidoses. Neurotox 20:457–460
14. Kruger R, Kuhn W, Muller T, Woitalla D, Graeber M, Kosel S, Przuntek H, Epplen JT, Schols L, Riess O (1998) Ala30Pro mutation in the gene encoding alpha-synuclein in Parkinson's disease. Nat Genet 18:106–108
15. Polymeropoulos MH, Lavedan C, Leroy E, Ide SE, Dehejia A, Dutra A, Pike B, Root H, Rubenstein J, Boyer R et al. (1997) Mutation in the alpha-synuclein gene identified in families with Parkinson's disease. Science 276:2045–2047
16. Zarranz JJ, Alegre J, Gomez-Esteban JC, Lezcano E, Ros R, Ampuero I, Vidal L, Hoenicka J, Rodriguez O, Atares B et al. (2004) The new mutation, E46K, of alpha-synuclein causes Parkinson and Lewy body dementia. Ann Neurol 55:164–173
17. Farrer M, Kachergus J, Forno L, Lincoln S, Wang DS, Hulihan M, Maraganore D, Gwinn-Hardy K, Wszolek Z, Dickson D et al. (2004) Comparison of kindreds with parkinsonism and alpha-synuclein genomic multiplications. Ann Neurol 55:174–179
18. Singleton AB, Farrer M, Johnson J, Singleton A, Hague S, Kachergus J, Hulihan M, Peuralinna T, Dutra A, Nussbaum R et al. (2003) alpha-Synuclein locus triplication causes Parkinson's disease. Science 302:841
19. Gai WP, Power JH, Blumbergs PC, Culvenor JG, Jensen PH (1999) Alpha-synuclein immunoisolation of glial inclusions from multiple system atrophy brain tissue reveals multiprotein components. J Neurochem 73:2093–2100
20. Spillantini MG, Crowther RA, Jakes R, Hasegawa M, Goedert M (1998) alpha-Synuclein in filamentous inclusions of Lewy bodies from Parkinson's disease and dementia with lewy bodies. Proc Natl Acad Sci USA 95:6469–6473
21. Conway KA, Lee SJ, Rochet JC, Ding TT, Harper JD, Williamson RE, Lansbury PT Jr. (2000) Accelerated oligomerization by Parkinson's disease linked alpha-synuclein mutants. Ann N Y Acad Sci 920:42–45
22. Greenbaum EA, Graves CL, Mishizen-Eberz AJ, Lupoli MA, Lynch DR, Englander SW, Axelsen PH, Giasson BI (2005) The E46K mutation in alpha-synuclein increases amyloid fibril formation. J Biol Chem 280:7800–7807
23. Narhi L, Wood SJ, Steavenson S, Jiang Y, Wu GM, Anafi D, Kaufman SA, Martin F, Sitney K, Denis P et al. (1999) Both familial Parkinson's disease mutations accelerate alpha-synuclein aggregation. J Biol Chem 274:9843–9846
24. Galpern WR, Lang AE (2006) Interface between tauopathies and synucleinopathies: a tale of two proteins. Ann Neurol 59:449–458
25. Chen L, Feany MB (2005) alpha-Synuclein phosphorylation controls neurotoxicity and inclusion formation in a *Drosophila* model of Parkinson disease. Nat Neurosci 8: 657–663
26. Fujiwara H, Hasegawa M, Dohmae N, Kawashima A, Masliah E, Goldberg MS, Shen J, Takio K, Iwatsubo T (2002) alpha-Synuclein is phosphorylated in synucleinopathy lesions. Nat Cell Biol 4:160–164
27. Hashimoto M, Hsu LJ, Xia Y, Takeda A, Sisk A, Sundsmo M, Masliah E (1999) Oxidative stress induces amyloid-like aggregate formation of NACP/alpha- synuclein in vitro. Neuroreport 10:717–721
28. McNaught KS, Olanow CW, Halliwell B, Isacson O, Jenner P (2001) Failure of the ubiquitin-proteasome system in Parkinson's disease. Nat Rev Neurosci 2:589–594
29. Buee L, Bussiere T, Buee-Scherrer V, Delacourte A, Hof PR (2000) Tau protein isoforms, phosphorylation and role in neurodegenerative disorders. Brain Res Rev 33: 95–130
30. Lindersson E, Lundvig D, Petersen C, Madsen P, Nyengaard JR, Hojrup P, Moos T, Otzen D, Gai WP, Blumbergs PC et al. (2005) p25alpha Stimulates alpha-synuclein aggregation and is co-localized with aggregated alpha-synuclein in alpha-synucleinopathies. J Biol Chem 280:5703–5715

31. Liu IH, Uversky VN, Munishkina LA, Fink AL, Halfter W, Cole GJ (2005) Agrin binds alpha-synuclein and modulates alpha-synuclein fibrillation. Glycobiology 15:1320–1331
32. Takeda A, Arai N, Komori T, Iseki E, Kato S, Oda M (1997) Tau immunoreactivity in glial cytoplasmic inclusions in multiple system atrophy. Neurosci Lett 234:63–66
33. Braak H, Braak E (2000) Pathoanatomy of Parkinson's disease. J Neurol 247 Suppl 2:II 3–10
34. Forno LS (1996) Neuropathology of Parkinson's disease. J Neuropathol Exp Neurol 55: 259–272
35. Braak H, Del Tredici K, Rub U, De Vos RAI, Steur ENHJ, Braak E (2003) Staging of brain pathology related to sporadic Parkinson's disease. Neurobiol Aging 24:197–211
36. Thomas B, Beal MF (2007) Parkinson's disease. Hum Mol Genet 16:R183–R194
37. Gasser T (2007) Update on the genetics of Parkinson's disease. Mov Disorders 22: S343–S350
38. Paisan-Ruiz C, Jain S, Evans EW, Gilks WP, Simon J, van der Brug M, de Munain AL, Aparicio S, Gil AM, Khan N et al. (2004) Cloning of the gene containing mutations that cause PARK8-linked Parkinson's disease. Neuron 44:595–600
39. Zimprich A, Biskup S, Leitner P, Lichtner P, Farrer M, Lincoln S, Kachergus J, Hulihan M, Uitti RJ, Calne DB et al. (2004) Mutations in LRRK2 cause autosomal-dominant Parkinsonism with pleomorphic pathology. Neuron 44:601–607
40. Hohl U, CoreyBloom J, Hansen LA, Thomas RG, Thal LJ (1997) Diagnostic accuracy of dementia with Lewy bodies: A prospective evaluation. Neurology 48:2032
41. Hansen LA, Samuel W (1997) Criteria for Alzheimer's disease and the nosology of dementia with Lewy bodies. Neurology 48:126–132
42. Burn DJ, Jaros E (2001) Multiple system atrophy: cellular and molecular pathology. J Clin Pathol Mol Pathol 54:419–426
43. Jellinger KA (2003) Neuropathological spectrum of synucleinopathies. Mov Disorders 18:S2–S12
44. Wakabayashi K, Takahashi H (2006) Cellular pathology in multiple system atrophy. Neuropathol 26:338–345
45. Nishie M, Mori F, Yoshimoto M, Takahashi H, Wakabayashi K (2004) A quantitative investigation of neuronal cytoplasmic and intranuclear inclusions in the pontine and inferior olivary nuclei in multiple system atrophy. Neuropathol App Neurobiol 30:546–554
46. Maroteaux L, Campanelli JT, Scheller RH (1988) Synuclein – A neuron-specific protein localized to the nucleus and presynaptic nerve-terminal. J Neurosci 8:2804–2815
47. Abeliovich A, Schmitz Y, Farinas I, Choi-Lundberg D, Ho WH, Castillo PE, Shinsky N, Verdugo JM, Armanini M, Ryan A et al. (2000) Mice lacking alpha-synuclein display functional deficits in the nigrostriatal dopamine system. Neuron 25:239–252
48. Clayton DF, George JM (1998) The synucleins: a family of proteins involved in synaptic function, plasticity, neurodegeneration and disease. Trends Neurosci 21:249–254
49. Weinreb PH, Zhen W, Poon AW, Conway KA, Lansbury PT Jr. (1996) NACP, a protein implicated in Alzheimer's disease and learning, is natively unfolded. Biochem 35: 13709–13715
50. Giasson BI, Uryu K, Trojanowski JQ, Lee VM (1999) Mutant and wild type human alpha-synucleins assemble into elongated filaments with distinct morphologies in vitro. J Biol Chem 274:7619–7622
51. Hashimoto M, Hsu LJ, Sisk A, Xia Y, Takeda A, Sundsmo M, Masliah E (1998) Human recombinant NACP/alpha-synuclein is aggregated and fibrillated in vitro: relevance for Lewy body disease. Brain Res 799:301–306
52. Conway KA, Lee SJ, Rochet JC, Ding TT, Williamson RE, Lansbury PT, Jr. (2000) Acceleration of oligomerization, not fibrillization, is a shared property of both alpha-synuclein mutations linked to early-onset Parkinson's disease: implications for pathogenesis and therapy. Proc Natl Acad Sci USA 97:571–576
53. Lashuel H, Petre B, Wall J, Simon M, Nowak R, Walz T, Lansbury P (2002) alpha-Synuclein, especially the Parkinson's disease-associated mutants, forms pore-like annular and tubular protofibrils. J Mol Biol 322:1089

54. Uversky VN (2007) Neuropathology, biochemistry, and biophysics of alpha-synuclein aggregation. J Neurochem 103:17–37
55. Volles MJ, Lansbury PT Jr. (2002) Vesicle permeabilization by protofibrillar alpha-synuclein is sensitive to Parkinson's disease-linked mutations and occurs by a pore-like mechanism. Biochem 41:4595–4602
56. Anderson JP, Walker DE, Goldstein JM, de Laat R, Banducci K, Caccavello RJ, Barbour R, Huang JP, Kling K, Lee M et al. (2006) Phosphorylation of Ser-129 is the dominant pathological modification of alpha-synuclein in familial and sporadic Lewy body disease. J Biol Chem 281:29739–29752
57. Nishie M, Mori F, Fujiwara H, Hasegawa M, Yoshimoto M, Iwatsubo T, Takahashi H, Wakabayashi K (2004) Accumulation of phosphorylated alpha-synuclein in the brain and peripheral ganglia of patients with multiple system atrophy. Acta Neuropathol 107: 292–298
58. Giasson BI, Duda JE, Murray IV, Chen Q, Souza JM, Hurtig HI, Ischiropoulos H, Trojanowski JQ, Lee VM (2000) Oxidative damage linked to neurodegeneration by selective alpha-synuclein nitration in synucleinopathy lesions. Science 290:985–989
59. Souza JM, Giasson BI, Chen QP, Lee VMY, Ischiropoulos H (2000) Dityrosine crosslinking promotes formation of stable alpha-synuclein polymers – Implication of nitrative and oxidative stress in the pathogenesis of neurodegenerative synucleinopathies. J Biol Chem 275:18344–18349
60. Conway KA, Rochet JC, Bieganski RM, Lansbury PT, Jr. (2001) Kinetic stabilization of the alpha-synuclein protofibril by a dopamine – alpha-synuclein adduct. Science 294: 1346–1349
61. Giasson BI, Forman MS, Higuchi M, Golbe LI, Graves CL, Kotzbauer PT, Trojanowski JQ, Lee VMY (2003) Initiation and synergistic fibrillization of tau and alpha-synuclein. Science 300:636–640
62. Dev KK, Hofele K, Barbieri S, Buchman VL, van der Putten H (2003) Part II: alpha-synuclein and its molecular pathophysiological role in neurodegenerative disease. Neuropharmacology 45:14–44
63. Takahashi M, Tomizawa K, Ishiguro K, Sato K, Omori A, Sato S, Shiratsuchi A, Uchida T, Imahori K (1991) A novel brain-specific 25 kDa protein (p25) is phosphorylated by a Ser Thr-Pro kinase (TPK-II) from tau protein-kinase fractions. FEBS Lett 289:37–43
64. Takahashi M, Tomizawa K, Fujita SC, Sato K, Uchida T, Imahori K (1993) A brain-specific protein p25 is localized and associated with oligodendrocytes, neuropil, and fiber-like structures of the Ca_3 hippocampal region in the rat-brain. J Neurochem 60:228–235
65. Kovacs GG, Gelpi E, Lehotzky A, Hoftberger R, Erdei A, Budka H, Ovádi J (2007) The brain-specific protein TPPP/p25 in pathological protein deposits of neurodegenerative diseases. Acta Neuropathol 113:153–161
66. Song YJC, Lundvig DMS, Huang Y, Gai WP, Blumbergs PC, Hojrup P, Otzen D, Halliday GM, Jensen PH (2007) P25 alpha relocalizes in oligodendroglia from myelin to cytoplasmic inclusions in multiple system atrophy. Am J Pathol 171:1291–1303
67. Skjoerringe T, Lundvig DMS, Jensen PH, Moos T (2006) P25 alpha/tubulin polymerization promoting protein expression by myelinating oligodendrocytes of the developing rat brain. J Neurochem 99:333–342
68. Otzen DE, Lundvig DMS, Wimmer R, Nielsen LH, Pedersen JR, Jensen PH (2005) p25 alpha is flexible but natively folded and binds tubulin with oligomeric stoichiometry. Prot Sci 14:1396–1409
69. Seki N, Hattori A, Sugano S, Suzuki Y, Nakagawara A, Muramatsu M, Hori T, Saito T (1999) A novel human gene whose product shares significant homology with the bovine brain-specific protein p25 on chromosome 5p15.3. J Hum Genet 44:121–122
70. Zhang Z, Wu CQ, Huang W, Wang S, Zhao EP, Huang QS, Xie Y, Mao YM (2002) A novel human gene whose product shares homology with bovine brain-specific protein p25 is expressed in fetal brain but not in adult brain. J Hum Genet 47:266–268

71. Hlavanda E, Kovács J, Oláh J, Orosz F, Medzihradszky KF, Ovádi J (2002) Brain-specific p25 protein binds to tubulin and microtubules and induces aberrant microtubule assemblies at substoichiometric concentrations. Biochem 41:8657–8664
72. Lehotzky A, Tirián L, Tökési N, Lénárt P, Szabó B, Kovács J, Ovádi J (2004) Dynamic targeting of microtubules by TPPP/p25 affects cell survival. J Cell Sci 117:6249–6259
73. Tirián L, Hlavanda E, Oláh J, Horváth I, Orosz F, Szabó B, Kovács J, Szabad J, Ovádi J (2003) TPPP/p25 promotes tubulin assemblies and blocks mitotic spindle formation. Proc. Natl. Acad. Sci USA 100:13976–13981.
74. Sirover MA (2005) New nuclear functions of the glycolytic protein, glyceraldehyde-3-phosphate dehydrogenase, in mammalian cells. J Cell Biochem 95:45–52
75. Chuang DM, Hough C, Senatorov VV (2005) Glyceraldehyde-3-phosphate dehydrogenase, apoptosis and neurodegenerative diseases. Annu Rev Pharmacol Toxicol 45:269–290
76. Oláh J, Tökési N, Vincze O, Horváth I, Lehotzky A, Erdei A, Szájli E, Katalin FM, Orosz F, Kovacs GG et al. (2006) Interaction of TPPP/p25 protein with glyceraldehyde-3-phosphate dehydrogenase and their co-localization in Lewy bodies. FEBS Lett 580:5807–5814
77. Nelson TJ, Backlund PS, Alkon DL (2004) Hippocampal protein-protein interactions in spatial memory. Hippocampus 14:46–57
78. Mcmahon HT, Missler M, Li C, Sudhof TC (1995) Complexins – Cytosolic Proteins That Regulate Snap Receptor Function. Cell 83:111–119
79. Hlavanda E, Klement E, Kókai E, Vincze O, Tökési N, Orosz F, Medzihradszky KF, Dombrádi V, Ovádi J (2007) Phosphorylation blocks the activity of tubulin polymerization-promoting protein (TPPP) – Identification of sites targeted by different kinases. J Biol Chem 282:29531–29539
80. Martin CP, Vazquez J, Avila J, Moreno FJ (2002) P24, a glycogen synthase kinase 3 (GSK 3) inhibitor. Biochim Biophys Acta-Mol Basis Dis 1586:113–122
81. Acevedo K, Li R, Soo P, Suryadinata R, Sarcevic B, Valova VA, Graham ME, Robinson PJ, Bernard O (2007) The phosphorylation of p25/TPPP by LIM kinase 1 inhibits its ability to assemble microtubules. Exp Cell Res 313:4091–4106
82. Yokozeki T, Homma K, Kuroda S, Kikkawa U, Ohno S, Takahashi M, Imahori K, Kanaho Y (1998) Phosphatidic acid-dependent phosphorylation of a 29-kDa protein by protein kinase C alpha in bovine brain cytosol. J Neurochem 71:410–417
83. Baker KG, Huang Y, McCann H, Gai WP, Jensen PH, Halliday GM (2006) P25 alpha immunoreactive but alpha-synuclein immunonegative neuronal inclusions in multiple system atrophy. Acta Neuropathol 111:193–195
84. Kovacs GG, László L, Kovács J, Jensen PH, Lindersson E, Botond G, Molnár T, Perczel A, Hudecz F, Mezö G et al. (2004) Natively unfolded tubulin polymerization promoting protein TPPP/p25 is a common marker of alpha-synucleinopathies. Neurobiol Dis 17:155–162
85. Jellinger KA (2006) P25 alpha immunoreactivity in multiple system atrophy and Parkinson disease. Acta Neuropathol 112:112
86. Goedert M, Spillantini MG, Jakes R, Rutherford D, Crowther RA (1989) Multiple isoforms of human microtubule-associated protein tau – sequences and localization in neurofibrillary tangles of Alzheimer's disease. Neuron 3:519–526
87. Goedert M, Spillantini MG, Potier MC, Ulrich J, Crowther RA (1989) Cloning and sequencing of the cDNA-encoding an isoform of microtubule-associated protein tau containing 4 tandem repeats: differential expression of tau protein messenger-RNAs in human-brain. EMBO J 8:393–399
88. Forman MS, Lee VMY, Trojanowski JQ (2000) New insights into genetic and molecular mechanisms of brain degeneration in tauopathies. J Chem Neuroanat 20:225–244
89. Litvan I (2003) Update on epidemiological aspects of progressive supranuclear palsy. Mov Disorders 18:S43–S50
90. Dickson DW, Bergeron C, Chin SS, Duyckaerts C, Horoupian D, Ikeda K, Jellinger K, Lantos PL, Lippa CF, Mirra SS et al. (2002) Office of rare diseases neuropathologic criteria for corticobasal degeneration. J Neuropathol Exp Neurol 61:935–946

91. Hutton M, Lendon CL, Rizzu P, Baker M, Froelich S, Houlden H, Pickering-Brown S, Chakraverty S, Isaacs A, Grover A et al. (1998) Association of missense and 5'-splice-site mutations in tau with the inherited dementia FTDP-17. Nature 393:702–705
92. Giasson BI, Lee VMY, Trojanowski JQ (2003) Interactions of amyloidogenic proteins. Neuromol Med 4:49–58
93. Conway KA, Harper JD, Lansbury PT (1998) Accelerated in vitro fibril formation by a mutant alpha-synuclein linked to early-onset Parkinson disease. Nat Med 4:1318–1320
94. Goedert M, Jakes R, Spillantini MG, Hasegawa M, Smith MJ, Crowther RA (1996) Assembly of microtubule-associated protein tau into Alzheimer-like filaments induced by sulphated glycosaminoglycans. Nature 383:550–553
95. Horiguchi T, Uryu K, Giasson BI, Ischiropoulos H, LightFoot R, Bellmann C, Richter-Landsberg C, Lee VMY, Trojanowski JQ (2003) Nitration of tau protein is linked to neurodegeneration in tauopathies. Am J Pathol 163:1021–1031
96. Hasegawa M, Fujiwara H, Nonaka T, Wakabayashi K, Takahashi H, Lee VMY, Trojanowski JQ, Mann D, Iwatsubo T (2002) Phosphorylated alpha-synuclein is ubiquitinated in alpha-synucleinopathy lesions. J Biol Chem 277:49071–49076
97. Lee VMY, Giasson BI, Trojanowski JQ (2004) More than just two peas in a pod: common amyloidogenic properties of tau and alpha-synuclein in neurodegenerative diseases. Trends Neurosci 27:129–134
98. Engelender S, Kaminsky Z, Guo X, Sharp AH, Amaravi RK, Kleiderlein JJ, Margolis RL, Troncoso JC, Lanahan AA, Worley PF et al. (1999) Synphilin-1 associates with alpha-synuclein and promotes the formation of cytosolic inclusions. Nat Genet 22:110–114
99. Kawamata H, Mclean PJ, Sharma N, Hyman BT (2001) Interaction of alpha-synuclein and synphilin-1: effect of Parkinson's disease-associated mutations. J Neurochem 77:929–934
100. Engelender S, Wakabayashi K, Wanner T, Kaminsky Z, Kleiderlein JJ, Margolis RL, Tsuji S, Takahashi H, Ross CA (1999) The alpha-synuclein-associated protein, synphilin-1: Gene structure and localization, and presence of synphilin-1 protein in Lewy bodies. Am J Hum Genet 65:A270
101. Ribeiro CS, Carneiro K, Ross CA, Menezes JRL, Engelender S (2002) Synphilin-1 is developmentally localized to synaptic terminals, and its association with synaptic vesicles is modulated by alpha-synuclein. J Biol Chem 277:23927–23933
102. Smith WW, Margolis RL, Li XJ, Troncoso JC, Lee MK, Dawson VL, Dawson TM, Iwatsubo T, Ross CA (2005) alpha-Synuclein phosphorylation enhances eosinophilic cytoplasmic inclusion formation in SH-SY5Y cells. J Neurosci 25:5544–5552
103. Chung KK, Zhang Y, Lim KL, Tanaka Y, Huang H, Gao J, Ross CA, Dawson VL, Dawson TM (2001) Parkin ubiquitinates the alpha-synuclein-interacting protein, synphilin-1: implications for Lewy-body formation in Parkinson disease. Nat Med 7:1144–1150
104. Ito T, Niwa J, Hishikawa N, Ishigaki S, Doyu M, Sobue G (2003) Dorfin localizes to Lewy bodies and ubiquitylates synphilin-1. J Biol Chem 278:29106–29114
105. Nagano Y, Yamashita H, Takahashi T, Kishida S, Nakamura T, Iseki E, Hattori N, Mizuno Y, Kikuchi A, Matsumoto M (2003) Siah-1 facilitates ubiquitination and degradation of synphilin-1. J Biol Chem 278:51504–51514
106. Liani E, Eyal A, Avraham E, Shemer R, Szargel R, Berg D, Bornemann A, Riess O, Ross CA, Rott R et al. (2004) Ubiquitylation of synphilin-1 and alpha-synuclein by SIAH and its presence in cellular inclusions and Lewy bodies imply a role in Parkinson's disease. Proc Natl Acad Sci USA 101:5500–5505
107. Wakabayashi K, Engelender S, Yoshimoto M, Tsuji S, Ross CA, Takahashi H (2000) Synphilin-1 is present in Lewy bodies in Parkinson's disease. Ann Neurol 47:521–523
108. Wakabayashi K, Engelender S, Tanaka Y, Yoshimoto M, Mori F, Tsuji S, Ross CA, Takahashi H (2002) Immunocytochemical localization of synphilin-1, an alpha-synuclein-associated protein, in neurodegenerative disorders. Acta Neuropathol 103:209–214
109. Eyal A, Szargel R, Avraham E, Liani E, Haskin J, Rott R, Engelender S: Synphilin-1A (2006) An aggregation-prone isoform of synphilin-1 that causes neuronal death and is

present in aggregates from alpha-synucleinopathy patients. Proc Natl Acad Sci USA 103: 5917–5922
110. Neumann M, Sampathu DM, Kwong LK, Truax AC, Micsenyi MC, Chou TT, Bruce J, Schuck T, Grossman M, Clark CM et al. (2006) Ubiquitinated TDP-43 in frontotemporal lobar degeneration and amyotrophic lateral sclerosis. Science 314:130–133
111. Amador-Ortiz C, Lin WL, Ahmed Z, Personett D, Davies P, Dara R, Graff-Radford NR, Hutton ML, Dickson DW (2007) TDP-43 immunoreactivity in hippocampal sclerosis and Alzheimer's disease. Ann Neurol 61:435–445
112. Higashi S, Iseki E, Yamamoto R, Minegishi M, Hino H, Fujisawa K, Togo T, Katsuse O, Uchikado H, Furukawa Y et al. (2007) Concurrence of TDP-43, tau and alpha-synuclein pathology in brains of Alzheimer's disease and dementia with Lewy bodies. Brain Res 1184:284–294
113. MacLeod D, Dowman J, Hammond R, Leete T, Inoue K, Abeliovich A (2006) The familial parkinsonism gene LRRK2 regulates neurite process morphology. Neuron 52:587–593
114. Perry G, Zhu X, Babar AK, Siedlak SL, Yang Q, Ito G, Iwatsubo T, Smith MA, Chen SG (2008) Leucine-rich repeat kinase 2 colocalizes with alpha-synuclein in parkinson's disease, but not tau-containing deposits in tauopathies. Neurodegener Dis 5, 222–224
115. Gerard M, Debyser Z, Kahle PJ, Baekeland V, Engelborghs Y (2006) The aggregation of alpha-synuclein is stimulated by FK506 binding proteins as shown by fluorescence correlation spectroscopy. FASEB J 20:A954
116. Galat A (2003) Peptidylprolyl cis/trans isomerases (immunophilins): biological diversity – targets – functions. Curr Top Med Chem 3:1315–1347
117. Lyons WE, George EB, Dawson TM, Steiner JP, Snyder SH (1994) Immunosuppressant FK506 promotes neurite outgrowth in cultures of PC12 cells and sensory ganglia. Proc Natl Acad Sci USA 91:3191–3195
118. Gold BG, Katoh K, Stormdickerson T (1995) The immunosuppressant FK506 increases the rate of axonal regeneration in rat sciatic-nerve. J Neurosci 15:7509–7516
119. Costantini LC, Chaturvedi P, Armistead DM, McCaffrey PG, Deacon TW, Isacson O (1998) A novel immunophilin ligand: distinct branching effects on dopaminergic neurons in culture and neurotrophic actions after oral administration in an animal model of Parkinson's disease. Exp Neurol 153:382
120. Avramut M, Achim CL (2002) Immunophilins and their ligands: insights into survival and growth of human neurons. Physiol Behav 77:463–468
121. Masliah E, Rockenstein E, Veinbergs I, Mallory M, Hashimoto M, Takeda A, Sagara Y, Sisk A, Mucke L (2000) Dopaminergic loss and inclusion body formation in alpha-synuclein mice: Implications for neurodegenerative disorders. Science 287:1265–1269
122. Manning-Bog AB, McCormack AL, Li J, Uversky VN, Fink AL, Di Monte DA (2002) The herbicide paraquat causes up-regulation and aggregation of alpha-synuclein in mice – Paraquat and alpha-synuclein. J Biol Chem 277:1641–1644
123. Goers J, Manning-Bog AB, McCormack AL, Millett IS, Doniach S, Di Monte DA, Uversky VN, Fink AL (2003) Nuclear localization of alpha-synuclein and its interaction with histones. Biochemistry 42:8465–8471
124. Kontopoulos E, Parvin JD, Feany MB (2006) Alpha-synuclein acts in the nucleus to inhibit histone acetylation and promote neurotoxicity. Hum Mol Genet 15:3012–3023
125. Outeiro TF, Kontopoulos E, Altmann SM, Kufareva I, Strathearn KE, Amore AM, Volk CB, Maxwell MM, Rochet JC, McLean PJ et al. (2007) Sirtuin 2 inhibitors rescue alpha-synuclein-mediated toxicity in models of Parkinson's disease. Science 317: 516–519
126. Lindersson EK, Hojrup P, Gai WP, Locker D, Martin D, Jensen PH (2004) alpha-Synuclein filaments bind the transcriptional regulator HMGB-1. Neuroreport 15:2735–2739
127. Alim MA, Hossain MS, Arima K, Takeda K, Izumiyama Y, Nakamura M, Kaji H, Shinoda T, Hisanaga S, Ueda K (2002) Tubulin seeds alpha-synuclein fibril formation. J Biol Chem 277:2112–2117

128. D'Andrea MR, Ilyin S, Plata-Salaman CR (2001) Abnormal patterns of microtubule-associated protein-2 (MAP-2) immunolabeling in neuronal nuclei and Lewy bodies in Parkinson's disease substantia nigra brain tissues. Neurosci Lett 306:137–140
129. Jensen PH, Hager H, Nielsen MS, Hojrup P, Gliemann J, Jakes R (1999) alpha-synuclein binds to Tau and stimulates the protein kinase A- catalyzed tau phosphorylation of serine residues 262 and 356. J Biol Chem 274:25481–25489
130. Jensen PH, Islam K, Kenney J, Nielsen MS, Power J, Gai WP (2000) Microtubule-associated protein 1B is a component of cortical Lewy bodies and binds alpha-synuclein filaments. J Biol Chem 275:21500–21507
131. Payton JE, Perrin RJ, Clayton DF, George JM (2001) Protein-protein interactions of alpha-synuclein in brain homogenates and transfected cells. Brain Res Mol Brain Res 95:138–145
132. Sharma N, Hewett J, Ozelius LJ, Ramesh V, McLean PJ, Breakefield XO, Hyman BT (2001) A close association of torsinA and alpha-synuclein in Lewy bodies – A fluorescence resonance energy transfer study. Am J Pathol 159:339–344
133. Lotharius J, Brundin P (2002) Impaired dopamine storage resulting from alpha-synuclein mutations may contribute to the pathogenesis of Parkinson's disease. Hum Mol Genet 11:2395–2407
134. Sidhu A, Wersinger C, Vernier P (2004) Does alpha-synuclein modulate dopaminergic synaptic content and tone at the synapse? FASEB J 18:637–647
135. Donahue JE, Berzin TM, Rafii MS, Glass DJ, Yancopoulos GD, Fallon JR, Stopa EG: Agrin in Alzheimer's disease (1999) Altered solubility and abnormal distribution within microvasculature and brain parenchyma. J Neuropathol Exp Neurol 58:534
136. Verbeek MM, Otte-Holler I, van den Born J, van den Heuvel LPWJ, David G, Wesseling P, de Waal RMW (1999) Agrin is a major heparan sulfate proteoglycan accumulating in Alzheimer's disease brain. Am J Pathol 155:2115–2125

Chapter 10
TPPP/p25: A New Unstructured Protein Hallmarking Synucleinopathies

Ferenc Orosz, Attila Lehotzky, Judit Oláh and Judit Ovádi

Abstract There is increasing evidence that unfolded and misfolded proteins initiate a cascade of pathogenic protein-protein interactions that culminate in neuronal dysfunction. This is a multistep process which results in toxic protein aggregates; thus they are potent targets for development of early diagnosis and of drugs to improve therapies of conformational diseases. The hallmark proteins of these diseases such as Parkinson's, Alzheimer's or Huntington's diseases, are α-synuclein, tau or mutant huntingtin, respectively, which do not have well-defined 3D structures and require protein partners to express their pathological functions. In this paper we review a new unstructured protein denoted Tubulin Polymerization Promoting Protein, TPPP/p25, from the discovery to its enrichment in human pathological inclusions characteristic for synucleinopathies with specific emphasis on its pursuits in single cells. There is a gappy area in the research of unfolded proteins referring to their structure-derived physiological and pathological functions. The studies of TPPP-homologous proteins at different levels of organization, molecular, cellular and tissue levels, rendered possible to reveal some TPPP/p25 specific structural and functional features, in addition to the general items for the role of the unfolded regions of the highly flexible proteins in their physiological and/or pathological functions.

10.1 Occurrence of TPPP Proteins

We isolated and identified a "new" protein which significantly enhanced the tubulin polymerization rate in the presence of brain but not in the presence of muscle extract [1]. We noticed by screening the protein databases that this protein, p25, had already been found as a contaminant of a tau kinase (cyclin-dependent kinase-5, Cdk5) fraction by a Japanese group [2]; and its complete sequence was determined

---

F. Orosz (✉)
Institute of Enzymology, Biological Research Center, Hungarian Academy of Sciences, Karolina út 29, Budapest, H-1113 Hungary
e-mail: orosz@enzim.hu

from bovine brain [3] and human neuroblastoma [4] cDNA libraries. As we have pointed out [5], TPPP/p25 differs completely from the extensively characterized protein p25, which is a truncated form of p35 that regulates Cdk5 activity by causing prolonged activation and mislocalization of the kinase [6]. To avoid further confusion about the name of this protein, we suggest using the term TPPP (**T**ubulin **P**olymerization **P**romoting **P**rotein), which highlights its characteristic action on tubulin [1, 5].

TPPP/p25 is the first member of a new protein family, the primary sequence of which differs from that of other known proteins. However, at **gene level** two paralogous human genes, p25beta and CGI-38, were identified that we termed *TPPP2* and *TPPP3*. The genes of *TPPP, TPPP2* and *TPPP3* included into the HUGO (http://www.genenames.org/data/hgnc_data.php?hgnc_id=24164; 24162; 19293) are located on the 5th, 14th and 16th chromosomes, respectively [4, 7]. BLAST analysis showed that orthologous TPPP genes can be found throughout the animal kingdom (and in the green algae) but not in prokaryotes, land plants, or fungi [8].

Human *TPPP2* was found at mRNA level as p25beta in fetal brain and cloned by Zhang et al. [7]. The transcript was highly expressed in liver and pancreas, and had a moderate expression level in heart, skeletal muscle and kidney, and it was not detected in lung, placenta or brain except in fetal brain. There is no evidence at present for the expression of the gene product, TPPP2/p18, at protein level. The *TPPP3* gene was found by comparative genomics searching for novel human genes evolutionarily conserved in *Caenorhabditis elegans* as CGI-38 [9]. Its ortholog was also identified in the mouse transcriptome (RIKEN cDNA 2700055K07) [10]. We provided unambiguous evidence that the corresponding protein, TPPP3/p20, is expressed in bovine brain, by its isolation from this tissue [8]. Its physiological function is unknown yet. The (hypothetic) human homologous gene products are shorter than TPPP/p25 and display 60% identity with it and each other.

In a recent study based upon homologous gene sequence analysis, in silico comparative genomic studies, and bioinformatics search, TPPP/p25 has been proposed to be a ciliary protein (Orosz unpublished data). Cilia and flagella, which are microtubule-filled, cellular extensions, are present on almost all human cells including brain neurons. We do not know the direct role of TPPP/p25 in the multiple functions of cilia, but it is likely that there is an intimate relationship between TPPP/p25 and cilia function. In these complex processes the TPPP/p25 may play an important role via its microtubule-stabilizing capability.

10.2 Unfolded Structural Features of TPPP Proteins

10.2.1 Prediction

The term "intrinsically unstructured" (IUPs) or "intrinsically disordered" designates proteins or protein domains that do not have well-defined 3D structure *in vitro* and under physiological conditions in absence of a binding partner [11–13]. They are flexible, display some "residual structure" [12] in the form of transient secondary structures, rather α-helix than β-sheet [14]. However, their folding can be promoted by their associations to target proteins, chaperons, lipid membranes, or by chemical agents (i.e. trifluoroethanol, TFE). There are common sequence features of the proteins belonging to this family, which make them distinct from globular proteins. (There are two excellent chapters (Sect. 1 and 2) in this issue highly related to the structural prediction of IUPs). For example, they are generally characterized by low hydrophobicity and by high net charge. They are generally depleted in the so-called "order-promoting" amino acids: W, C, F, I, Y, L, N, V and enriched in "disorder-promoting" amino acids: Q, S, P, E, K, A, R, G ([11, 13] and Sect. 2). Here we explore how the properties of TPPP/p25 correspond to these criteria in comparison with its paralogs. The low aromaticity (5.5%) and high pI value (9.5) of TPPP/p25 indicate its disorder. Figure 10.1 shows that TPPP/p25 is depleted in all "order-promoting" amino acids without exception, while in the case of the two shorter homologues this tendency is weaker but still exists. On the contrary, TPPP/p25 is enriched in "disorder-promoting" residues, except glutamine and proline. These two amino acids, in addition to alanine and arginine, are underrepresented in TPPP3/p20 and TPPP2/p18 as well. These data suggest that TPPP/p25 is an unstructured IUP, while the shorter homologues, especially TPPP2/p18 are much less unfolded.

Numerous disorder predictors were developed to search naturally disordered regions of proteins as PONDR®, a neural network-based algorithm [15, 16]; DisoPred2 [17], which based on a support vector machine algorithm trained on a specific

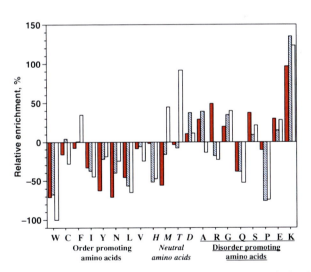

Fig. 10.1 Amino acid composition profiles of human TPPP proteins: deviation from globular proteins. The composition of ordered proteins was taken from [13]. Calculations were done as described in Sect. 2.3. Enrichment or depletion relative to globular proteins in each amino acid type appears as a positive or negative bar, respectively. TPPP/p25, TPPP/p20 and TPPP/p18 are represented by black, striped and empty bars, respectively

dataset of high resolution X-ray structures; IUPred [18, 19], predicting unstructured regions from amino acid sequences by estimating their total pair-wise inter-residue interaction energy; and others. These methods render possible prediction of long disordered and ordered regions (>30–40 amino acids) of proteins with good confidence. Using these predictors, we found that TPPP/p25 is supposed to contain one long disordered region of about 50 amino acids, namely the N-terminal part (Fig. 10.2). It is in accordance with the fact that about 80% of its amino acid residues belongs to the disorder-promoting ones. The majority of proline and alanine residues are located in this part of the protein. In the middle part of the protein an ordered region of similar length was predicted. The C-terminal part contains a pattern of alternating ordered and disordered short sections, however, the various predictors did not give unambiguous result for this part of the protein, except for a short (11 amino acid) disordered sequence. The overall percent of disordered residues was 43–46% depending on the predictor used. We also used a neural network based program, SCPRED, for prediction of amino acid residues involved in strong long-range interactions, denoted as "stabilization centers" [21]. The program can help in prediction of the folding class of a protein with unknown structure. We found practically no stabilization centers in the first 77 amino acids of TPPP/p25 (Fig. 10.2). Stabilization center residues are supposed to be responsible for the prevention of the decay of the folded structure thus lack of them can be indicative for the unfolded structure of the N-terminal end of TPPP/p25 [22]. On the contrary, most of the stabilization centers were predicted to be in the middle, ordered region. Interestingly, even the last 13 amino acids of this 50 amino acid ordered region lack stabilization centers, which may be indicative for the decreased stability of the secondary structures if they exist.

We have predicted the disordered and ordered regions of the homologues by this approach [8] (cf. Fig. 10.3). The disordered regions of TPPP/p25 appear to be the most extended as compared to those of the two shorter forms. Quantitative evaluation of the prediction data unambiguously indicated the distinct conformational state of TPPPs even if the disordered N-terminal segment of TPPP/p25 is disregarded. The overall percent of disorder is the highest for TPPP/p25 and the least for TPPP2/p18; this establishment is valid when the predicted disorder values of the N-terminal-free TPPP/p25 and the two homologues are compared in spite of their high sequence homologies. This amazing throughput was supported by experimental data.

10.2.2 Experimental

While the prediction methods used for unstructured proteins supply information for the unfolded regions of the proteins, the biochemical and physico-chemical methods can provide qualitative, sometimes quantitative data about their secondary structures such as α-helix, β-sheet or random coil reinforcing the lack of the well-defined tertiary structure. Extensive structural studies on TPPP/p25 protein in our laboratory showed that TPPP/p25 may have some secondary structural elements but not 3D

	1	M	A	D	K	A	K	P	A	K	A	A	N	R	T	P	P	K	S	P	G	D	P	S	K	D	R	A	A	K	R	L	S	E	G	A	G	E	G	A	A	A	S	P	E	L	S	A	L	E	E	A
PONDR®		*	*	*	*	*	*	*	*	*	*	*	*	*	*	*	*	*	*	*	*	*	*	*	*	*	*	*	*	*	*	*	*	*	*	*	*	*	*	*	*	*	*	*	*	*	*	*	*	*	*	
DisoPred2		*	*	*	*	*	*	*	*	*	*	*	*	*	*	*	*	*	*	*	*	*	*	*	*	*	*	*	*	*	*	*	*	*	*	*	*	*	*	*	*	*	*	*	*	*	*	*	*	*	*	*
IUPred		*	*	*	*	*	*	*	*	*	*	*	*	*	*	*	*	*	*	*	*	*	*	*	*	*	*	*	*	*	*	*	*	*	*	*	*	*	*	*	*	*	*	*	*	*	*	*	*	*	*	
SCPRED		*	*	*	*	*	*	*	*	*	*	*	*	*	*	*	*	*	*	*	*	*	*	*	*	*	*	*	*	*	*	*	*	*	*	*	*	*	*	*	*	*	*	*	*	*	*	*	*			

	55	F	R	R	F	A	V	H	G	D	A	R	A	T	G	R	E	M	H	G	K	N	W	S	K	L	C	K	D	C	Q	V	I	D	G	R	N	V	T	V	T	D	V	D	I	V	F	S	K	I	K	G	K	S	C	R
PONDR®						*																																																		
DisoPred2													*	*																																										
IUPred																																																								
SCPRED																																																								

	110	T	I	T	F	E	Q	F	Q	E	A	L	E	E	L	A	K	K	R	F	K	D	K	S	S	E	E	A	V	R	E	V	H	R	L	I	E	G	K	A	P	I	I	S	G	V	T	K	A	I	S	P	T	V	S
PONDR®																					*	*	*	*	*	*	*		*	*	*	*	*	*	*							*	*	*	*	*								
DisoPred2																																																							
IUPred																		*	*	*	*	*	*	*	*	*	*																												
SCPRED																																																							

	165	R	L	T	D	T	T	K	F	T	G	S	H	K	E	R	F	D	P	S	G	K	G	K	G	K	A	G	R	V	D	L	V	D	E	S	G	Y	V	S	G	Y	K	H	A	G	T	Y	D	Q	K	V	Q	G	K
PONDR®		*	*	*	*	*	*	*	*	*	*	*	*	*	*	*	*	*	*	*	*	*	*	*	*	*	*	*	*	*	*	*	*	*	*	*	*	*	*	*	*	*	*	*	*	*	*	*	*	*	*	*	*	*	*	*
DisoPred2																*	*	*	*	*	*	*	*	*	*	*	*																						*	*	*	*	*	*	*
IUPred		*	*	*	*	*	*																																																	
SCPRED																																																							

Fig. 10.2 Prediction of structural order/disorder of TPPP/p25 using various disorder predictors. In the case of disorder predictors PONDR® [15], DisoPred2 [17] and IUPred [19] the asterisks note disordered amino acid residues. In the case of SCPRED [21] the asterisks label the lack of stabilization centers. Black, dark gray and light gray background indicate if the amino acid residue was predicted to be disordered by 3, 2 or 1 disorder predictor, respectively. In the case of PONDR® and IUPred the threshold, above which residues are considered to be disordered, was 0.5. In DisoPred2 the false positive rate threshold was set at 5%, from which follows that at least 57% of the disordered residues was recovered [17]. Amino acids from 55 to 219 represent the Pfam05517 domain, which is common in TPPP proteins

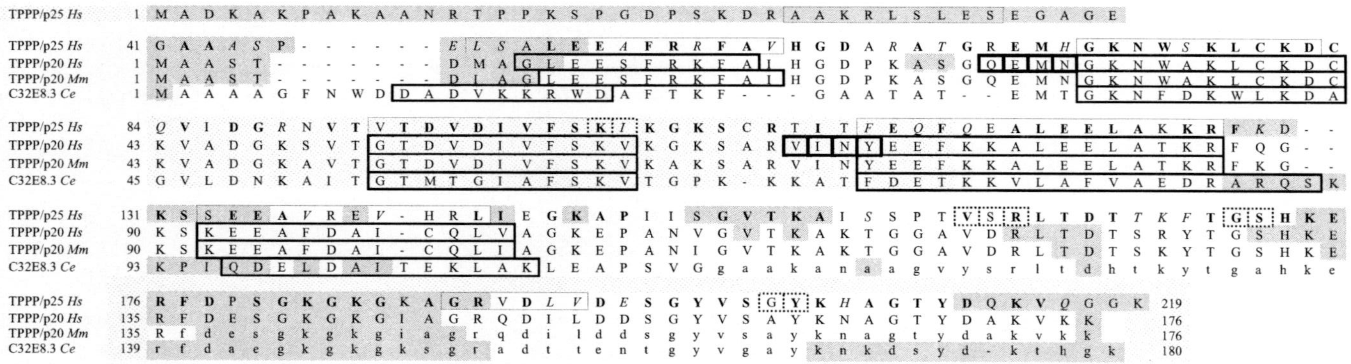

Fig. 10.3 Multiple sequence alignment and detected and predicted structural features of some TPPP proteins. Grey background indicates disordered amino acid residues predicted by at least two disorder predictors. Amino acids in rectangles and squares label α-helices and β-sheets, respectively, determined by NMR spectroscopy [25, 27, 28]. Amino acids in rectangles and squares bordered by dotted lines label tentative α-helices and β-sheets, respectively, in human TPPP/p25 predicted by using PSIPRED [30]. In the sequence of human TPPP/p25 amino acid residues identical with and similar to those of the TPPP/p20 proteins are labeled by bold and italic letters. In the sequences of TPPP/p20 *Mm* and C32E8.3 *Ce* the small letters indicate amino acids which were absent in these proteins when the NMR structures were determined. *Hs, Homo sapiens, Mm, Mus musculus, Ce, Caenorhabditis elegans*

structure [1, 8, 22], therefore it belongs to the IUPs. It is important to emphasize that IUPs may have transient structural elements due to their flexible character; however, they bear at least 30–40 amino acid segment without any structure [12]. The unfolded character of TPPP/p25 was found to be more obvious when its features were compared to those of its homologues [8].

Limited proteolysis is a sensitive method to test the structural integrity of proteins. Our data using trypsin IV, a human brain-specific protease, and chymotrypsin showed that the TPPP proteins display distinct resistance against the proteolytic digestions: while TPPP/p25 was immediately digested into small fragments, its homologues display less vulnerability toward both proteases; in the case of TPPP2/p18 significant amount of intact protein was detected even after 4 h of digestion [8]. These data reveal that the structural integrity of TPPP/p25 (if there is any) is much lower than that of the shorter homologues, especially that of TPPP2/p18, and that the lack of integrity is not restricted to the N-terminal segment of TPPP/p25 but protease-sensitive bonds are exposed in additional parts of the protein as well.

The unfolding of proteins by elevation of the temperature is accompanied by a positive heat capacity change, which can be quantified by differential scanning calorimetry. Concerning the features of the homologues, there are significant differences in their molar heat capacity functions. The transition temperature was the lowest (56.4°C) for TPPP/p25, then for TPPP3/p20 (59.7°C); both values are significantly lower as compared to that for TPPP2/p18 (65.1°C), which was still lower than for the control, lysozyme (72.1°C). The calorimetric enthalpy change is associated with the disruption of intramolecular interactions and the concomitant formation of interactions between water and unburied groups because of the unfolding process. The calorimetric enthalpy change values of all TPPPs evaluated were much lower than that for lysozyme. The values for TPPP/p25 and TPPP3/p20 were similar to each other but lower than that of p18. Significant temperature-induced structural alterations occurred only in TPPP2/p18. These data indicate that the stabilizing elements occur at a very low level in the cases of the two former homologues [8].

The circular dichroism (CD) spectra of the human recombinant homologues were measured in the far-UV range to obtain comparative data for the secondary structures of TPPPs [8]. The spectrum of TPPP/p25 with a minimum around 205 nm was described first as „a not typical for globular proteins" [1]. This first report was followed by several studies in our and other laboratories using a couple of additional biophysical methods such as fluorescence spectroscopy, gelfiltration or proton NMR [22–24]. All these investigations show that TPPP/p25 is a flexible unstructured protein which has some secondary structure; no data is available at present that this protein is folded forming 3D structure under physiological conditions. In contrast to TPPP/p25, we reported that TPPP2/p18, rather than TPPP3/p20, is significantly less unfolded; the former may have 3D structure [8]. The structure of TPPP3/p20 determined by multinuclear NMR is now available [25] which help us to envisage the "structure" of the more unfolded TPPP/p25. According to these data this homologue has a core with five α-helices and two β-sheets and about half of this molecule has no structural elements in spite of the fact that the N-terminal segment existing in TPPP/p25 is missing from TPPP3/p20. The N-terminal part of TPPP/p25 is of special interest since this part was predicted unambiguously to be unstructured (1–47 amino acid segment showed disorder values above 0.8 by PONDR® prediction) [22]. The small difference (0.48 kcal/mol) in structural stability between TPPP/p25 and its truncated form, TPPP/p25Δ3–43, [24] supports the view that the deleted N-terminal region is largely unstructured since it would not contribute to the stabilization of the protein by, e.g., side-chain contacts and docking of secondary structure elements.

α-helix forming potential of TPPP/p25 was revealed [22] by the use of TFE, which mimics the hydrophobic environment experienced by proteins in protein–protein interactions and is therefore widely used as a probe to discover regions that have a propensity to undergo an induced folding [26]. CD spectrum of TPPP/p25 showed an increased α-helicity (from 4 to 43%) upon the addition of TFE [22], which is comparable with the α-helix content predicted by various methods (30–43%) [1].

In the light of all these results it is not surprising that the crystallization of TPPP/p25 for X-ray analysis has been unsuccessful to date. This finding could be interpreted by the prediction data. TPPP proteins, especially TPPP/p25, are depleted in bulky hydrophobic (Ile, Leu, and Val) and aromatic amino acid residues (Trp, Tyr, and Phe), which would normally form the hydrophobic core of a folded globular protein. NMR studies of the mouse and human recombinant TPPP3/p20 homologues and the *Caenorhabditis elegans* TPPP homologue suggested a central core with secondary structural elements flanked by unfolded short N- and long C-terminal domains [25, 27, 28]. Comparing the sequence of these proteins we can conclude that their primary structures seem to be less conserved than the secondary ones. The secondary structural elements fell into the predicted central ordered region of these proteins except some amino acids of the fifth helix in the *C. elegans* ortholog (Fig. 10.3). Moreover, the five helices correspond to the helices predicted by PSIPRED [29, 30], except the third helix of the *C. elegans* protein, which is predicted to be a β-strand (data not shown). There are several amino acid changes in the corresponding sequences in TPPP/p25 in comparison with TPPP3/p20, however, the homology is still high enough, and all the five helices are predicted for TPPP/p25 as well (Fig. 10.3). However, only three of the predicted helices fell into the "core" region of TPPP/p25. Interestingly, one of them (FEQFQEALEELAKKR) does not contain even a single predicted stabilization centre, and another one (GKNWSKL-CKD) mostly lacks them as well. There is only one predicted helix (VTDVDIVFS) where all the amino acid residues are stabilization centers. The last two helices mentioned show the highest identity with the corresponding TPPP3/p20 sequences. *Regular secondary structure elements as helix and sheet can be disordered as a result of flexibility in the intervening regions of globally disordered proteins.* The extra flexibility of TPPP/p25, which is not surprising in the light of the predictive data, can explain why the physico-chemical and spectroscopic methods did not show significant structure for TPPP/p25.

10.3 Interacting Partners of TPPP/p25

A characteristic feature of unstructured proteins is formation of protein-protein interactions promoted by the crowded intracellular milieu, thus failing their destined degradation by the proteolytic systems such as proteasome machinery. The fate of unstructured proteins is accelerated in lack of interacting partners. Additionally, they have an intrinsic ability to step into non-physiological interactions due to their relative structural flexibility. We and other laboratories have identified potential interacting partners of TPPP/p25 from brain tissues by affinity chromatography and immunoprecipitation coupled with mass spectrometry or proteomics. These data are listed in Table 10.1. In a couple of cases we do not know the possible significance of these interactions since no functions have been determined. However, there are a few interactions which have physiological and/or pathological relevance shortly described below.

The primary target of TPPP/p25 is tubulin/microtubule under *in vitro* and *in vivo* conditions. Their *in vitro* interaction was shown by CD, surface plasmon resonance, co-polymerization and pelleting experiments [1, 5]. *In vitro* TPPP/p25 induces tubulin assembly into intact-like microtubules and aberrant forms depending on the circumstances as shown by electron microscopy and atomic force microscopy [1, 31]. TPPP/p25 displays very extensive microtubule bundling activity independently of whether the microtubules were preformed by paclitaxel or produced by TPPP/p25. These properties are likely prerequisite for this unstructured protein to fulfill its physiological and pathological functions (Scheme 10.1).

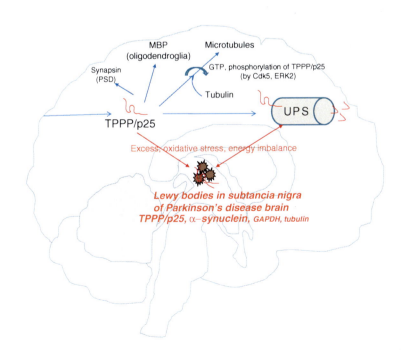

Table 10.1 Interacting partners of TPPP/p25

Interacting partner	Source	Method	Reference	Role of the interaction
[I] Proteins involved in signal transduction				
ERK2	bovine	AF	[31]	phosphorylation, which inhibits MT-formation
Glycogen synthase kinase 3 (GSK-3)	rat	IP	[33]	TPPP/p25 inhibits GSK-3 activity
Cdk5	in vitro	phosphorylation	[2, 31, 33]	phosphorylation, which inhibits MT-formation
Protein kinase A	in vitro	phosphorylation	[31, 33]	phosphorylation
Protein kinase C	bovine	IP	[34]	phosphatidic acid-dependent phosphorylation
LIM Kinase	sheep	AF, IP	[35]	phosphorylation, which inhibits MT-formation
[II] Structural proteins				
Tubulin β	bovine	co-polymerization, AF, CD, SPR, AFM	[1, 5, 45]	polymerization of tubulin, bundling of MTs
[III] Proteins involved in cellular metabolism				
GAPDH	bovine	AF, IP	[45]	co-occurrence in *pathological inclusions*
[IV] (Pre)synaptic proteins				
Complexin	rat	in vivo	[50]	the interaction decreased in trained animals; Ca^{2+}-dependent interaction
α-synuclein	in vitro	co-sedimentation	[40]	co-occurrence in *pathological inclusions*
[V] Glial proteins				
Myelin basic protein (MBP)	porcine bovine	AF, IP, CD, FL, SPR	[37]	TPPP/p25 and MBP is expressed in differentiating oligodendrocytes when they reach the myelinating state
[VI] Translation/ribosomal proteins				
Elongation factor α	bovine	AF	[45]	

AF – affinity chromatography, AFM, atomic force microscopy, FL – fluorescence spectroscopy, IP – immunoprecipitation, SPR – surface plasmon resonance, MT – microtubule.

The interaction of TPPP/p25 with microtubules is highly dynamic as demonstrated by fluorescence recovery after photobleaching experiment [32]; and can be affected by a couple of factors such as GTP, a tubulin binding nucleotide, the hydrolysis of which is coupled with the *in vivo* microtubule assembly; the truncation or the specific phosphorylation of the highly unfolded N-terminal segment of the protein [5, 31].

TPPP/p25 is a phosphoprotein which has been supported by *in vitro* and *in vivo* data. *In vitro* phosphorylation of TPPP/p25 or its fragments by tau kinase II (Cdk5) [2, 31, 33], protein kinase A [33], extracellular signal-regulated protein kinase 2 (ERK2) [31], protein kinase C [34] and LIM kinase [35] has been demonstrated. The phosphorylation sites, Thr14, Ser18 and Ser160 for Cdk5, Ser18 and Ser160 for ERK2, and Ser32 for protein kinase A, were identified by mass spectrometry, and were consistent with the bioinformatic predictions. These sites were also found to be phosphorylated *in vivo* in TPPP/p25 isolated from bovine or porcine brain [31, 36]. Affinity binding experiments provided evidence for the direct interaction between TPPP and ERK2 [31]. The phosphorylation by ERK2 or Cdk5 resulted in the loss of microtubule assembling activity of TPPP/p25. The fact that the specific phosphorylation of TPPP/p25 on the N-terminal tail can regulate the function of this unfolded protein suggests that this post-translational modification at the "signaling sequence" might have pathological relevance in addition to its physiological role.

Myelin basic protein (MBP), a marker protein of differentiated oligodendrocytes, has been isolated by affinity chromatography using TPPP/p25 column from a crude cytosolic extract of porcine brain [37], and also found to be co-purified with TPPP/p25 from bovine brain extract (our unpublished result). The tight interaction of these two basic, oligodendroglial proteins was characterized *in vitro* by different biophysical methods [37]. This finding concerns with the observation that these two proteins co-localize in differentiated oligodendrocyte [38, 39]. (See more details in Sect. 9.4.1 in this issue).

α-synuclein, a prototype of the IUPs, is a characteristic marker of synucleinopathies like Parkinson's disease (PD). TPPP/p25 displays a number of common features with α-synuclein in spite of the fact that the net charge for TPPP/p25 and α-synuclein are $+10$ and -10, respectively: low molecular mass, low aromaticity, long predicted disordered region, increased α-helix content from 2–4 to 39–43% by addition of TFE, lack of 3D structure, heat stability. All these features are hallmarks of unstructured proteins. These two unfolded proteins interact with each other, and TPPP/p25 promotes the fibrillization of α-synuclein *in vitro* [40]. However, it has been suggested that α-synuclein itself is not sufficient to cause aggregation leading to Lewy body (LB)-like inclusions, but additional factors such as oxidative stress, mitochondrial dysfunction and macromolecular interactions probably play role in the pathogenesis [41].

In contrast to the early view that glyceraldehyde-3-phosphate dehydrogenase (GAPDH) is a classical glycolytic housekeeping enzyme with its glycolytic activity, significant evidence has accumulated that GAPDH is a multifunctional enzyme displaying a number of additional activities due to its multiple interactions with different cytosolic and nuclear proteins, nucleic acids and subcellular particles ([42, 43] and references therein). GAPDH interacts with different proteins involved in age-related neurodegenerative disorders such as Huntington's disease, Alzheimer's disease and ataxias [44]. GAPDH has been found to be a stimulator of α-synuclein aggregation in PD; indeed, the over-expression of both GAPDH and α-synuclein in COS-7 cells induced LB-like cytoplasmic inclusions [41].

Our recent immunoprecipitation and affinity chromatography experiments with bovine brain cell-free extract revealed salt- and NAD^+-sensitive interaction between TPPP/p25 and GAPDH. Interestingly, TPPP/p25 aligned with microtubule network did not co-localize with GAPDH in HeLa cells expressing TPPP/p25 at low level, the glycolytic enzyme was distributed uniformly in the cytosol. At high expression levels, however, GAPDH co-localized with TPPP/p25 in the aggresome-like aggregate [45]. The pathological relevance of this observation was supported by immunohistochemistry which showed enrichment of TPPP/p25 and GAPDH within the α-synuclein positive LB [45]. Recently, proteomic analysis suggested the co-occurrence of TPPP/p25 and GAPDH in rodent brain postsynaptic density (PSD) [46–49].

10.4 TPPP Expression at Cell Level

The prerequisite to understand the role of unstructured proteins in neurodegenerative processes is to characterize their ultrastructure-related intracellular functions under physiological and pathological conditions. While the characterization of the unstructured proteins at molecular level has been coming on, which is represented well in two chapters of this issue too (Sect. 1 and 2), there is a gap concerning the studies at cellular level. The accumulation of unfolded proteins causes aberrant protein aggregation unless they are degraded by proteasome and/or phagosome/lysosome machineries ensuring the correct protein turnover. These processes result in pathological alterations such as oxidative and photolytic stress, mitochondrial and synaptic dysfunctions [51]; in addition, the impairment of energy metabolism as a cause or effect of the unwanted processes has been also suggested [52]. Obviously, therefore, the adequate level of the unstructured proteins with physiological function is critical for the cell life, otherwise toxic protein aggregates are formed.

We generated human cell models to explore the behavior of TPPP proteins under conditions which might mimic physiological and pathological situations (Scheme 10.1). One of these models is transfected living HeLa cells which express human recombinant TPPP/p25 fusing with enhanced green fluorescent protein (EGFP). The expression of the fusion protein at low and high levels produced distinct microtubule ultrastructures [32]. The other cell model, a stable neuroblastoma cell line, K4, expressing EGFP-TPPP/p25, was established to elaborate the effect of TPPP/p25 on energy metabolism at system level.

10.4.1 Structures and Effects in Living Cells

Immunofluorescence studies with living HeLa cells expressing EGFP-TPPP/p25 showed that at low expression level the green fluorescent fusion protein is perfectly aligned with the filament network during the interphase [32]. Double immunostaining procedure indicated that TPPP/p25 co-localized exclusively with the interphase microtubule network but not with actin or intermediate filaments. This phenomenon was not specific for HeLa cells; similar behavior was observed with rat kidney (NRK) cells as well as in SK-N-MC neuroblastoma cells [22, 32]. During the cell cycle, when the microtubule system is characterized by its extensive reorganization, specific accumulation of TPPP/p25 was detected in the centrosome region at interphase, on the spindle microtubules at metaphase and the furrow region at cytokinesis. The intracellular visualization of TPPP/p25 in single cells rendered it possible to evaluate some key characteristics of this protein at its low intracellular level as follows: i) the microtubule network was resistant to the depolymerizing effect of vinblastine, an anti-microtubule agent, suggesting the stabilization effect of TPPP/p25 by its bundling activity; ii) TPPP/p25 does not cause energy deficit, the energy state-dependent polarization of the mitochondrial membrane is normal; iii) TPPP/p25 does not arrest the cell cycle, however, it can slow down it at cytokinesis by its stabilizing effect as demonstrated by the enhanced number of the cells in cytokinesis phase with highly fluorescent furrow; iv) the association of TPPP/p25 to microtubule network is highly dynamic, fast exchange (with less than 5 second half-time recovery rate) of the microtubule-bound TPPP/p25 was demonstrated by time-lapse experiments [32]. These data suit to our idea that TPPP/p25 stabilizes the microtubule network in a dynamic manner, which could be physiologically relevant. The dynamism could be intracellularly modulated in various manners such as by a specific ligand or by post-translational modification.

The former one is supported by our experiments with *Drosophila* embryo [5]. We found that TPPP/p25 inhibited the mitosis in *Drosophila* embryos expressing green fluorescent protein-tubulin. The purified TPPP/p25 microinjected into the posterior region of the embryo arrested the mitosis along the diffusion of the protein; during the mitotic wave mitotic spindles were not formed at the posterior region, and the normal size of the embryo nuclei were multiplied without breakdown of nuclear envelope or inhibition of centrosome replication [5]. The inhibitory effect of TPPP/p25 on the mitosis was suspended when the protein was microinjected together with GTP, a nucleotide, the hydrolysis of which is tightly coupled with *in vivo* microtubule assembly. Therefore GTP could be a potential modulator of the microtubule-related physiological function.

The specific phosphorylation of the unfolded N-terminal tail of TPPP/p25 affects the tubulin/microtubule-TPPP/p25 interaction as we demonstrated *in vitro* (see above). The facts that TPPP/p25 is able to interact with ERK2 in brain cytosolic extract, and the endogenous TPPP/p25 was found to be phosphorylated in the N-terminal tail [31] prompted us to speculate about the role of phosphorylation of TPPP/p25 in synaptic plasticity since its occurrence as a phospho-protein in the postsynaptic density was reported [49, 53, 54].

10.4.2 Energy State

Mitochondrial membrane potential is highly related to the energy state of the cells; it is an important parameter determining the fate of the cells [55]. The accumulation of a fluorescent dye, tetramethylrhodamine ethyl ester visualized by fluorescence microscopy, signed the polarized state of the mitochondrial membrane related to the energy state of the living cell [56]. We found that K4 cells, stably expressing EGFP-TPPP/p25, showed strikingly high fluorescence intensity as compared to the control SK-N-MC cells suggesting that the TPPP/p25 expression did not cause energy impairment but enhanced the membrane polarization state.

In agreement with this effect of TPPP/p25 detected at cell level, the ATP concentration measured experimentally was 1.5-fold higher in the extract of the K4 cells as compared to that of the control cells. The flux analysis of the glucose metabolism based upon direct measurement and computation using the V_{max} values of the individual enzymes in the extracts of SK-N-MC and K4 cells revealed that the enhanced ATP concentration was, at least partly, due to the activation of some key glycolytic enzymes [22]. These data suggest that the stable expression of TPPP/p25 at level which is well tolerated by K4 cells sets up an increased energy state.

10.4.3 Physiological Function

Oligodendrocytes such as CG-4 cells express endogenous TPPP/p25 [38]. CG-4 is a stable cell line derived from primary cultures of bipotential oligodendrocyte-type 2-astrocyte (O-2A) progenitor cells of rat central nervous system glial precursors which could be differentiated into oligodendrocytes *in vitro* [57]. Our experiments unambiguously indicated that the expressions of TPPP/p25 and MBP increased concomitantly during the differentiation of the oligodendrocytes. The two proteins showed similar subcellular distribution at the periphery of the cells; their extensive co-localization was visualized especially at the long projections of prematured oligodendrocytes [38]. The microtubules are the key structural elements of the diversified projections, and we detected co-immunopositivity of TPPP/p25 and class IV β-tubulin in differentiating oligodendrocytes. The immunofluorescence studies, therefore, provided evidence for the co-localization of TPPP/p25, MBP and tubulin on the long projections at the cellular periphery of the oligodendrocytes suggesting key role of the two unstructured proteins in the stabilization of the microtubule-based projections [37].

Additional data with the CG-4 cells revealed that TPPP/p25 was not, or only in very small amounts, present in progenitor cells, however, it was strongly up-regulated in differentiating oligodendrocytes [38]. Recently, a microarray analysis of gene expression of oligodendrocyte progenitor cells expressing the A2B5+ marker and differentiating O4+ oligodendrocytes revealed that the TPPP/p25 transcript level increased about 30-fold during the progression from A2B5+ to the O4+ stage [59]. Very recently genome-wide transcriptional profiling comparing oligodendrocytes from developing and mature mouse forebrain have revealed gene expression changes during oligodendrocyte specification and differentiation [60]. *TPPP* gene was found among the genes whose expression was significantly (10.5-times) enriched in myelinating oligodendrocytes compared to oligodendrocyte progenitor cells.

ATP promotes oligodendrocyte differentiation which is prerequisite of the myelination of axons of neuronal cells, and it might regulate remyelination processes in the central nervous system diseases such as multiple sclerosis [61]. Since the differentiation of oligodendrocytes is coupled with TPPP/p25 expression [38], and TPPP/p25 expression in non-differentiated K4 cells resulted in enhancement of ATP level, one can hypothesize indirect role of TPPP/p25 in the ATP-producing energy metabolism.

10.4.4 Human Cell Model for Aggresome Development

The over-expression of TPPP/p25 in transiently transfected HeLa cells promotes aggresome formation at the centrosome region visualized by fluorescence microscopy [32]. The centrosome, a perinuclear microtubule-organizing center, is the main site of γ-tubulin; its primary function is the assembly and organization of microtubules during cell division [62]. The formation of the intracellular aggresomes requires the transport of the regionally accumulated proteins along microtubule network by the motor protein, dynein, to the centrosome. In fact, the aggresome represents expanded centrosome, which was identified within TPPP/p25-expressing HeLa cells by positive immunostaining for the centrosome-related tubulin.

The electron microscopic studies revealed that one of the characteristic features of these cells was the clustering of cell components in the area of the centrosome identified by the presence of the centriole [32]. This process leaves the periphery of the cytoplasm free of mitochondria and cisternae of the rough endoplasmic reticulum. In the center of the centrosomal area a distinct round body can frequently be observed that consists of a dense network of filamentous material composed of microtubules and 10 nm intermediate filaments. Inside its core the fibers run randomly, however, in the periphery they tend to be arranged in parallel bundles. Within this network ribosomes and small cisternae of the rough endoplasmic reticulum and annulate lamellae are dispersed, however, mitochondria and Golgi membranes are excluded. Formation of pericentriolar membrane-free, cytoplasmic inclusions containing filamentous proteins was first described in cells expressing misfolded membrane cystic fibrosis transmembrane conductance regulator protein [63]. This inclusion body-like protein aggregate, which was formed as a consequence of proteasome inhibition, was denoted the aggresome. Based on the intracellular and ultrastructural features of the bright fluorescent aggregate, we identified the round juxtanuclear bodies as aggresomes.

Formation of aggresomes in tissue culture and formation of inclusions in tissue are considered as the same process [62, 63]. LBs similarly to aggresome are developed in response to proteolytic stress since the capacity to clear aggregated proteins lacks due to the overwhelming production of the unwanted proteins (Scheme 10.2).

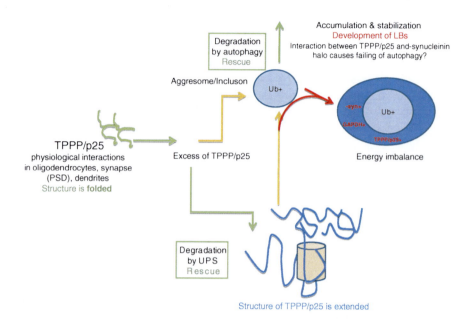

We noticed that the expression level of TPPP/p25 in transfected HeLa cells was affected by the activity of the ubiquitin proteasome system (UPS). The inhibition of the proteolysis with carbobenzoxy-Leu-Leu-Leu-al caused about twofold increase of the fluorescent signal of microscopic field as compared to that measured in non-treated transfected cells. In addition, under this condition more aggresome-bearing fluorescent cells were visualized. All these data proposed that the TPPP/p25 level was controlled by the proteasome machinery, the transiently expressed fusion TPPP/p25 was quickly degraded in HeLa cells.

Aggresomes appear to be less toxic than some other pathological protein aggregates such as amorphous ones. TPPP/p25 at high expression level promotes the formation of another microtubule ultrastructure, the perinuclear cage, which is distinct from the aggresome [32] (Fig. 10.4). Tetramethylrhodamine ethyl ester staining of EGFP-TPPP/p25 expressing cells was visualized by fluorescence microscopy to establish the energy state of the cells owing different microtubule ultrastructures. While the low expression did not cause damage in the polarization of the mitochondrial membrane, the high expression of the unstructured protein resulted in partial and complete mitochondrial dysfunction of the aggresome and cage containing HeLa cells, respectively [32] (Fig. 10.4). This finding is consistent with the view that the aggresome, which mimics the pathological inclusions, can protect intracellular organelles and molecules by sequestration of unwanted proteins.

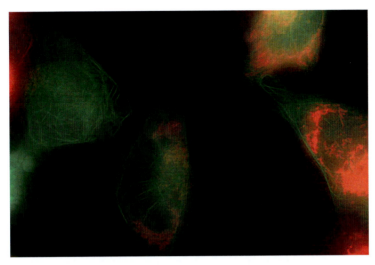

Fig. 10.4 Effect of the TPPP/p25 expression on the energy state of HeLa cells. Live HeLa cells transfected with pEGFP-TPPP/p25 are stained with rhodamine 123 (red), a marker of the hyperpolarization of mitochondrial membrane. The green fluorescent fusion protein is aligned along the microtubule network. Different cells show different phenotypes corresponding to the expressed level of TPPP/p25 (expression level is related to the green fluorescence intensity). The mitochondria of the cells expressing TPPP/p25 at low level are intensively stained (red signal) indicating normal membrane potential; the cell displaying aggresome formation at high expression level of TPPP/p25 (green at centrosome region) shows shortened, fragmented mitochondria with modest red staining suggesting impairment in energy state (at the center of the image). TPPP/p25-induced microtubule cage formed at high expression level shows mitochondrial dysfunction, depolarization of mitochondrial membrane (no red staining)

10.5 TPPP/p25 in Brain Tissues

10.5.1 Normal Brain

TPPP/p25 was identified as a brain-specific protein occurring mainly in oligodendrocytes, in neuropil and fiber-like structures of the CA3 hippocampal region in the rat brain [64]. Later on, TPPP/p25 was found to be present in oligodendroglial cells of the white matter and perineuronal oligodendroglial cells in the cortex of the normal human brain, in addition to faint immunostaining of the neuropil [65]. The pre-myelinating cells are formed from oligodendrocyte progenitor cells during the brain development and differentiate into myelin-forming oligodendrocytes. This differentiation process is accompanied by the appearance of long, arborized projections as well as the expression of MBP [66]. The physiological connection between MBP and TPPP/p25 was suggested first by Colello and co-workers [67], who searched for proteins associated with myelination in oligodendrocytes during normal development or with remyelination after insult or disease, using a proteomic approach. The occurrence of TPPP/p25 and MBP was detected in oligodendrocytes of normal rat brain tissue as well [39]. Their extensive co-localization was visualized at cell level predominantly at the long projections of pre-matured oligodendrocytes suggesting that TPPP/p25 might be partially membrane associated protein in oligodendrocytes [38]. These observations concern with our recent data which revealed that the expression of TPPP/p25 at the protein level is extensively up-regulated in differentiating oligodendrocytes and co-localizes with microtubule network ensuring the stabilization of the projection structure [38]. This finding is of special importance in the light of our recent recognition that TPPP/p25 expression is absent in neoplastic cells such an oligodendrogliomas [68]. At the same time lack of immunodetection of TPPP/p25 was found in normal astrocytes of human brain tissue, contrary to the positivity of granular TPPP/p25 immunoreactivity in reactive astrocytes [65]. Very recently a genome-wide transcriptional profiling of mouse brain confirmed cell-type specific expression of the *TPPP* gene: its transcript was 8.4-times enriched in oligodendrocytes as compared to the other cell types, it was practically absent in astrocytes, while its moderate level was detected in neurons [60].

The detection of TPPP/p25 immunoreactivity in neurons bearing lipofuscin and granulovacuolar degeneration may be due to the degeneration or breakdown of the microtubule system of the neuronal cytoskeleton. This suggests that neurons also possibly harbor low levels of TPPP/p25, at least in specified parts of the brain. Using Allen Brain Atlas, (http://www.brain-map.org.welcome.do/) TPPP/p25 transcript was identified in hippocampal pyramidal layers in the highest level. Subcellular proteomics revealed TPPP/p25 as a component of neuromelanin granules, thereby it is not surprising to find TPPP/p25 immunoreactivity in neuromelanin granules of substantia nigra neurons [65]. Neuromelanin is considered to be of lysosomal origin [69], containing both TPPP/p25 and α-synuclein [69, 70]. Proteomics methods identified TPPP/p25 in various synaptic preparations as a component of the postsynaptic density [46–48, 54], or transiently associating with synaptic [71] and chlatrin coated [72] vesicles. Moreover, interactions of TPPP/p25 with the synaptic proteins complexin [50] and synapsin (cf. Table 10.1) were found. Complexin binds to the synaptic SNARE (soluble N-ethylmaleinimide-sensitive fusion-attachment protein receptor) complex the assembly of which is triggered by Ca^{2+}-influx [73]. The SNARE complex initiates the fusion of synaptic vesicles to the plasma membrane. Thus the Ca^{2+}-dependent interaction of TPPP/p25 with complexin [50] suggests again its occurrence within neuronal cells even at modest level (cf. Table 10.2).

10.5.2 Pathological Brain

Unfolded/misfolded proteins like α-synuclein, tau or mutant huntingtin protein are hallmarks of conformational diseases. They form fibrils or protein aggregates with distinct ultrastructures; in pathological brain tissues they appear as intracellular inclusions which are tightly linked to the development of PD, Alzheimer's or Huntington's diseases. Immunohistochemistry and confocal microscopy demonstrates that TPPP/p25 is enriched in filamentous α-synuclein bearing LBs of PD and diffuse Lewy body disease, as well as glial inclusions of multiple system atrophy; in addition, the immunoreactivity of the two unstructured proteins display

Table 10.2 Occurrence of TPPP/p25 in normal and diseased brain tissue

	Reference	Remarks
Proteomics		
Hippocampus	[50]	in connection with spatial memory; rat
Neuromelanin	[70]	human
Postsynaptic density	[46–48, 54, 74]	identified as phospho-protein in proteomics studies; rodents
Synaptosomal preparation	[49, 53]	mouse, human
Synaptic vesicle	[71]	rat
Clathrin coated vesicle	[72]	rat
Photoreceptor	[78]	mouse
Immunohistochemistry in normal brain		
Hippocampus	[64]	neuropil in first and second layer, fibers in CA3; rat
Nucleus supraopticus	[39]	rat
Oligodendrocytes	[39, 40, 64, 65, 67, 75]	in perykarya and in myelin, rat and human
Astrocytes	[65]	in reactive astrocytes only, granular; human
Microglia		not detected
Immunohistochemistry in human pathological brain		
Multiple system atrophy	[23, 65, 76, 77]	glial cytoplasmic inclusions in astrocytes and oligodendrocytes
Parkinson's disease	[23, 65]	LBs and Lewy neurites
Diffuse LB disease	[65]	LBs and additional inclusions
Huntington's disease	[65]	some extracellular inclusions
Alzheimer's disease	[23, 65]	dot-like, along hyper-phosphorylated tau, only pretangles
granulovacuolar degeneration	[65]	
Oligodendrogliomas	[65, 68]	cancerous cells did not stained compared with normal oligodendrocytes

correlation shown by Western-blot and fluorescent images obtained by confocal microscopy [23] (Table 10.2). The co-localization of TPPP/p25 and α-synuclein in synucleinopathies was also supported by Jensen and co-workers [40]. TPPP/p25 appears as a specific immunomarker of synucleinopathies since it is not associated with abnormally phosphorylated tau in the inclusions of Pick's disease, progressive supranuclear palsy, and corticobasal degeneration. However, immunoelectron microscopy confirms clusters of TPPP/p25 immunoreactivity along filaments of unstructured (pretangle) but not compact neurofibrillary tangles in Alzheimer's disease [23]. This peculiar finding might label a common origin of synucleinopathies and tauopathies (Table 10.2). On the basis of our extensive immunohistochemical studies performed in various human pathological brain tissues we proposed TPPP/p25 to be a novel marker of synucleinopathies [22].

It should be, however, noted that the specificity of the immunohistochemical labeling of TPPP/p25 in human brain tissues was dependent on the polyclonal antisera used [65]. We demonstrate that one antibody, raised against the amino acid segment 184–200 of TPPP/p25 is better in immunolabeling the majority of α-synuclein immunopositive neuronal and glial pathological profiles detectable in PD, diffuse Lewy body disease, and multiple system atrophy; while another, raised against full-length human recombinant TPPP/p25, is more suitable to immunodetect normal oligodendrocytes [65]. This difference could be due to the distinct ultrastructural states of this unstructured protein in the myelinated oligodendrocytes and in aggregated form within inclusions.

One can hypothesize that this accumulation of the TPPP/p25 results in unbalanced homeostasis of neurons and/or glial cells similarly as described for tau isoforms as well as for α-synuclein [79–84]. Nevertheless, it is debated whether generation of protein aggregates is a cytoprotective or destructive process in cells [83–87]; in the case of aggresome or fibril formation the former possibility is more likely [84, 88]. The stabilization of protein aggregates by inclusion formation, however, counteracts with the protecting mechanisms such as proteasomal degradation or autophagy. The fine balance of these processes is a key step in the etiology of neurodegeneration. TPPP/p25 accumulation causes the rearrangement of the cytoskeleton and formation of aberrant microtubule ultrastructures, however, the mechanism which leads to cell death is unclear yet; we assume that accumulation of the unstructured protein followed by formation of stable proteinaceous aggregates near the centrosomes could interfere with intracellular transport processes and could alter other cytoskeletal functions. This could explain the death of the transfected cells at high TPPP/p25 expression level, and suggests the involvement of TPPP/p25 in the pathomechanism of synucleinopathies. Therefore, we propose that the intracellular enrichment of TPPP/p25 coupled with the development of pathological inclusions in brain leads to neurodegenerative disorders.

10.6 Impact of Unfolded Structures on Physiological and Pathological Functions

This review reflects the recent advance and growth in the field of TPPP/25 research. This unstructured protein has become the target of many structural, functional and pathological studies due to its potential involvement in neurodegenerative diseases. TPPP/p25 was discovered as the first member of a new protein family and characterized as an unstructured protein. On one hand, it displays MAP-like function like the also unstructured tau protein; on the other hand, it is involved in the pathomechanism of certain neurodegenerative disorders. One could, therefore, propose that the structural and functional features of TPPP/p25 are determined by its extended N-terminal segment which does not have any structural element. However, our comparative studies with its homologous proteins, as well as with its truncated mutant provided astonishing information against this assumption.

Two shorter paralogs of TPPP/p25, TPPP2/p18 and TPPP3/p20, were identified [7, 8] and cloned. The homologous proteins were used for comparative studies [8] in order to evaluate the role of the highly unstructured N-terminal tail, which is missing from the TPPP3/p20 and TPPP2/p18, and of other sequence variances in its physiological and pathological functions. These studies showed that in spite of the high homology of the TPPP proteins, their structures were different [8]; in agreement with the prediction data (cf. Figs. 10.1, 10.2 and 10.3), TPPP3/p20 but rather TPPP2/p18 behaved much less as an unfolded protein than TPPP/p25; TPPP2/p18 likely possesses tertiary structure [8] while TPPP3/p20 disposes a core structure with five α-helices and two β-sheets flanked by unfolded N- and C-terminal domains [25] (Fig. 10.3). The structural differences of the homologues manifested themselves in their abilities to associate to and to re-organize microtubule structures: TPPP3/p20 and especially TPPP2/p18 showed significantly less or no affinity to tubulin in vitro and in vivo compared to TPPP/p25 [8]. The most significant difference was found between the two shorter homologues, although none of them has the extended 40 amino acid segment [8]; while the truncated homologue of TPPP/p25 (Δ3–43) with amino acid length similar to that of the shorter homologues, shares a number of structural and functional characteristics with its wild type form ([24] and our unpublished data). The fact that significant structural and functional differences were found between the two shorter homologues, but not between TPPP/p25 and its truncated form is against the paradigm regarding the dominant role of the intrinsically unfolded region (IUR) in the functional properties of this IUP. The differences should be related to the amino acid substitutions in the C-terminal part of the proteins. However, the situation is likely different if the role of phosphorylation in the impairment of TPPP/p25 functions [31] is considered. The highly disordered N-terminal segment phosphorylated by ERK2 and protein kinase A, as signal sequence, could ensure crucial role for TPPP/p25 in *pathological* processes. This idea is further proposed by the finding that TPPP3/p20 protein, in contrast to TPPP/p25, did not induce formation of aggresome/inclusion at cell level [8]. Studies to evaluate the interrelationship of the unfolded segment and pathological effects of TPPP/p25 are in progress in our laboratory.

Acknowledgments This work was supported by EU FP6–2003-LIFESCIHEALTH-I *BioSim* to J. Ovádi and by Hungarian National Scientific Research Fund Grants OTKA T-67963 to J. Ovádi and T-049247 to F. Orosz.

Abbreviations

CD, circular dichroism; Cdk5, cyclin-dependent kinase-5; EGFP, enhanced green fluorescent protein; ERK2, extracellular signal-regulated protein kinase 2; GAPDH, glyceraldehyde-3-phosphate dehydrogenase; IUP, intrinsically unstructured protein; LB, Lewy body; MBP, myelin basic protein; PD, Parkinson's disease; TFE, trifluoroethanol; TPPP, Tubulin Polymerization Promoting Protein

References

1. Hlavanda E, Kovács J, Oláh J, Orosz F, Medzihradszky KF, Ovádi J (2002) Brain-specific p25 protein binds to tubulin and microtubules and induces aberrant microtubule assemblies at substoichiometric concentrations. Biochemistry 41:8657–8664
2. Takahashi M, Tomizawa K, Ishiguro K, Sato K, Omori A, Sato S, Shiratsuchi A, Uchida T, Imahori K (1991) A novel brain-specific 25 kDa protein (p25) is phosphorylated by a Ser Thr-Pro kinase (TPK-II) from tau protein-kinase fractions. FEBS Lett 289:37–43
3. Shiratsuchi A, Sato S, Oomori A, Ishiguro K, Uchida T, Imahori K (1995) cDNA cloning of a novel brain-specific protein p25. Biochim Biophys Acta 1251:66–68
4. Seki N, Hattori A, Sugano S, Suzuki Y, Nakagawara A, Muramatsu M, Hori T, Saito T (1999) A novel human gene whose product shares significant homology with the bovine brain-specific protein p25 on chromosome 5p15.3. J Hum Genet 44:121–122
5. Tirián L, Hlavanda E, Oláh J, Horváth I, Orosz F, Szabó B, Kovács J, Szabad J, Ovádi J (2003) TPPP/p25 promotes tubulin assemblies and blocks mitotic spindle formation. Proc Natl Acad Sci USA 100:13976–13981
6. Patrick GN, Zukerberg L, Nikolic M, de la Monte S, Dikkes P, Tsai L-H (1999) Conversion of p35 to p25 deregulates Cdk5 activity and promotes neurodegeneration. Nature 402:615–622
7. Zhang Z, Wu CQ, Huang W, Wang S, Zhao EP, Huang QS, Xie Y, Mao YM (2002) A novel human gene whose product shares homology with the brain-specific protein p25 is expressed in fetal brain but not in adult brain. J Hum Genet 47:266–268
8. Vincze O, Tőkési N, Oláh J, Hlavanda E, Zotter Á, Horváth I, Lehotzky A, Tirián L, Medzihradszky KF, Kovács J et al. (2006) Tubulin polymerization promoting proteins (TPPPs): members of a new family with distinct structures and functions. Biochemistry 45:13818–13826
9. Lai CH CC, Ch'ang LY, Liu CS, Lin W (2000) Identification of novel human genes evolutionarily conserved in *Caenorhabditis elegans* by comparative proteomics. Genome Res 10:703–713
10. Strausberg RL, Feingold EA, Grouse LH, Derge JG, Klausner RD, Collins FS, Wagner L, Shenmen CM, Schuler GD, Altschul SF et al. (2002) Generation and initial analysis of more than 15,000 full-length human and mouse cDNA sequences. Proc Natl Acad Sci USA 99:16899–16903
11. Dunker AK, Lawson JD, Brown CJ, Williams RM, Romero P, Oh JS, Oldfield CJ, Campen AM, Ratliff CM, Hipps KW et al. (2001) Intrinsically disordered protein. J Mol Graph Model 19:26–59
12. Uversky VN (2002) Natively unfolded proteins: a point where biology waits for physics. Protein Sci 11:739–756
13. Tompa P (2002) Intrinsically unstructured proteins. Trends Biochem Sci 27:527–533
14. Fuxreiter M, Simon I, Friedrich P, Tompa P (2004) Preformed structural elements feature in partner recognition by intrinsically unstructured proteins. J Mol Biol 338:1015–1026
15. Li X, Romero P, Rani M, Dunker AK, Obradovic Z (1999) Predicting protein disorder for N-, C-, and internal regions. Genome Inform Ser Workshop Genome Inform 10:30–40

16. Romero P, Obradovic Z, Li X, Garner EC, Brown CJ, Dunker AK (2001) Sequence complexity of disordered protein. Proteins 42:38–48
17. Ward JJ, Sodhi JS, McGuffin LJ, Buxton BF, Jones DT (2004) Prediction and functional analysis of native disorder in proteins from the three kingdoms of life. J Mol Biol 337: 635–645
18. Dosztányi Z, Csizmók V, Tompa P, Simon I (2005) The pairwise energy content estimated from amino acid composition discriminates between folded and intrinsically unstructured proteins. J Mol Biol 347:827–839.
19. Dosztányi Z, Csizmók V, Tompa P, Simon I (2005) IUPred: web server for the prediction of intrinsically unstructured regions of proteins based on estimated energy content. Bioinformatics 21:3433–3434
20. Dunker AK, Brown CJ, Obradovic Z (2002) Identification and functions of usefully disordered proteins. Adv Protein Chem 62:25–49
21. Dosztányi Z, Fiser A, Simon I, (1997) Stabilization centers in proteins: identification, characterization and predictions. J Mol Biol 272:597–612
22. Orosz F, Kovács GG, Lehotzky A, Oláh J, Vincze O, Ovádi J (2004) TPPP/p25: from unfolded protein to misfolding disease: prediction and experiments. Biol Cell 96:701–711
23. Kovacs GG, László L, Kovács J, Jensen PH, Lindersson E, Botond G, Molnár T, Perczel A, Hudecz F, Mező G et al. (2004) Natively unfolded tubulin polymerization promoting protein TPPP/p25 is a common marker of alpha-synucleinopathies. Neurobiol Dis 17:155–162
24. Otzen DE, Lundvig DMS, Wimmer R, Nielsen LH, Pedersen JR, Jensen PH (2005) p25 alpha is flexible but natively folded and binds tubulin with oligomeric stoichiometry. Prot Sci 14:1396–1409
25. Aramini JM, Rossi P, Shastry R, Nwosu C, Cunningham K, Xiao R, Liu J, Baran MC, Rajan PK, Acton TB et al. (2007) Solution NMR structure of Tubulin polymerization-promoting protein family member 3 from Homo sapiens. http://www.pdb.org/pdb/explore/explore.do?structureId=2JRF
26. Hua QX, Jia WH, Bullock BP, Habener JF, Weiss MA (1998) Transcriptional activator–coactivator recognition: nascent folding of a kinase-inducible transactivation domain predicts its structure on coactivator binding. Biochemistry 37:5858–5866
27. Monleon D, Chiang Y, Aramini JM, Swapna GV, Macapagal D, Gunsalus KC, Kim S, Szyperski T, Montelione GT (2004) Backbone 1H, 15N and 13C assignments for the 21 kDa Caenorhabditis elegans homologue of "brain-specific" protein. J Biomol NMR 28: 91–92
28. Kobayashi N, Koshiba S, Inoue M, Kigawa T, Yokoyama S (2005) Solution structure of mouse CGI-38 protein. http://www.pdb.org/pdb/explore/explore.do?structureId=1WLM
29. Jones DT. (1999) Protein secondary structure prediction based on position-specific scoring matrices. J Mol Biol 292: 195–202
30. Bryson K, McGuffin LJ, Marsden RL, Ward JJ, Sodhi JS, Jones DT (2005) Protein structure prediction servers at University College London. Nucl Acids Res 33:W36–38
31. Hlavanda E, Klement E, Kókai E, Vincze O, Tőkési N, Orosz F, Medzihradszky KF, Dombrádi V, Ovádi J (2007) Phosphorylation blocks the activity of tubulin polymerization-promoting protein (TPPP) – Identification of sites targeted by different kinases. J Biol Chem 282: 29531–29539
32. Lehotzky A, Tirián L, Tőkési N, Lénárt P, Szabó B, Kovács J, Ovádi J (2004) Dynamic targeting of microtubules by TPPP/p25 affects cell survival. J Cell Sci 117:6249–6259
33. Martin CP, Vazquez J, Avila J, Moreno FJ (2002) P24, a glycogen synthase kinase 3 (GSK 3) inhibitor. Biochim Biophys Acta-Mol Basis Dis 1586:113–122
34. Yokozeki T, Homma K, Kuroda S, Kikkawa U, Ohno S, Takahashi M, Imahori K, Kanaho Y (1998) Phosphatidic acid-dependent phosphorylation of a 29-kDa protein by protein kinase C alpha in bovine brain cytosol. J Neurochem 71:410–417
35. Acevedo K, Li R, Soo P, Suryadinata R, Sarcevic B, Valova VA, Graham ME, Robinson PJ, Bernard O (2007) The phosphorylation of p25/TPPP by LIM kinase 1 inhibits its ability to assemble microtubules. Exp Cell Res 313:4091–4106

36. Kleinnijenhuis AJ, Hedegaard C, Lundvig D, Sundbye S, Issinger OG, Jensen ON, Jensen PH (2008) Identification of multiple post-translational modifications in the porcine brain specific p25alpha. J Neurochem 106:925–933
37. Song YJC, Lundvig DMS, Huang Y, Gai WP, Blumbergs PC, Hojrup P, Otzen D, Halliday GM, Jensen PH (2007) P25 alpha relocalizes in oligodendroglia from myelin to cytoplasmic inclusions in multiple system atrophy. Am J Pathol 171:1291–1303
38. Lehotzky A, Tőkési N, Gonzalez-Alvarez I, Merino V, Bermejo M, Orosz F, Lau P, Kovacs GG, Ovádi J (2008) Progress in the development of early diagnosis and a drug with unique pharmacology to improve cancer therapy. Philos Transact A Math Phys Eng Sci 366:3599–3617
39. Skjoerringe T, Lundvig DMS, Jensen PH, Moos T (2006) P25 alpha/tubulin polymerization promoting protein expression by myelinating oligodendrocytes of the developing rat brain. J Neurochem 99:333–342
40. Lindersson E, Lundvig D, Petersen C, Madsen P, Nyengaard JR, Hojrup P, Moos T, Otzen D, Gai WP, Blumbergs PC, Jensen PH (2005) p25alpha Stimulates alpha-synuclein aggregation and is co-localized with aggregated alpha-synuclein in alpha-synucleinopathies. J Biol Chem 280:5703–5715
41. Tsuchiya K, Tajima H, Kuwae T, Takeshima T, Nakano T, Tanaka M, Sunaga K, Fukuhara Y, Nakashima K, Ohama E, Mochizuki H, Mizuno Y, Katsube N, Ishitani R (2005) Pro-apoptotic protein glyceraldehyde-3-phosphate dehydrogenase promotes the formation of Lewy body-like inclusions. Eur J Neurosci 21:317–326
42. Ovádi J, Orosz F, Hollán S (2004) Functional aspects of cellular microcompartmentation in the development of neurodegeneration: mutation induced aberrant protein-protein associations. Mol Cell Biochem 256–257:83–93
43. Sirover MA (2005) New nuclear functions of the glycolytic protein, glyceraldehyde-3-phosphate dehydrogenase, in mammalian cells. J Cell Biochem 95:45–52
44. Sirover MA (1999) New insights into an old protein: the functional diversity of mammalian glyceraldehyde-3-phosphate dehydrogenase. Biochim Biophys Acta 1432:159–184
45. Oláh J, Tőkési N, Vincze O, Horváth I, Lehotzky A, Erdei A, Szájli E, Katalin FM, Orosz F, Kovács GG et al. (2006) Interaction of TPPP/p25 protein with glyceraldehyde-3-phosphate dehydrogenase and their co-localization in Lewy bodies. FEBS Lett 580:5807–5814
46. Jordan BA, Fernholz BD, Boussac M, Xu C, Grigorean G, Ziff EB, Neubert TA (2004) Identification and verification of novel rodent postsynaptic density proteins. Mol Cell Proteomics 3:857–871
47. Li KW, Hornshaw MP, Van der Schors RC, Watson R, Tate S, Casetta B, Jimenez CR, Gouwenberg Y, Gundelfinger ED, Smalla KH et al. (2004) Proteomics analysis of rat brain postsynaptic density. Implications of the diverse protein functional groups for the integration of synaptic physiology. J Biol Chem 279:987–1002
48. Yoshimura Y, Yamauchi Y, Shinkawa T, Taoka M, Donai H, Takahashi N, Isobe T, Yamauchi T (2004) Molecular constituents of the postsynaptic density fraction revealed by proteomic analysis using multidimensional liquid chromatography–tandem mass spectrometry. J Neurochem 88:759–768
49. Collins MO, Yu L, Coba MP, Husi H, Campuzano I, Blackstock WP, Choudhary JS, Grant SG (2005) Proteomic analysis of in vivo phosphorylated synaptic proteins. J Biol Chem 280:5972–5982
50. Nelson TJ, Backlund PS, Alkon DL (2004) Hippocampal protein-protein interactions in spatial memory. Hippocampus 14:46–57
51. Lee VM, Trojanowski JQ (2006) Mechanisms of Parkinson's disease linked to pathological alpha-synuclein: new targets for drug discovery. Neuron 53:33–38
52. Ovádi J, Orosz F (2007) Energy metabolism in conformational diseases. In: Bertau M, Mosekilde E, Westerhoff H (eds.) Biosimulation in drug development. Wiley-VCH, Weinheim, Germany. pp. 233–257
53. DeGiorgis JA, Jaffe H, Moreira JE, Carlotti CG, Leite JP, Pant HC, Dosemeci A (2005) Phosphoproteomic analysis of synaptosomes from human cerebral cortex. J Proteome Res 4:306–315

54. Trinidad JC, Specht CG, Thalhammer A, Schoepfer R, Burlingame AL (2006) Comprehensive identification of phosphorylation sites in postsynaptic density preparations.mol Cell Proteomics 5:914–922
55. Iijima T, Mishima T, Akagawa K, Iwao Y (2003) Mitochondrial hyperpolarization after transient oxygen-glucose deprivation and subsequent apoptosis in cultured rat hippocampal neurons. Brain Res 993:140–145
56. Collins TJ, Bootman MD (2003) Mitochondria are morphologically heterogeneous within cells. J Exp Biol 206:1993–2000
57. Louis JC, Magal E, Muir D, Manthorpe M, Varon S (1992). CG-4, a new bipotential glial cell line from rat brain, is capable of differentiating in vitro into either mature oligodendrocytes or type-2 astrocytes. J Neurosci Res 31:193–204
58. Harauz G, Ishiyama N, Hill CM, Bates IR, Libich DS, Farès C. (2004) Myelin basic protein-diverse conformational states of an intrinsically unstructured protein and its roles in myelin assembly and multiple sclerosis. Micron 35:503–542
59. Dugas JC, Tai YC, Speed TP, Ngai J, Barres BA (2006) Functional genomic analysis of oligodendrocyte differentiation. J Neurosci 26:10967–10983
60. Cahoy JD, Emery B, Kaushal A, Foo LC, Zamanian JL, Christopherson KS, Xing Y, Lubischer JL, Krieg PA, Krupenko SA et al. (2008) A transcriptome database for astrocytes, neurons, and oligodendrocytes: a new resource for understanding brain development and function. J Neurosci 28:264–278
61. Agresti C, Meomartini ME, Amadio S, Ambrosini E, Volonté C, Aloisi F, Visentin S (2005) ATP regulates oligodendrocyte progenitor migration, proliferation, and differentiation: involvement of metabotropic P2 receptors. Brain Res Brain Res Rev 48:157–165.
62. Olanow CW, Perl DP, DeMartino GN, McNaught KS (2004) Lewy-body formation is an aggresome-related process: a hypothesis. Lancet Neurol 3:496–503.
63. Johnston JA, Ward CL, Kopito RR (1998) Aggresomes: A cellular response to misfolded proteins. J Cell Biol 143:1883–1898
64. Takahashi M, Tomizawa K, Fujita SC, Sato K, Uchida T, Imahori K (1993) A brain-specific protein p25 is localized and associated with oligodendrocytes, neuropil, and fiber-like structures of the CA hippocampal region in the rat brain. J Neurochem 60:228–235
65. Kovacs GG, Gelpi E, Lehotzky A, Hoftberger R, Erdei A, Budka H, Ovádi J (2007) The brain-specific protein TPPP/p25 in pathological protein deposits of neurodegenerative diseases. Acta Neuropathol 113:153–161
66. Akiyama K, Ichinose S, Omori A, Sakurai Y, Asou H (2002). Study of expression of myelin basic proteins (MBPs) in developing rat brain using a novel antibody reacting with four major isoforms of MBP. J Neurosci Res 68:19–28
67. Colello RJ, Fuss B, Fox MA, Alberti J (2002) A proteomic approach to rapidly elucidate oligodendrocyte-associated proteins expressed in the myelinating rat optic nerve. Electrophoresis 23:144–151.
68. Preusser M, Lehotzky A, Budka H, Ovádi J, Kovacs GG (2007) TPPP/p25 in brain tumours: expression in non-neoplastic oligodendrocytes but not in oligodendroglioma cells. Acta Neuropathol 113:213–215
69. Tribl F, Marcus K, Meyer HE, Bringmann G, Gerlach M, Riederer P (2006) Subcellular proteomics reveals neuromelanin granules to be a lysosome-related organelle. J Neural Transm 113:741–749
70. Tribl F, Gerlach M, Marcus K, Asan E, Tatschner T, Arzberger T, Meyer HE, Bringmann G, Riederer P (2005) "Subcellular proteomics" of neuromelanin granules isolated from the human brain. Mol Cell Proteomics 4:945–957
71. Takamori S, Holt M, Stenius K, Lemke EA, Grønborg M, Riedel D, Urlaub H, Schenck S, Brügger B, Ringler P et al. (2006) Molecular anatomy of a trafficking organelle. Cell 127: 831–846
72. Blondeau F, Ritter B, Allaire PD, Wasiak S, Girard M, Hussain NK, Angers A, Legendre-Guillemin V, Roy L, Boismenu D et al. (2004) Tandem MS analysis of brain clathrin-coated vesicles reveals their critical involvement in synaptic vesicle recycling. Proc. Natl Acad Sci. USA 101:3833–3838

73. Pabst S, Margittai M, Vainius D, Langen R, Jahn R, Fasshauer D (2002) Rapid and selective binding to the synaptic SNARE complex suggests a modulatory role of complexins in neuroexocytosis. J Biol Chem 277:7838–7848
74. Vosseller K, Trinidad JC, Chalkley R.J, Specht CG, Thalhammer A, Lynn A.J, Snedecor JO, Guan S, Medzihradszky KF, Maltby DA et al. (2006) O-linked N-acetylglucosamine proteomics of postsynaptic density preparations using lectin weak affinity chromatography and mass spectrometry Mol Cell Proteomics 5:923–934
75. Lyck L, Dalmau I, Chemnitz J, Finsen B, Schrøder HD (2008) Immunohistochemical markers for quantitative studies of neurons and glia in human neocortex. J Histochem Cytochem 56:201–221
76. Baker KG, Huang Y, McCann H, Gai WP, Jensen PH, Halliday GM (2006) P25alpha immunoreactive but alpha-synuclein immunonegative neuronal inclusions in multiple system atrophy. Acta Neuropathol 111:193–195
77. Jellinger KA (2006) P25alpha immunoreactivity in multiple system atrophy and Parkinson disease. Acta Neuropathol 112:112
78. Liu Q, Tan G, Levenkova N, Li T, Pugh EN Jr, Rux JJ, Speicher DW, Pierce EA (2007) The proteome of the mouse photoreceptor sensory cilium complex Mol Cell Proteomics 6: 1299–1317
79. Mattson MP (1995) Degenerative and protective signaling mechanisms in the neurofibrillary pathology of AD. Neurobiol Aging 16:447–457
80. Hutton M, Lewis J, Dickson D, Yen SH, McGowan E (2001) Analysis of tauopathies with transgenic mice. Trends Mol Med 7:467–470
81. Yu S, Uéda K, Chan P (2005) Alpha-synuclein and dopamine metabolism. Mol Neurobiol 31:243–254
82. Mattson MP (2006) Neuronal life-and-death signaling, apoptosis, and neurodegenerative disorders. Antioxid Redox Signal. 8:1997–2006
83. Winklhofer KF, Tatzelt J (2006) The role of chaperones in Parkinson's disease and prion diseases. Handb Exp Pharmacol 172:221–258
84. Agorogiannis EI, Agorogiannis GI, Papadimitriou A, Hadjigeorgiou GM (2004) Protein misfolding in neurodegenerative diseases. Neuropathol Appl Neurobiol 30:215–224
85. Lee HG, Perry G, Moreira PI, Garrett MR, Liu Q, Zhu X, Takeda A, Nunomura A, Smith MA (2005) Tau phosphorylation in Alzheimer's disease: pathogen or protector? Trends Mol Med 11:164–169
86. Scheibel T, Buchner J (2006) Protein aggregation as a cause for disease. Handb Exp Pharmacol 172:199–219
87. Hol EM, Scheper W (2008) Protein quality control in neurodegeneration: walking the tight rope between health and disease. J Mol Neurosci. 34:23–33
88. Tanaka M, Kim YM, Lee G, Junn E, Iwatsubo T, Mouradian MM (2004) Aggresomes formed by alpha-synuclein and synphilin-1 are cytoprotective. J Biol Chem 279:4625–4631

Chapter 11
Protein-Based Neuropathology and Molecular Classification of Human Neurodegenerative Diseases

Gabor G. Kovacs and Herbert Budka

Abstract Neurodegenerative diseases are characterized by death and progressive loss of neurons in distinct areas of the central nervous system. Classification is based on clinical presentation, anatomical regions affected, inclusion bearing cell-types and conformationally altered proteins involved in the process. In this chapter, the current molecular pathological classification of neurodegenerative diseases is reviewed by summarizing the proteins relevant for neurodegenerative diseases and their morphological types as extra- and intracellular deposits.

11.1 Introduction

Death and progressive loss of neurons in distinct areas of the central nervous system (CNS) characterize neurodegenerative diseases. The selective loss of specific population of neurons determines clinical presentation. Complex pathways are involved in neurodegeneration, and the most important ones are summarized below.

A. Programmed cell death is one mechanism involved in selective neuronal vulnerability. In principle, cells die from one of two distinct processes [1].

   (1) Necrotic cell death: This is typically characterized by cell swelling and breakdown of the cell membrane, and requires no active participation of the degenerating cell. This is mainly observed in ischemia and not in neurodegenerative diseases.
   (2) Non-necrotic cell death: This is regulated by cell autonomous processes and typically produces distinctive ultrastructural changes (seen by electron microscopy). Two major forms are apoptotic and autophagic.
      a. Apoptosis: Specific cytological features including chromatin condensation and margination, nuclear fragmentation and cytoplasmic blebbing. Several proteins regulate this process.

G.G. Kovacs (✉)
Institute of Neurology, Medical University of Vienna, AKH 4J, POB 48, A-1097 Wien, Austria
e-mail: gabor.kovacs@meduniwien.ac.at

b. Autophagy: This is a degradative mechanism involved in the recycling and turnover of cytoplasmic components.
B. Synaptic degeneration: this implies that synapses, the important connection between neurons, degenerate before the neuron itself dies [2]. Since complex connections between neurons are one major feature of the CNS, the loss of synaptic structures leads to clinical symptoms, while the homeostasis of neurons may also be damaged.
C. Oxidative mechanisms: Cells fail to compensate for oxidative stress and cannot cope with accumulation of reactive oxygen species [3]. These are byproducts of cellular oxidative metabolism. Important components include superoxide, hydrogen peroxide, hydroxyl radicals, nitric oxide, and peroxynitrite.
D. Abnormal protein processing: This includes abnormal interactions between proteins that result in aggregation of protein in fibrillar structure. Important systems involved in these processes include the *ubiquitin-proteasome* system and the *endosomal-lysosomal* system [4].

11.2 Classification of Neurodegenerative Disease: Basic Concepts

A nosological classification of neurodegenderative diseases is based on clinical presentation, anatomical regions and cell types affected, conformationally altered proteins involved in the pathogenetic process, and etiology if known, e.g. genetic aberrations. Involvement of proteins defines the concept of *conformational diseases*. Their basis is that the structural conformation of a physiological protein changes, resulting in an altered function or potentially toxic intra- or extracellular accumulation. Mutations in the encoding genes are linked to hereditary forms of disease. While a plethora of descriptive data on definitely or potentially involved proteins has accumulated and has revolutionized our knowledge on these disorders, much less is understood with regard to their function and the pathophysiological role of their misfolded conformers.

The clinical classification is based on which anatomical regions or functional systems are affected by neuronal damage. The major clinical features of neurodegenerative disorders are the following:

1. Cognitive decline, dementia, alteration in high-order brain functions. The most important anatomical regions involved are the hippocampus and limbic system. A subtype of dementia is frontotemporal dementia, which is associated with degeneration of the frontal and temporal lobes (Frontotemporal lobar degeneration, FTLD). These patients present with either behavioural or speech disorders.
2. Movement disorder:
   a. Hypokinetic: Parkinson syndrome
   b. Hyperkinetic: Chorea
   c. Other: Ataxia, upper and lower motor neuron symptoms
3. Combination of these

The neuropathological classification is based on the following: (1) Evaluation of the anatomical distribution of neuronal loss and reactive astrogliosis, and additional histological features like spongiform change of the neuropil in prion disease, or microvacuolation in frontal and temporal cortical areas in many neurodegenerative disorders, swollen neurons or, furthermore, damage to the white matter. These changes are seen with conventional staining techniques. (2) Evaluation of protein deposits in the nervous system: these can be deposited intracellularly and extracellularly and are analyzed by immunohistochemistry.

One major difference between intracellular inclusion bodies and extracellular aggregates is that the first have been assumed to arise from hydrophobic aggregation of proteins in unfolded or denatured states, whereas the latter could arise from the polymerization of specific partially folded intermediates that are unstable in an aqueous environment [5, 6]. Intracellular protein deposits may be seen mainly in the cytoplasm of neurons, astrocytes and oligodendroglial cells. Less frequently intranuclear deposits may be observed, mostly in neurons. Such protein deposits are also called inclusions, sometimes with a specific name (e.g. Lewy body). Extracellular deposits may, or may not, show amyloid characteristics. The term "amyloid", meaning starch-like, was coined by Virchow in 1851 based on its mis-identification as carbohydrate. A defining feature of amyloid is apple-green birefringence when viewed through a polarising light microscope after tissue staining with Congo red.

Consistent and sensitive immunohistochemical methods are now used in practice and elucidate the underlying molecular pathology [7]. The neuropathological diagnosis considers the anatomical distribution of neuronal loss and protein deposition patterns in correlation with the clinical syndrome. Biochemical and genetic examinations are complementary to this.

11.3 Proteins with Relevance for the Classification of Neurodegenerative Diseases

11.3.1 Microtubule-Associated Protein Tau

The **microtubule-associated protein tau** (MAPT) is important for the assembly and stabilization of microtubules [8]. In nerve cells tau is found in axons, but in disease circumstances it may redistribute to dendrites and to the cell body. The normal human brain has six isoforms of tau, produced from a single gene by alternative mRNA splicing [8].

Biochemically, there are four main patterns of insoluble tau as observed on Western blotting. This includes (1) major bands at 60, 64, and 68 kDa; (2) bands at 64 and 68 kDa; (3) bands at 60 and 64 kDa; and (4) a minor band at 72 kDa that is usually associated with the first pattern [8]. It is also important to distinguish different isofoms of tau in diseases. The isoforms differ from one another by the presence or absence of a 29- or 58-amino acid insert in the amino-terminal half of the protein, and by the inclusion, or not, of a 31-amino acid repeat encoded by exon 10 of tau, in the carboxy-terminal half of the protein Three isoforms with 0, 1, or 2 inserts contain three microtubule-binding repeats (R) and are designated as 3R tau, and three isoforms, also with 0, 1, or 2 inserts containing four microtubule-binding repeats, are designated as 4R tau. Thus, diseases with pathological accumulation of tau show a biochemical signature characterized by the pattern of insoluble tau and further by the apperance of tau isoforms. 3R and 4R isoforms may be equally or unequally present in distinct disorders with altered tau [8, 9]. Although the brain tau isoform composition has been extensively analysed by Western blotting, availability of monoclonal antibodies detecting 3R and 4R tau opened the way for inclusion-specific mapping of isoforms [10]. In addition, the abnormal hyperphosphorylation of tau is a common feature to all disease with abnormal tau filaments. The site of phoshorylation differs between diseases; however, this seems to be less important in distingushing disease forms as compared to the patterns of tau bands described above.

The *Tau* (*MAPT*) gene maps to chromosome 17q21.2. Mutations lead to hereditary diseases that are associated with progressive neurodegenerative syndromes and accumulation of intracellular deposits of soluble and insoluble hyperphosphorylated tau protein [11]. Genetic variability in *MAPT*, in particularly a dinucleotide repeat polymorphism in intron 9 defined as H1 and H2 haplotypes, may contribute to risk of sporadic tau diseases [12].

11.3.2 β-Amyloid

β-**amyloid** (or Aβ) derives from the amyloid precursor protein (APP). APP has the structural characteristics of a cell surface receptor and undergoes proteolytic cleavages by β and γ secretases to generate the β-amyloid fragment of 40 or 42 residues [13]. Since secretases are linked to presenilin genes, presenilins are implicated in the trafficking and proteolytic cleavage of APP. APP fragments may undergo degradation or oligomerisation and extracellular plaque formation, which is a characteristic feature of the neuropathology of Alzheimer's disease (AD). In addition to 1–40 and 1–42, carboxy-terminally truncated Aβ peptides (1–37, 1–38, 1–39) may be found, and the Aβ peptide pattern may well reflect disease-specific changes [14].

The *APP* gene has been mapped to the centromeric region of chromosome 21. Mutations in the *APP* gene or the presenilin genes (*Ps1* located on chromosome 14 and *Ps2* located on chromosome (1) may lead to early onset AD; furthermore, several genes implicated in the processing of Aβ may influence late-onset AD [15, 16].

11.3.3 α-Synuclein

α-**Synuclein** is a 140-aa protein that belongs to a family of abundant brain proteins (α-, β-, and γ-synuclein). Until recently the function of α-synuclein remained unclear, and indeed mice deficient in either α- and β-synuclein, or both, survive without obvious brain defects [17]. α-Synuclein lacks secondary or tertiary structures, so it belongs to the family of natively unfolded proteins, many of which act as chaperones. It is known to interact with several other proteins [18].

It is mainly expressed in presynaptic terminals [19] and may play a role in the stabilization of the cytoskeleton, synaptic vesicular transport and maintenance of synaptic plasticity [19, 20]. It modulates the functional activity of dopamine transporter and presynaptic dopamine release [19–22]. In normal nerve terminals, wild type α-synuclein interacts directly with dopamine transporter, trafficking the latter from the membrane into the cytoplasm [19, 21–23]. It also attenuates the activity of dopamine transporter [23].

Western blot shows a band at 19 kDa and further ones at 29–36 and 45–55 kDa that represent complex α-synuclein forms [18, 24, 25]. Moreover, a proteinase-K resistant subset was demonstrated in diseased brains and correlated well with pathology [26]. Another study found high levels of α-synuclein in the detergent-soluble fraction of brain samples with multiple system atrophy. It was concluded that the solubility of α-synuclein differs between multiple system atrophy and Parkinson's disease (PD) [27].

The α-synuclein gene locates to chromosome 4. Mutations in the gene may associate with PD and/or Lewy body dementia (LBD) phenotypes [28, 29]. It must be noted that several other genes are implicated in PD and might be related to the formation of α-synuclein harbouring cellular inclusions [30].

11.3.4 Prion Protein

Prion protein (PrP) is predominantly expressed in neurons. PrP is central to prion diseases or transmissible spongiform encephalopathies. The normal (cellular) form is designated as PrP^C. Conformational change of this protein is central to the "prion" hypothesis and distinguishes PrP^C from the disease-associated form (PrP^{Sc}, where Sc indicates scrapie, a prion disease of sheep and goats) that seems to be infectious as "prion" [31, 32]. PrP^{Sc} exists as a predominantly ß-pleated sheet, while PrP^C is α-helix dominant [31]. PrP^{Sc} can be differentiated from PrP^C in Western blots after protease treatment, because the pathological form is not digestable [31], although a protease–sensitive but disease associated transitional form was also described [33].

The significance of PrP^C extends beyond prion diseases. Possible functions comprise a role in neurogenesis and differentiation of neural stem cells, neuritogenesis, synaptogenesis, neuronal survival via anti or pro-apoptotic functions, copper binding, redox homeostasis, long term renewal of haematopoetic stem cells, activation and development of T cells, differentiation, modulation of phagocytosis of leukocytes, and altering leukocyte recruitment to site of inflammation [34]. Thus, its upregulation appears important in inflammatory conditions that raise the possibility to have an increased substrate available for the PrP^C-PrP^{Sc} conversion [35, 36].

The encoding gene of PrP (*PRNP*) locates to chromosome 20. Reported point and insertional mutations number more than 30. A polymorphism at codon 129 has a clear influence on the phenotype in all (sporadic, genetic, acquired) subtypes [37]. Sporadic Creutzfeldt-Jakob disease (CJD) is classified according to the constellation of codon 129 (valine or methionine allele) of *PRNP*, and the Western blot signature of protease-resistant PrP. At least three patterns may be distinguished (although some authors observe even more [38]), which, in combination with the codon 129 polymorphism, define at least six phenotypic groups [39].

11.3.5 TAR-DNA-Binding Protein 43 (TDP-43)

TDP-43 or transactive response (TAR) DNA-binding protein 43 is a 414-amino-acid nuclear protein. It is a highly conserved protein expressed in non CNS tissues as well. It contains 2 RNA recognition motifs as well as a glycine-rich C-terminal sequence [40, 41]. This C-terminal domain of TDP-43 binds to several proteins of the heterogeneous nuclear ribonucleoprotein family, which are involved with the biogenesis of mRNA. TDP-43 may also act as scaffold for nuclear bodies [42].

TDP-43 is encoded by the *TDP* (*TARDBP*) gene on chromosome 1. Mutations in this gene have been linked to amyotrophic lateral sclerosis, both in familial and sporadic forms [43, 44]. Interestingly, mutations in *progranulin* (*PGRN*) and *valosin-containing protein* (*VCP*) genes are also characterized by TDP-43 inclusions, suggesting multiple and complex pathways of pathological alterations linked to this protein [45].

11.3.6 α-Internexin

α-**Internexin** is expressed predominantly by neurons. It belongs to the family of intermediate filaments including also three neurofilament proteins (light, medium, heavy) and peripherin. It was recently shown to be a sensitive marker of most of the inclusions of neurofilament inclusion body disease [46]. The gene for α-internexin is located on chromosome 10 and transcibes to a 499-amino acid protein [47]. It must be noted that α-internexin is not the structural component of inclusion in this disease and possibly accumulates secondarily [48].

11.3.7 Other Proteins

There are more forms of genetic neurodegenerative disorders with abnormal protein inclusions, comprising proteins encoded by genes linked to neurological trinucleotide repeat disorders. Their basis is the expansion of unstable trinucleotide repeats that account for at least 16 neurological disorders, ranging from developmental childhood disorders to late onset neurodegenerative diseases such as Huntington's disease and inherited ataxias [49]. The gene product of these, such as huntingtin in Huntington's disease, atrophin in dentatorubral-pallidoluysian atrophy, or the ataxins in spinocerebellar ataxias, may be deposited in cells in the form of inclusions.

Moreover, rare inherited disorders are characterized by cellular neuroserpin or ferritin inclusions. Neuroserpin is a serine protease inhibitor mainly expressed in the CNS. It is a regulatory element of extracellular proteolytic events and inhibits the activity of a tissue plasminogen activator [50]. The gene encoding neuroserpin is mapped to chromosome 3q26 and is also designated as *PI12* [51]. Its open reading frame encodes 410 amino acids [50]. Inclusion bodies (Collins bodies) are composed of neuroserpin and are deposited both in the neuropil and within neuronal perikarya and processes [52, 53]. In ferritin-related neurodegenerative diseases, the molecular genetic defect resides in the *ferritin light polypeptide* gene located on chromosome 19 [54]. Inclusions may be found in neurons and glial cells both in the cytoplasm and nucleus. In familial British and Danish dementias with deposition of amyloid proteins in the extracellular spaces of the brain and in blood vessels, the molecular genetic defect is a mutation in the BRI_2 gene [55]. Both the ferritinopathies and the latter amyloidoses accumulate proteins with a carboxyl-terminus that is abnormal in length and primary sequence [54].

In addition to the aforementioned proteins that are linked to specific diseases, it is important to mention that many other proteins may be found within inclusions and proteinaceous deposits; however, these are not structural elements of abnormal fibrils, but are proteins binding to them. Such proteins comprise components of the ubiquitin-proteasome system, proteins implicated in cellular responses or associated with phosphorylation and signal transduction, cytoskeletal proteins, cell cycle proteins, or even cytosolic or serum proteins that passively diffuse and bind to the protein deposits. Thus immunohistochemistry may demonstrate such components in proteinopathy inclusions such as tau-associated neurofibrillary tangles and α-synuclein-associated Lewy bodies. Recently, the tubulin polymerization promoting protein (TPPP/p25) was demonstrated to be associated selectively with the formation of α-synuclein inclusions, in particularly those of multiple system atrophy [56, 57].

The role of ubiquitin immunohistochemistry is of particular interest in diagnostic neuropathology. Ubiquitin is a small stress-induced protein, which participates in the degradation of short-lived and damaged proteins, and is found in diverse filamentous inclusions of neurodegenerative disorders [58]. Several studies indicate that the ubiquitin-binding protein 62/sequestosome 1, a cytosolic 62-kd protein (p62) is also a common component of various inclusions [59, 60]. The formation of inclusions follows a certain kind of "maturation" pathway and involves ubiquitination only later [61]. While early protein deposits may be detectable only with specific antibodies raised against the original protein, late aggregates or fully developed inclusion bodies may be nicely visualized by anti-ubiquitin or anti-p62 immunostaining [62].

Based on the most important proteins related to neurodegenerative disorders listed above, diseases are classsified also as tauopathies, synucleinopathies, prion diseases, trinucleotide repeat disorders, or TDP-43 proteinopathies. Some forms do not fit strictly into these categories and are referred to e. g. as FTLD with ubiquitin-positive TDP-43-negative neuronal inclusions [63], including cases with mutation in the *charged multivesicular body protein 2B (CHMP2B)* gene [45, 64], others as neuronal intermediate filament inclusion disease, and others as cerebral amyloidoses. Major disease groups are summarized in Table 11.1.

Table 11.1 Major conformational neurodegenerative diseases

Tauopathy	Synucleino-pathy	Trinucleotide-repeat disorder	Prion disease	TDP-43 proteinopathy	Other
• AD	• PD	• Huntington's disease	• CJD	• FTLD-U (type 1–4)	NIFID
• FTLD with Pick bodies	• LBD	• Huntington's disease-like 2	• GSS	• FTLD-U with MND	• Neuroserpinopathy
• FTLD with MAPT mutation	• MSA	• SCA 1, 2, 3, 6, 7, 8, 10, 12, 17	• FFI	• MND (ALS)	• Ferritinopathy
• CBD		• DRPLA Fragile X and XE syndrome	• Kuru		• FTLD-U lacking TDP-43 IR
• AGD		• Friedreich ataxia			• British and Danish familial dementia with amyloidosis
• PSP		• Myotonic dystrophy			
• NFT dementia		• Fragile X Tremor-Ataxia syndrome			
• Unclassifiable		• Spinobulbar muscular atrophy			

AD shows features of tauopathies associated with extracellular deposition of Aβ. Abbreviations: PSP: progressive supranuclear palsy, CBD: corticobasal degeneration, AGD: argyrophilic grain disease, NFT: neurofibrillary tangle, MSA: multiple system atrophy, SCA: spinocerebellar atrophy, DRPLA: dentatorubral-pallidoluysian atrophy, GSS: Gerstmann-Sträussler-Scheinker disease, FFI: fatal familial insomnia, FTLD-U: FTLD with ubiquitin immunoreactive neuronal inclusions, MND: motor neuron disease, ALS: amyotrophic lateral sclerosis, NIFID: neuronal intermediate filament inclusion disease.

11.4 Morphological Types of Extra- and Intracellular Protein Deposits

Localization of deposits related to different proteins is summarized in Table 11.2 and Fig. 11.1

11.4.1 Extracellular Protein Deposition

Aβ of Alzheimer type pathology may show several morphological forms of deposits. The qualification as amyloid may be inappropriate for some of these since they are not visualized by Congo red or thioflavine stains. Immunostaining for Aβ reveals compact and non-compact deposits. Focal immunodeposits with amyloid cores are referred to as "classic" plaques [65]. Further plaques have less compact amyloid deposits and lack a definite core, thus are referred to as primitive plaques. Non-compact plaques are referred to as preamyloid or diffuse deposits with irregular contours. Moreover, lake-like and fleecy types of deposits are also seen. Cotton wool plaques are usually associated with familial AD. In addition, deposition of Aβ in vessel walls is also noted. The biochemical composition of Aβ differs between vascular and parenchymal deposits. Except for cotton wool plaques, the aforementioned morphological subtypes do not specify any disease form.

In the prion diseased human brain, PrP^{Sc} is deposited in diffuse/synaptic, patchy/perivacuolar, perineuronal, and plaque-like patterns [66]. Plaques may have, or may not have, amyloid characteristics. PrP amyloid plaques are further classified according to the morphology of the core, such as unicentric (Kuru type), multicentric, or florid plaques that are surrounded by vacuoles. PrP^{Sc} may also accumulate in astrocytes and microglia and also may co-localize with both chemical and electric

Table 11.2 Extra- and intracellular distribution of deposits composed of different proteins in neurodegenerative disorders

Protein	Localization of deposits in CNS			
	Intracellular			Extracellular*
	Neuron		Glia	
	Cytoplasm	Nucleus		
Tau	+	−	+	−
α-Synuclein	+	+	+	−
TDP-43	+	+	+	−
α-Internexin	+	−	−	−
Huntingtin	−	+	−	−
Neuroserpin	+	−	−	+
PrP**	−	−	−	+
Aβ**	−	−	−	+

**Intracellular protein deposits are described but are not the major components of histopathological alterations and are not classified as inclusions.
*Some proteins may seem to be extracellular after death of the cells.

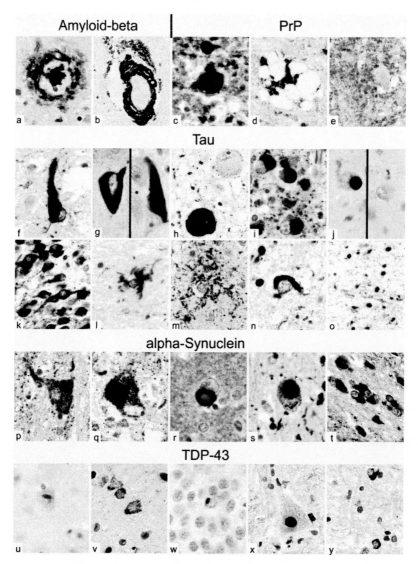

Fig. 11.1 Examples of extra- and intracellular protein deposits. **a**. Cored amyloid plaque showing immunoreactivity for Aβ in AD brain. **b**. Vascular amyloid and perivascular fibrillar Aβ immunodeposition. **c**. Unicentric amyloid plaque (Kuru plaque) in CJD demonstrated by anti-PrP immunostaining. **d**. Patchy/perivacuolar PrP deposition in CJD. **e**. Diffuse/synaptic PrP deposits in CJD. **f**. Neurofibrillary tangle immunostained by anti-hyperphosphorylated tau antibody (clone AT8) in AD. **g**. Neurofibrillary tangles are demonstrated by anti-3R (left side) and anti-4R (right side) Tau immunostaining. **h**. Globose tangle in the locus coeruleus in progressive supranuclear palsy (AT8). **i**. Pick bodies in frontotemporal lobar degeneration with Pick bodies are demonstrated by anti-hyperphosphorylated tau immunostaining (AT8). **j**. Pick bodies are immunoreactive for anti-3R tau (left side) but not for anti 4R tau (right side). **k**. Abundant neuronal tau pathology in the dentate gyrus in an individual with mutation in the *MAPT* gene. **l**. Tufted astrocyte in the putamen in progressive supranuclear palsy (AT8). **m**. Astrocytic plaque in the cingular cortex in

synapses, in the neuronal cell body and dendrites, thus both post- and presynaptically, furthermore, in intra- and adaxonal localisations [67]. PrP^{Sc} was identified within macrophages and vascular associated dendritic cells in the vessel wall and perivascular area in sporadic CJD [68].

In contrast to extracellular Aβ deposits, PrP immunostaining patterns may associate with certain molecular subtypes of prion disease, and thus are helpful to characterize these disorders. The immunomorphology of PrP^{Sc} deposits may differ according to the protease-resistant PrP type. Fine deposits like diffuse/synaptic is usually observed in combination with the presence of type 1 protease-resistant PrP, while more coarse deposits like amyloid plaques, plaque-like formation, and patchy/perivacuolar appearance is usually associated with type 2 protease-resistant PrP. PrP typing relies on the size of the unglycosylated PrP fragment on Western blots [66]. The special morphological type of perineuronal PrP^{Sc} immunodeposits is extensively seen in sporadic CJD with the presence of valine in at least one of the alleles encoded at codon 129 and type 2 protease-resistant PrP [39, 69].

Further extracellular amyloid deposition is seen in famical British and Danish dementia. In neuroserpinopathies Collins bodies may be observed in the neuropil. It must be noted that several inclusion bodies may become extracellular after the death of the cell harbouring that inclusion, exemplified by extracellular Lewy bodies or so called ghost tangles.

11.4.2 Intracellular Protein Deposition

Neuron and glia-related protein-deposits include cytoplasmic, intranuclear, or cell process (axonal or dendritic) related deposits. When examined by electron microscopy, the filaments of inclusions differ in structure. Their morphology is distinct; for example tau immunoreactive oligodendroglial coiled bodies are distinguishable from α-synuclein bearing oligodendroglial Papp-Lantos bodies. Also within the

---

Fig. 11.1 corticobasal degeneration (AT8). **n**. Oligodendroglial coiled-body in the frontal white matter in corticobasal degeneration (AT8). **o**. Hyperphosphorylated-tau (AT8) immunoreactive grains in the entorhinal cortex in argyrophilic grain disease. **p**. Small cytoplasmic spherical α-synuclein immunoreactivity lacking the typical appearance of a Lewy body in a neuron in the substantia nigra of an individual with PD. **q**. Typical brainstem-type α-synuclein immunoreactive Lewy body in the substantia nigra of an individual with PD. **r**. α-Synuclein immunoreactive cortical Lewy body in the cingular cortex of an individual with Lewy body dementia. **s**. Cytoplasmic and nuclear α-synuclein immunoreactive inclusions in a pontine base neuron in an individual with multiple system atrophy. **t**. Typical α-synuclein immunoreactive oligodendroglial cytoplasmic inclusions (Papp-Lantos bodies) in multiple system atrophy. **u**. Intranuclear TDP-43 immunopositive inclusion in the putamen in an indvidual with familial TDP-43 proteinopathy. **v**. Granular cytoplasmic TDP-43 immunoreactivity in an individual with frontotemporal lobar degeneration. **w**. Cytoplasmic TDP-43 immunopositive inclusion in a granular layer neuron in the dentate gyrus in an individual with frontotemporal lobar degeneration. **x**. Compact spherical TDP-43 immunoreactive cytoplasmic inclusion in a lower motor neuron in a patient with amyotrophic lateralsclerosis. **y**. Oligodendroglial TDP-43 immunopositive inclusions in the white matter of the spinal cord in a patient with amyotrophic lateralsclerosis

same proteinopathy group, the filaments may vary in structure, such as paired helical filaments, straight filaments, or twisted ribbons in tauopathies. At the light microscopic level, the morphological appearance of inclusions is distinguishable as well within the same proteinopathy group. In tauopathies, inclusions may differ whether they are composed of 3R or 4R tau isoforms or both. This renders classification schemes rather complex. Moreover, there are several conditions where inclusions are demonstrated by conventional stains or ubiquitin immunohistochemistry, while the essential structural component of the particular inclusion has not been clarified. These include basophilic inclusion body disease, neuronal intranuclear (hyaline) inclusion disease, or FTLD cases with ubiquitin immunoreactive inclusions that are TDP-43, tau, α-synuclein, PrP, or α-internexin immunonegative. If the detailed histological, immunohistochemical and biochemical analysis fails to reveal signature lesions the remaining diagnosis may be dementia lacking distinctive histological features. Inclusions in different proteinopathies are listed in Table 11.3, while extra- and intracellular protein deposits are demonstrated in Fig. 11.1.

11.5 Other Distinguishing Morphological Features

Some morphological alterations are also detectable with immunostaining methods related to proteins listed above, but are not specific for any disorder.

Perisomatic granules in FTLD with Pick bodies (previously called Pick's disease) and AD predominantly appear in the hippocampal CA1 region. They are ubiquitin and glutamate receptor immunoreactive and preferentially associate with tau immunopositive pretangles [70]. They can be differentiated from granulovacuolar degeneration of neurons that are usually observed in the same area. Interestingly, granulovacuolar degeneration that is a feature seen in ageing, AD and FTLD with Pick bodies, is not associated with tau.

Argyrophilic grains are oval, spindle, or comma shaped abnormally phosphorylated tau immunoreactive structures that are usually located around apical dendrites or tiny axons of neurons [71] and also follow a maturation process from ubiquitin-negative tau-positive deposits to ubiquitin and tau positive argyrophilic structures [72]. A recent study suggested that it is an age-related process that can be divided into stages [73].

Tangle-associated neuritic clusters were first described in AD and ageing and were shown to be not necessarily associated with Aβ deposition. Recently, they were characterized in progressive supranuclear palsy brains as well and were demonstrated as distinct from ghost tangles although both structures relate to the degradation of neurofibrillary tangles [74].

11.6 Synthesis: Classification of Neurodegenerative Diseases

An algorithm for the molecular pathological classification of neurodegenerative diseases is proposed in Fig. 11.2.

Table 11.3 A list of inclusions in tauopathies, synucleinopathies, and TDP-43 proteinopathies

Protein	Neuronal cytoplasm	Neuronal process	Cellular pathology Neuronal nucleus	Astrocyte	Oligodendrocyte
Tau	Pretangle	Dystrophic neurite	–	Tufted	Coiled body
	Neurofibrillary tangle	Grain		Thorn-shaped	Globular inclusions
	Pick body	Thread		Astrocytic plaque	
	Spherical inclusions			Ramified astrocyte	
	Dots				
	Perinuclear ring				
	Corticobasal body				
α-**Synuclein**	Lewy body	Lewy neurite	Nuclear inclusion	Coil-like, star, crescent shaped	Nuclear inclusion
	Pale body				Cytoplasmic inclusion (Papp-Lantos)
	Cytoplasmic inclusion (other)				Cytoplasmic inclusion (circular, coil-shaped)
TDP-43	Perinuclear granules	Dystrophic neurite (long or dot-like)	Nuclear inclusion (round, rod, and lentiform: "Cat-eye")	–	Cytoplasmic inclusions: (round, triangular, flame-shaped, coiled-body-like)
	Cytoplasmic inclusions (compact round; skein-like)				

The starting point for neuropathological diagnosis and disease classification includes collection of clinical information, such as age at onset, duration of disease, and in particular a family history of neuropsychiatric disease. Evaluation of atrophy during neuroimaging investigations and *post mortem* macroscopical inspection of the brain may suggest vulnerable areas where neuronal loss, the major feature of neurodegeneration, predominates. The next step is the detailed anatomical mapping of neuronal loss and reactive astrogliosis to further define the selective vulnerability pattern of the particular case. At this level, morphological alterations may be detected such as eosinophilic or argyrophilic neuronal or glial inclusions, as well as extracellular congophilic amyloid deposits with or without surrounding degeneration of neural processes ("neuritic" plaques) that are well visualized by silver stains.

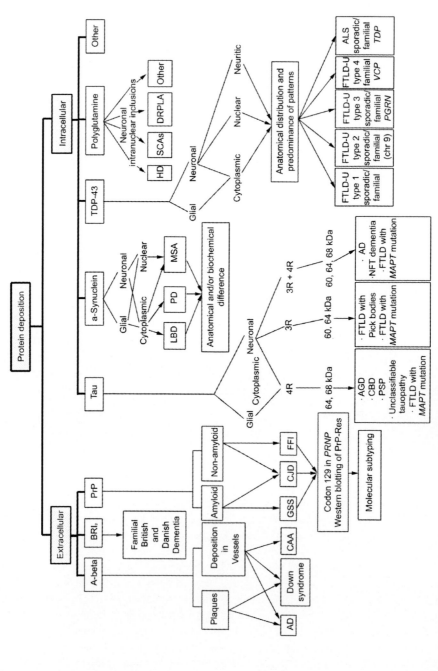

Fig. 11.2 Algorithm for the classification of neurodegenerative diseases

Presence of intra- and extracellular protein deposition requires immunohistochemistry for further classification [7]. Extracellular deposits may comprise a variety of plaques and vascular deposits with immunoreactivity for Aβ, PrP, or BRI$_2$ proteins.

Care is warranted when interpreting Aβ deposition as AD, since it also appears in ageing. Indeed, there is a spectrum of neurodegenerative features between neurodegenerative disease and nomal ageing, and it is sometimes more quantity than quality that differentiates pathology from "normal". Intraneuronal hyperphosphorylated tau inclusions, neurofibrillary tangles, must be considered as well. AD is a conformational disease associated with deposition of two proteins, Aβ extracellularly and tau intracellularly, the latter principally intraneuronally and not in glia. At a biochemical level, insoluble tau shows three major bands at 60, 64 and 68 kDa together with a minor band at 72 kDa, while both 3R and 4R tau isoforms are found in the diseased brains (3R + 4R tauopathy). In cases where there is abundance of neuritic plaques and lack of neocortical neurofibrillary tangles, plaque-predominant AD may be diagnosed. In addition, abundance of Aβ deposits, in particularly cotton wool plaques should raise the possibility of mutations in genes related to the processing or Aβ. On the other hand, relative paucity of Aβ deposits together with abundance of neurofibrillary tangles and neuropil threads mainly in the allocortex argue for neurofibrillary tangle dementia, which shows a similar biochemical pattern of tau as in AD. In summary, AD may be diagnosed as combination of Aβ plaques, including neuritic forms, together with neurofibrillary tangles and a documented clinical neuropsychiatric syndrome. However, quantity and distribution are relevant: Neurofibrillary degeneration follows an anatomical route described by Braak and Braak (I to VI stages) [75]. Semiquantitative scoring of neuritic plaques was the basis of neuropathological diagnostic criteria suggested by the Consortium to Establish a Registry for AD (CERAD) [76], while both lesions are considered in the NIA-Reagan Insitute criteria that are usually better suited for clinico-neuropathological correlation [77].

In prion disease extracellular PrP deposits may have, or have not, amyloid characteristics. PrP plaque amyloidosis is a major feature of Gerstmann-Sträussler-Scheinker disease, but may be seen in CJD as well. Further molecular classification of prion diseases involves Western blot examination of the protease resistant PrP and constellation of the codon 129 polymorphism in *PRNP*. These molecular subtypes have prognostic relevance and associate with distinct phenotypes [78].

Amyloid peptides in familial British and Danish dementia are cleavage products of mutated forms of the BRI$_2$ protein. In patients with familial British dementia, a single nucleotide substitution, and in patients with familial Danish dementia a 10-nucleotide duplication, in BRI_2 is the genetic background leading to amyloid formation in the brain [54].

In cases where intracellular protein deposits are suspected, not only the structural protein but the cell types and subcellular localization of inclusion should be defined as well. Proteins include tau, α-synuclein, TDP-43, or proteins linked to trinucleotide repeat diseases. In cases where inclusions are not immunoreactive for these, neurofilaments along with α-internexin, neuroserpin, and ferritin may be examined. Moreover, there are conditions where only ubiquitin immunoreactivity in inclusions suggests the presence of aggregated proteins.

When hyperphosphorylated tau is the major constituent of cellular inclusions, it is important to evaluate which cell types harbour inclusions and what is the morphology of these inclusions (see Table 11.3). In AD, neuronal inclusions predominate (neurofibrillary tangles), in FTLD with Pick bodies typical neuronal inclusions are seen together with less abundant astrocytic and oligodendroglial tau inclusions [45]. Insoluble tau in FTLD with Pick bodies is clearly distinct with bands at 60 and 64 kDa, while neuronal inclusions are composed of 3R tau. Ramified astrocytes and sparse oligodendroglial inclusions are observed as well. In progressive supranuclear palsy and corticobasal degeneration, features may overlap since both diseases contain glial and neuronal tau inclusions with a predominance of 4R isoforms and 64, 68 kDa bands of the insoluble tau in Western blotting. Here the anatomical distribution and morphological types of cellular inclusions distinguish the disorders [79]. While oligodendroglial coiled bodies are seen in both, astrocytic plaques are associated mainly with corticobasal degeneration, and tufted astrocytes with progressive supranuclear palsy. Distribution of threads in the white and gray matter is also helpful to discriminate these disorders. Globose neurofibrillary tangles are seen in progressive supranuclear palsy. Argyrophilic grain disease is another 4R tauopathy where oligodendroglial coiled bodies, pretangles and tau immunoreactive argyrophilic grains are restricted to limbic areas [80]. Mutations in *MAPT* may be associated with a variety of inclusions, partly overlapping with aforementioned sporadic diseases and partly unique in morphology. All types of patterns of insoluble tau, 3R, 4R or 3R + 4R tau isoforms may be detected [45].

When α-synuclein is the predominating protein of inclusions, three major disease forms should be considered. Although the exact differentiation of PD and LBD has debatable issues, PD features mainly intraneuronal cytoplasmic and neuritic deposits (Lewy bodies and Lewy neurites) following a defined anatomical pathway from lower brainstem regions to neocortical areas as suggested by Braak and coworkers (stages I to VI) [81]. Diagnostic criteria for LBD are also available [82]. Both Lewy body disease forms may show astrocytic (coil-like, star, crescent shaped) and oligodendroglial inclusions (Table 11.3) [83, 84]. In contrast to these, multiple system atrophy is a disease dominated by glial cytoplasmic inclusions (Papp-Lantos bodies [85]) that are rarely seen in PD and LBD; also the biochemical pattern of α-synuclein may differ [27]. Neuronal cytoplasmic inclusions are also distinguishable from Lewy bodies. In addition, neuronal or rarely glial nuclear inclusions may be seen in multiple system atrophy, but are not observed in Lewy body disorders. Interestingly, some neuronal intranuclear inclusions are immunoreactive for TPPP/p25 but not for α-synuclein [86].

TDP-43 has been identified as the disease protein in cases of sporadic and familial FTLD with ubiquitin immunoreactive inclusions with or without motor neuron disease, and in most cases of sporadic and familial amyotrophic lateral sclerosis. Depending on the subcellular (neuronal, glial, cytoplasmic, nuclear, neuritic, see Table 11.3) and anatomical (layer) distribution of TDP-43 inclusions, four subtypes are suggested for cases with FTLD with ubiquitin immunoreactive inclusions [45].

The genetic background links to chromosome 9, or to the *progranulin* gene, or to the *valosin containing protein* gene [64]. In familial forms of amyotrophic lateral sclerosis mutations in the *TDP* gene, but not *SOD1* (superoxide dismutase1) gene, may be associated with the presence of TDP-43 inclusions [43, 44].

The exact classification of further disease forms lacking relevant amount of aforementioned protein deposits depends on the evaluation of ubiquitin immunoreactivity. Further investigations here may include immunostaining for neurofilaments along with α-internexin, neuroserpin, or ferritin. In the presence of ubiquitin immunoreactive neuronal intranuclear inclusions, detailed genetic analyses may be needed, although the anatomical distribution of alterations may help in choosing which gene might be examined. Still cases will emerge in practice, where one cannot reach an exact classification. Here an interdisciplinary approach involving molecular biologists, biochemists, and neuropathologists seems most promising.

11.7 Concluding Remarks

There is considerable overlap in the accumulation of different aggregated proteins. Tau, Aβ, α-synuclein immunoreactive structures may be found in diseases defined by other neuropathological alterations. Recently, TDP-43 immunopositive neuronal inclusions were described in AD and LBD cases, but await clarification of their significance. Conditions associated with tau, α-synuclein or TDP-43 inclusions are listed in Table 11.4. Moreover, several pathologies may co-exist in the same brain

Table 11.4 Conditions associated with tau, α-synuclein, or TDP-43 immunoreactive deposits

Tau
Ageing
Down's syndrome
Gerstmann-Sträussler-Scheinker disease
Myotonic dystrophy
Neurodegeneration with brain iron accumulation type I
Familial British dementia
Niemann-Pick disease type C
PrP cerebral amyloid angiopathy
Subacute sclerosing panencephalitis
Dementia pugilistica
Postencephalitic Parkinsonism

α-Synuclein
Ageing
Down's syndrome
Neurodegeneration with brain iron accumulation type I
Subacute sclerosing panencephalitis
Neuroaxonal dystrophy
Ataxia teleangiectasia

TDP-43
AD
LBD

and may confound disease classification. Notorious examples include AD with Lewy body pathology. In fact, overlap neurodegeneration may be more the rule than the exception. While improved neuropathologic diagnosis greatly contributes to classification schemes of neurodegenerative diseases, the pathogenetic role of many misfolded and aggregated proteins still needs to be further elucidated. All this emphasizes the role of neuropathological evaluation in research studies.

In conclusion, molecular-neuropathologic diagnosis is based on the evaluation of the anatomical distribution of extra- and intracellular deposits composed of abnormal protein conformers. Biochemical examination of these proteins provides further information on isoforms that may specify morphologically distinct disease forms. Such a complex biochemical-morphological approach, in particularly the definition of protein forms related to specific clinical phenotypes, may serve as a basis for developing body-fluid based *in vivo* diagnostic biomarkers.

Acknowledgments This work was performed in the frame of EU Grant FP6, BNEII No LSHM-CT-2004-503039.

Notes

Since the submission of the manuscript the nomenclature of FTLD-U changed and now includes those cases, which do not show tau, α-synuclein, or TDP-43 immunoreactive inclusions. Former FTLD-U cases with TDP-43 immunopositive structures should be designated as FTLD-TDP.

Abbreviations

Aβ: β-amyloid; AD: Alzheimer's disease; APP: amyloid precursor protein; CJD: Creutzfeldt-Jakob disease; CNS: central nervous system; FTLD: Frontotemporal lobar degeneration; LBD: Lewy body dementia; MAPT: microtubule-associated protein tau; PD: Parkinson's disease; PrP: prion protein; TDP-43: transactive response (TAR) DNA-binding protein 43.

References

1. Bredesen DE, Rao RV, Mehlen P (2006) Cell death in the nervous system. Nature 443:796–802
2. Wishart TM, Parson SH, Gillingwater TH (2006) Synaptic vulnerability in neurodegenerative disease. J Neuropathol Exp Neurol 65:733–739
3. Sayre LM, Perry G, Smith MA (2008) Oxidative stress and neurotoxicity. Chem Res Toxicol 21:172–188
4. Mayer RJ, Tipler C, Arnold J, Laszlo L, Al-Khedhairy A, Lowe J, Landon M (1996) Endosome-lysosomes, ubiquitin and neurodegeneration. Adv Exp Med Biol 389:261–269
5. Fink AL (1998) Protein aggregation: folding aggregates, inclusion bodies and amyloid. Fold Des 3:R9–R23

6. Dobson CM (2003) Protein folding and misfolding. Nature 426:884–890
7. Dickson DW (2005) Required techniques and useful molecular markers in the neuropathologic diagnosis of neurodegenerative diseases. Acta Neuropathol (Berl) 109:14–24
8. Lee VMY, Goedert M, Trojanowski J (2001) Neurodegenerative tauopathies. Annu Rev Neurosci 24:1121–1159
9. Goedert M, Klug A, Crowther RA (2006) Tau protein, the paired helical filament and Alzheimer's disease. J Alzheimers Dis 9:195–207
10. de Silva R, Lashley T, Strand C, Shiarli AM, Shi J, Tian J, Bailey KL, Davies P, Bigio EH, Arima K et al. (2006) An immunohistochemical study of cases of sporadic and inherited frontotemporal lobar degeneration using 3R- and 4R-specific tau monoclonal antibodies. Acta Neuropathol (Berl) 111:329–340
11. Goedert M, Jakes R (2005) Mutations causing neurodegenerative tauopathies. Biochim Biophys Acta 1739:240–250
12. Dickson DW, Rademakers R, Hutton ML (2007) Progressive supranuclear palsy: pathology and genetics. Brain Pathol 17:74–82
13. Masters CL, Beyreuther K (2003) Molecular pathogenesis of Alzheimer's disease. In: Dickson D (ed) Neurodegeneration: The molecular pathology of dementia and movement disorders. ISN Neuropath Press, Basel, pp 69–73
14. Zetterberg H, Ruetschi U, Portelius E, Brinkmalm G, Andreasson U, Blennow K, Brinkmalm A (2008) Clinical proteomics in neurodegenerative disorders. Acta Neurol Scand 118: 1–11
15. Bertram L, Tanzi R (2003) Genetics of Alzheimer's disease. In: Dickson D (ed) Neurodegeneration: The molecular pathology of dementia and movement disorders. ISN Neuropath Press, Basel, pp 40–46
16. Chai CK (2007) The genetics of Alzheimer's disease. Am J Alzheimers Dis Other Demen 22:37–41
17. Chandra S, Fornai F, Kwon HB, Yazdani U, Atasoy D, Liu X, Hammer RE, Battaglia G, German DC, Castillo PE et al. (2004) Double-knockout mice for alpha- and beta-synucleins: effect on synaptic functions. Proc Natl Acad Sci U S A 101:14966–14971
18. Clayton DF, George JM (1998) The synucleins: a family of proteins involved in synaptic function, plasticity, neurodegeneration and disease. Trends Neurosci 21:249–254
19. Sidhu A, Wersinger C, Vernier P (2004) alpha-Synuclein regulation of the dopaminergic transporter: a possible role in the pathogenesis of Parkinson's disease. FEBS Lett 565: 1–5
20. Cabin DE, Shimazu K, Murphy D, Cole NB, Gottschalk W, McIlwain KL, Orrison B, Chen A, Ellis CE, Paylor R et al. (2002) Synaptic vesicle depletion correlates with attenuated synaptic responses to prolonged repetitive stimulation in mice lacking alpha-synuclein. J Neurosci 22:8797–8807
21. Sidhu A, Wersinger C, Vernier P (2004) Does alpha-synuclein modulate dopaminergic synaptic content and tone at the synapse? FASEB J 18:637–647
22. Wersinger C, Sidhu A (2003) Attenuation of dopamine transporter activity by alpha-synuclein. Neurosci Lett 340:189–192
23. Wersinger C, Sidhu A (2005) Disruption of the interaction of alpha-synuclein with microtubules enhances cell surface recruitment of the dopamine transporter. Biochemistry 44:13612–13624
24. Goedert M (2001) Alpha-synuclein and neurodegenerative diseases. Nat Rev Neurosci 2: 492–501
25. Dickson DW (2001) Alpha-synuclein and the Lewy body disorders. Curr Opin Neurol 14:423–432
26. Neumann M, Muller V, Kretzschmar HA, Haass C, Kahle PJ (2004) Regional distribution of proteinase K-resistant alpha-synuclein correlates with Lewy body disease stage. J Neuropathol Exp Neurol 63:1225–1235
27. Campbell BC, McLean CA, Culvenor JG, Gai WP, Blumbergs PC, Jakala P, Beyreuther K, Masters CL, Li QX (2001) The solubility of alpha-synuclein in multiple system atrophy differs from that of dementia with Lewy bodies and Parkinson's disease. J Neurochem 76:87–96

28. Polymeropoulos MH, Lavedan C, Leroy E, Ide SE, Dehejia A, Dutra A, Pike B, Root H, Rubenstein J, Boyer R et al. (1997) Mutation in the alpha-synuclein gene identified in families with Parkinson's disease. Science 276:2045–2047
29. Gwinn-Hardy K, Singleton AA (2002) Familial Lewy body diseases. J Geriatr Psychiatry Neurol 15:217–223
30. Gasser T (2007) Update on the genetics of Parkinson's disease. Mov Disord 22(Suppl 17):S343–S350
31. Prusiner SB (1998) Prions. Proc Natl Acad Sci U S A 95:13363–13383
32. Legname G, Baskakov IV, Nguyen HO, Riesner D, Cohen FE, DeArmond SJ, Prusiner SB (2004) Synthetic mammalian prions. Science 305:673–676
33. Safar J, Wille H, Itri V, Groth D, Serban H, Torchia M, Cohen FE, Prusiner SB (1998) Eight prion strains have PrP(Sc) molecules with different conformations. Nat Med 4: 1157–1165
34. Caughey B, Baron GS (2006) Prions and their partners in crime. Nature 443:803–810
35. Kovacs GG, Kalev O, Gelpi E, Haberler C, Wanschitz J, Strohschneider M, Molnár MJ, László L, Budka H (2004) The prion protein in human neuromuscular diseases. J Pathol 204:241–247
36. Aguzzi A, Heikenwalder M (2006) Pathogenesis of prion diseases: current status and future outlook. Nat Rev Microbiol 4:765–775
37. Kovacs GG, Trabattoni G, Hainfellner JA, Ironside JW, Knight RS, Budka H (2002) Mutations of the prion protein gene phenotypic spectrum. J Neurol 249:1567–1582
38. Hill AF, Joiner S, Wadsworth JD, Sidle KC, Bell JE, Budka H, Ironside JW, Collinge J (2003) Molecular classification of sporadic Creutzfeldt-Jakob disease. . Brain 126:1333–1346
39. Parchi P, Giese A, Capellari S, Brown P, Schulz-Schaeffer W, Windl O, Zerr I, Budka H, Kopp N, Piccardo P et al. (1999) Classification of sporadic Creutzfeldt-Jakob disease based on molecular and phenotypic analysis of 300 subjects. Ann Neurol 46:224–233
40. Ayala YM, Pagani F, Baralle FE (2006) TDP43 depletion rescues aberrant CFTR exon 9 skipping. FEBS Lett 580:1339–1344
41. Buratti E, Dörk T, Zuccato E, Pagani F, Romano M, Baralle FE (2001) Nuclear factor TDP-43 and SR proteins promote in vitro and in vivo CFTR exon 9 skipping. EMBO J 20: 1774–1784
42. Wang IF, Reddy NM, Shen CK (2002) Higher order arrangement of the eukaryotic nuclear bodies. Proc Natl Acad Sci U S A 99:13583–13588
43. Sreedharan J, Blair IP, Tripathi VB, Hu X, Vance C, Rogelj B, Ackerley S, Durnall JC, Williams KL, Buratti E et al. (2008) TDP-43 mutations in familial and sporadic amyotrophic lateral sclerosis. Science 319:1668–1672
44. Gitcho MA, Baloh RH, Chakraverty S, Mayo K, Norton JB, Levitch D, Hatanpaa KJ, White CL III, Bigio EH, Caselli R et al. (2008) TDP-43 A315T mutation in familial motor neuron disease. Ann Neurol 63:535–538
45. Cairns NJ, Bigio EH, Mackenzie IR, Neumann M, Lee VM, Hatanpaa KJ, White CL III, Schneider JA, Grinberg LT, Halliday G et al. (2007) Neuropathologic diagnostic and nosologic criteria for frontotemporal lobar degeneration: consensus of the Consortium for Frontotemporal Lobar Degeneration. Acta Neuropathol (Berl) 114:5–22
46. Cairns NJ, Uryu K, Bigio EH, Mackenzie IR, Gearing M, Duyckaerts C, Yokoo H, Nakazato Y, Jaros E, Perry RH et al. (2004) alpha-Internexin aggregates are abundant in neuronal intermediate filament inclusion disease (NIFID) but rare in other neurodegenerative diseases. Acta Neuropathol (Berl) 108:213–223
47. Cairns NJ, Lee VM, Trojanowski JQ (2004) The cytoskeleton in neurodegenerative diseases. J Pathol 204:438–449
48. Yokota O, Tsuchiya K, Terada S, Ishizu H, Uchikado H, Ikeda M, Oyanagi K, Nakano I, Murayama S, Kuroda S et al. (2008) Basophilic inclusion body disease and neuronal intermediate filament inclusion disease: a comparative clinicopathological study. Acta Neuropathol (Berl) 115:561–575

49. Orr HT, Zoghbi HY (2007) Trinucleotide repeat disorders. Annu Rev Neurosci 30:575–621
50. Takao M, Benson MD, Murrell JR, Yazaki M, Piccardo P, Unverzagt FW, Davis RL, Holohan PD, Lawrence DA, Richardson R et al. (2000) Neuroserpin mutation S52R causes neuroserpin accumulation in neurons and is associated with progressive myoclonus epilepsy. J Neuropathol Exp Neurol 59:1070–1086
51. Schrimpf SP, Bleiker AJ, Brecevic L, Kozlov SV, Berger P, Osterwalder T, Krueger SR, Schinzel A, Sonderegger P (1997) Human neuroserpin (PI12): cDNA cloning and chromosomal localization to 3q26. Genomics 40:55–62
52. Davis RL, Holohan PD, Shrimpton AE, Tatum AH, Daucher J, Collins GH, Todd R, Bradshaw C, Kent P, Feiglin D et al. (1999) Familial encephalopathy with neuroserpin inclusion bodies. Am J Pathol 155:1901–1913
53. Davis RL, Shrimpton AE, Holohan PD, Bradshaw C, Feiglin D, Collins GH, Sonderegger P, Kinter J, Becker LM, Lacbawan F et al. (1999) Familial dementia caused by polymerization of mutant neuroserpin. Nature 401:376–379
54. Vidal R, Delisle MB, Ghetti B (2004) Neurodegeneration caused by proteins with an aberrant carboxyl-terminus. J Neuropathol Exp Neurol 63:787–800
55. Vidal R, Ghiso J, Frangione B (2000) New familial forms of cerebral amyloid and dementia. Mol Psychiatry 5:575–576
56. Kovacs GG, Gelpi E, Lehotzky A, Höftberger R, Erdei A, Budka H, Ovádi J (2007) The brain-specific protein TPPP/p25 in pathological protein deposits of neurodegenerative diseases. Acta Neuropathol (Berl) 113:153–161
57. Kovacs GG, László L, Kovács J, Jensen PH, Lindersson E, Botond G, Molnár T, Perczel A, Hudecz F, Mező G et al. (2004) Natively unfolded tubulin polymerization promoting protein TPPP/p25 is a common marker of alpha-synucleinopathies. Neurobiol Dis 17: 155–162
58. Kovacs GG, László L (2001) The message of ubiquitin immunohistochemistry in conformational neurodegenerative diseases. In: Solomon B, Taraboulos A, Katchalski-Katzir E (eds) Conformational diseases – A compendium. Bialik Institute, Jerusalem, pp 249–258.
59. Wooten MW, Hu X, Babu JR, Seibenhener ML, Geetha T, Paine MG, Wooten MC (2006) Signaling, polyubiquitination, trafficking, and inclusions: Sequestosome 1/p62's role in neurodegenerative disease. J Biomed Biotechnol 2006:62079
60. Kuusisto E, Kauppinen T, Alafuzoff I (2008) Use of p62/SQSTM1 antibodies for neuropathological diagnosis. Neuropathol Appl Neurobiol 34:169–180
61. Bancher C, Brunner C, Lassmann H, Budka H, Jellinger K, Wiche G, Seitelberger F, Grundke-Iqbal I, Iqbal K, Wisniewski HM (1989) Accumulation of abnormally phosphorylated tau precedes the formation of neurofibrillary tangles in Alzheimer's disease. Brain Res 477: 90–99
62. Kuusisto E, Parkkinen L, Alafuzoff I (2003) Morphogenesis of Lewy bodies: dissimilar incorporation of alpha-synuclein, ubiquitin, and p62. J Neuropathol Exp Neurol 62:1241–1253
63. Mackenzie IR, Foti D, Woulfe J, Hurwitz TA (2008) Atypical frontotemporal lobar degeneration with ubiquitin-positive, TDP-43-negative neuronal inclusions. Brain 131:1282–1293
64. Pickering-Brown SM (2007) The complex aetiology of frontotemporal lobar degeneration. Exp Neurol 206:1–10
65. Duyckaerts C, Dickson DW (2003) Neuropathology of Alzheimer's disease. In: Dickson D (ed) Neurodegeneration: The molecular pathology of dementia and movement disorders. ISN Neuropath Press, Basel, pp 47–65
66. Kovacs GG, Head MW, Hegyi I, Bunn TJ, Flicker H, Hainfellner JA, McCardle L, László L, Jarius C, Ironside JW et al. (2002) Immunohistochemistry for the prion protein: comparison of different monoclonal antibodies in human prion disease subtypes. Brain Pathol 12:1–11
67. Kovacs GG, Preusser M, Strohschneider M, Budka H (2005) Subcellular localization of disease-associated prion protein in the human brain. Am J Pathol 166:287–294
68. Koperek O, Kovacs GG, Ritchie D, Ironside JW, Budka H, Wick G (2002) Disease-associated prion protein in vessel walls. Am J Pathol 161:1979–1984

69. Kovacs GG, Head MW, Bunn T, Laszlo L, Will RG, Ironside JW (2000) Clinicopathological phenotype of codon 129 valine homozygote sporadic Creutzfeldt-Jakob disease. Neuropathol Appl Neurobiol 26:463–472
70. Probst A, Herzig MC, Mistl C, Ipsen S, Tolnay M (2001) Perisomatic granules (non-plaque dystrophic dendrites) of hippocampal CA1 neurons in Alzheimer's disease and Pick's disease: a lesion distinct from granulovacuolar degeneration. Acta Neuropathol (Berl) 102:636–644
71. Tolnay M, Clavaguera F (2004) Argyrophilic grain disease: a late-onset dementia with distinctive features among tauopathies. Neuropathology 24:269–283
72. Kovacs GG, Pittman A, Revesz T, Luk C, Lees A, Kiss E, Tariska P, Laszlo L, Molnár K, Molnar MJ et al. (2008) MAPT S305I mutation: implications for argyrophilic grain disease. Acta Neuropathol (Berl). 116:103–118
73. Saito Y, Ruberu NN, Sawabe M, Arai T, Tanaka N, Kakuta Y, Yamanouchi H, Murayama S (2004) Staging of argyrophilic grains: an age-associated tauopathy. J Neuropathol Exp Neurol 63:911–918
74. Arima K, Nakamura M, Sunohara N, Nishio T, Ogawa M, Hirai S, Kawai M, Ikeda K (1999) Immunohistochemical and ultrastructural characterization of neuritic clusters around ghost tangles in the hippocampal formation in progressive supranuclear palsy brains. Acta Neuropathol (Berl) 97:565–576
75. Braak H, Braak E (1991) Neuropathological stageing of Alzheimer-related changes. Acta Neuropathol (Berl) 82:239–259
76. Mirra SS, Heyman A, McKeel D, Sumi SM, Crain BJ, Brownlee LM, Vogel FS, Hughes JP, van Belle G, Berg L (1991) The Consortium to Establish a Registry for Alzheimer's Disease (CERAD). Part II. Standardization of the neuropathologic assessment of Alzheimer's disease. Neurology 41:479–486
77. Group W (1997) Consensus recommendations for the postmortem diagnosis of Alzheimer's disease. The National Institute on Aging, and Reagan Institute Working Group on Diagnostic Criteria for the Neuropathological Assessment of Alzheimer's Disease. Neurobiol Aging 18:S1–S2
78. Pocchiari M, Puopolo M, Croes EA, Budka H, Gelpi E, Collins S, Lewis V, Sutcliffe T, Guilivi A, Delasnerie-Laupretre N et al. (2004) Predictors of survival in sporadic Creutzfeldt-Jakob disease and other human transmissible spongiform encephalopathies. Brain 127:2348–2359
79. Dickson DW (1999) Neuropathologic differentiation of progressive supranuclear palsy and corticobasal degeneration. J Neurol 246 Suppl 2:II6–II15
80. Ferrer I, Santpere G, van Leeuwen FW (2008) Argyrophilic grain disease. Brain 131: 1416–1432
81. Braak H, Del Tredici K, Rub U, de Vos RA, Jansen Steur EN, Braak E (2003) Staging of brain pathology related to sporadic Parkinson's disease. Neurobiol Aging 24:197–211
82. McKeith IG, Dickson DW, Lowe J, Emre M, O'Brien JT, Feldman H, Cummings J, Duda JE, Lippa C, Perry EK et al. (2005) Diagnosis and management of dementia with Lewy bodies: third report of the DLB Consortium. Neurology 65:1863–1872
83. Terada S, Ishizu H, Yokota O, Tsuchiya K, Nakashima H, Ishihara T, Fujita D, Ueda K, Ikeda K, Kuroda S (2003) Glial involvement in diffuse Lewy body disease. Acta Neuropathol (Berl) 105:163–169
84. Piao YS, Wakabayashi K, Hayashi S, Yoshimoto M, Takahashi H (2000) Aggregation of alpha-synuclein/NACP in the neuronal and glial cells in diffuse Lewy body disease: a survey of six patients. Clin Neuropathol 19:163–169
85. Papp MI, Kahn JE, Lantos PL (1989) Glial cytoplasmic inclusions in the CNS of patients with multiple system atrophy (striatonigral degeneration, olivopontocerebellar atrophy and Shy-Drager syndrome). J Neurol Sci 94:79–100
86. Baker KG, Huang Y, McCann H, Gai WP, Jensen PH, Halliday GM (2006) P25alpha immunoreactive but alpha-synuclein immunonegative neuronal inclusions in multiple system atrophy. Acta Neuropathol (Berl) 111:193–195

Index

A

Aβ, *see* Amyloid β-protein (Aβ)
ABri peptide, 24, 31, 52, 53
AD, *see* Alzheimer's disease (AD)
ADan peptide, 24, 53
Adaptor protein-4, *see* AP-4
Aggregation, 1, 5–7, 25, 34–37, 46, 53, 77, 80, 81, 83–85, 87–90, 97, 99, 100, 111, 112, 116, 117, 120, 133, 134, 136, 137, 139–142, 144–146, 148, 151, 159, 162, 163, 165, 167, 175–194, 207–216, 235, 236, 252, 253
Aggresome
  formation, 78, 84–86, 212, 236, 239–241, 244
  clearance, 84–86
Alexander's disease, 24, 53
Alpers disease, 24, 54
Alpha-synuclein, *see* α-Synuclein
Alzheimer's disease (AD), 1, 7, 8, 13, 22, 24, 27, 32–35, 38, 40, 42, 52, 53, 79, 97, 111, 116–119, 121, 125, 133, 134, 140, 144, 148, 160, 176, 179, 207, 213–216, 225, 236, 242–244, 254, 258–260, 262, 265–268
  familial, 117, 259
  sporadic, 32, 33, 42
ALS, *see* Amyotrophic lateral sclerosis (ALS)
AMPA receptor, 101
Amyloid, 6–13, 27, 32–36, 44–46, 48, 52, 53, 134–136, 139–142, 144, 147, 150, 151, 159–164, 176, 177, 194, 208–210, 213, 214, 253, 254, 257, 259, 261, 263, 265, 267
  the structure of, 11, 12
Amyloid β-protein (Aβ), 27, 32, 33, 254
Amyloid β-protein precursor (APP), 7, 8, 32, 33, 38, 119, 208, 254
Amyloidosis, 7, 22, 34, 134, 258, 265

Amyotrophic lateral sclerosis (ALS), 39, 39, 41, 80, 83, 98–100, 105, 111, 116, 117, 122, 123, 214, 256, 258, 266, 267
Androgen receptor, 23, 31, 46, 47, 49
Antioxidant, 175, 176, 182–188, 192–194
Anti-prion compounds, 136, 149, 150
Anti-Wallerian degeneration, 97, 103, 106
AP-4, 100, 101
Argyrophilic grain disease, 27, 39, 258, 261, 266
Ataxin-1, 23, 31, 46
Ataxin-2, 23, 31, 49, 50
Ataxin-3, 23, 31, 46, 50
Ataxin-7, 31, 51
Atrophin-1, 22, 31, 46–49
Autophagy, 77, 85–89, 97, 100, 105, 106, 112, 114, 115, 183, 189, 244, 252
  chaperone-mediated, 178–183
Axonal protection, 97–106
Axonal regeneration, 97, 101–103, 106

B

Baicalein, 185–187
Bovine spongiform encephalopathy, 8, 22, 36, 134

C

CAG repeat, 8, 46, 48, 50
  diseases, 8
Cdk5, *see* Cyclin-dependent kinase-5
Cellular prion protein, 120, 133, 140, 141, 146
  See also Prion protein (Prp), native (Prpc)
Central nervous system (CNS), 36, 41, 45, 53, 99, 105, 123, 139, 167, 210, 238, 239, 251, 252, 256, 257, 259
Centripetal degeneration, 99
Cerebral amyloid angiopathy, 34, 53, 267
Chaperones, 2, 25

Chaperone-mediated autophagy (CMA), *see* Autophagy, chaperone-mediated
Charged multivesicular body protein 2B, 258
Cockayne syndrome, 24, 54
Collins bodies, 257, 261
Complexin, 161, 168, 212, 234, 242
Corticobasal body, 263
Corticobasal degeneration, 22, 39, 121, 213, 244, 258, 261, 266
Creutzfeldt-Jakob disease, 8, 36, 39, 120, 134, 256, 258, 260, 261, 265
Cross-β structure, 10, 13, 33, 164, 177
Crowding, 5, 6, 44
αB-crystallin, 53, 183, 189–191, 194
Curcumin, 186, 187
Cyclin-dependent kinase-5, 225

D

D^2 concept, 56
Dementia with Lewy bodies, 22, 27, 38, 40, 45, 179, 207, 261, 268
Dentatorubral-pallidoluysian atrophy (DRPLA), 22, 31, 46–48, 122, 257, 258
Diffuse Lewy body disease (DLBD), 22, 38, 97, 99, 242, 244
Disorder predictors, *see* Predictors of intrinsic disorder
Disorder-to-order transition, 165
DisoPred, 4, 26, 229
DJ-1, 183, 189, 192–194
DNA excision repair protein ERCC-6, 24, 31, 54, 55
Dopamine, 40, 98, 160, 164, 175, 176, 180–183, 211, 214, 216, 255
Dopaminergic neurons, 24, 40, 98, 99, 01, 103, 118–120, 175, 176, 184, 185, 189, 193
Down's syndrome, 27, 33, 38, 42, 213, 267
Dystrophic neurite, 35, 39, 43, 119, 263

E

Electron spin resonance (ESR), 164
Endoplasmic reticulum (ER), 50, 111–125, 135, 136, 239
  associated degradation (ERAD), 112–114, 116
  stress, 111–125
  stress in neurodegenerative disorders, 116, 125
  stress induced cell death, 116, 124, 125
ERAD, *see* Endoplasmic reticulum (ER), associated degradation (ERAD)
Extracellular signal-regulated protein kinase 2 (ERK2), 235

F

Familial British dementia (FBD), 24, 38, 52, 53, 257, 261, 265, 267
Familial Danish dementia (FDD), 24, 53, 257, 261, 265
Fatal familial insomnia, 22, 36, 39, 120, 258
Ferritin, 257, 265, 267
Fibril, 11–13, 25, 52, 53, 134, 140, 142, 150, 151, 159, 160, 162–164, 180, 188–190, 213, 244
FK506-binding proteins, 215
Flavonoids, 185
Fragile X tremor-ataxia syndrome, 258
Friedreich ataxia, 258
Frontotemporal dementia, 35, 38, 39, 80, 83, 97, 111, 116, 121, 213, 252
Frontotemporal lobar degeneration (FTLD), 214, 252, 258, 261, 262, 266, 268
  with Pick bodies, *see* Pick's disease

G

Geldanamycin, 183, 189
Gerstmann-Sträussler-Scheinker disease, 22, 36, 37, 120, 134, 142, 144, 145, 258, 265, 267
Glial cytoplasmic inclusion (GCI), 39, 42, 43, 207, 210, 212, 214, 243, 266
Glial fibrillary acidic protein (GFAP), 24, 31, 53, 54
GlobPlot, 26, 49
Glutathione, 183, 185, 192, 193
Glyceraldehyde-3-phosphate dehydrogenase (GAPDH), 212, 234–236

H

HD, *see* Huntington's disease (HD)
Heat shock proteins, 53, 112, 124, 189–191, 193
α-Helix, 4, 8, 34, 37, 44, 50, 133, 138, 140–143, 145–151, 168, 227, 228, 230, 232, 235, 245
Hemorrhage with amyloidosis-Dutch type (HCHWA-D), 34
Hippocampus, 32, 33, 35, 42, 45, 98, 117, 118, 121, 212, 243
  CA1 region of, 105
Histone, 51, 208, 215, 216
Hsp27, 53, 189–191
Hsp70, 112, 115, 120, 183, 189–191, 193, 194
Hsp90, 183, 190, 191
Huntingtin, 1, 7, 8, 31, 46–48, 85, 97, 98, 105, 225, 242, 257, 259
Huntingtin yeast-two hybrid protein K (HYPK), 48

Index

Huntington's disease (HD), 1, 7, 22, 24, 46–48, 84, 85, 97, 98, 105, 111, 116, 122, 134, 225, 236, 242, 243, 257, 258
6-Hydroxydopamine (6-OHDA), 101, 105, 119, 120

I

Immunohistochemistry, 35, 118, 124, 189, 236, 242, 243, 253, 257, 262, 265
Inclusion
  biogenesis, 82, 87
  clearance, 77, 78, 82, 85, 86
  cytoplasmic, 39, 42, 43, 84, 207, 208, 210, 212–214, 236, 239, 243, 261, 263, 266
  neuronal intermediate filament, 258
  nuclear, 210, 263, 266
  Papp-Lantos, 43, 261, 263, 266
Inclusion bodies, 38, 53, 77, 80, 82–85, 87, 100, 214, 239, 253, 256, 257, 261, 262
α-Internexin, 256, 259, 262, 265, 267
Intrinsically disordered protein (IDP), 1–7, 10, 22, 25–29, 32, 43–45, 48, 49, 56
  amino acid composition, 5
Intrinsically disordered region (IDR), 25–27, 51, 56
Intrinsically unstructured protein (IUP), 1, 5, 25, 227, 245
IDP, see Intrinsically disordered protein (IDP)
IDR, see Intrinsically disordered region (IDR)
IUP, see Intrinsically unstructured protein (IUP)
IUPred, 5, 26, 228, 229

K

K63 ubiquitination, 83, 84, 86–88
Kennedy's disease, 23, 46, 47
Kuru, 8, 22, 36, 39, 120, 258–260

L

Leucine-rich repeat Ig-containing protein, see LINGO-1
Leucine-rich repeat kinase 2 (LRRK2), 215
Leukoencephalopathy with vanishing white matter, 123
Lewy body (LB), 22, 27, 28, 38–45, 80, 82, 83, 97, 99, 119, 120, 160, 161, 175, 176, 179, 182, 190–192, 207–210, 212–216, 235, 236, 239, 240, 242–244, 253, 255, 257, 261, 263, 266, 268
Lewy neurites (LNs), 28–42, 45, 209, 210, 215, 243, 263, 266
LINGO-1, 101–103
Lysosome, 85, 88, 115, 135, 236

M

Metal binding properties, 142
Methionine sulfoxide reductase A (MsrA), 175, 183
N-methyl-4-phenyl-1,2,3,6-tetrahydropyridine, see MPTP
Microtubule
  stabilizing drugs, 106
Microtubule-associated protein (MAP), 8, 115, 216, 245
Microtubule-associated protein tau, see Tau
Misfolding, 1, 6, 8, 10, 11, 24, 25, 77, 79, 97–106, 112, 123, 133–135, 140, 143, 146, 148, 150, 151, 189
Misfolding diseases, 1, 8, 11, 24
Mitochondrial DNA polymerase γ 24, 31, 54
Molten globule, 2, 23, 25, 49, 136
Motor neuron disease, 27, 239, 258, 266
MPTP, 98, 101, 179, 215
MSA, see Multiple system atrophy (MSA)
MsrA, see Methionine sulfoxide reductase A (MsrA)
Multiple sclerosis, 117, 123, 239
Multiple system atrophy (MSA), 22, 27, 28, 39, 42, 43, 179, 191, 207–210, 212–216, 242–244, 255, 257, 258, 261, 266
Myelin basic protein, 212, 234, 235, 238, 241
Myotonic dystrophy, 27, 258, 267

N

Natively unfolded protein, 46, 53, 175, 176, 210, 255
Neurodegenerative diseases, 1, 7, 13, 21, 22, 27–29, 32, 34, 36, 50, 56, 77–80, 82, 85, 97–100, 103–106, 111, 116, 117, 122, 124, 125, 136, 186, 188, 208, 213, 251–253, 257, 258, 262, 264, 265, 268
  classification of, 251–253, 262, 264
Neurodegenerative disorders, 7, 21, 24, 25, 28, 35, 37, 38, 40–43, 46, 47, 49, 53, 79, 81, 85, 99, 111, 113, 116, 117, 120, 121, 123–125, 134, 165, 175, 207, 209, 213, 214, 216, 236, 244, 245, 252, 253, 256–259
Neurodegeneration, 7, 21, 31, 32, 37, 43, 46, 52, 56, 80, 81, 85, 97, 98, 100, 101, 105, 106, 111, 112, 115, 120, 121, 124, 125, 135, 136, 139, 140, 181, 189, 193, 194, 207, 208, 244, 251, 263, 268
Neurodegeneration with brain iron accumulation type 1 (NBIA1), 22, 39, 43, 207, 267

Neurofibrillary tangle (NFT), 27, 33–35, 52, 53, 79, 117–119, 121, 213, 214, 216, 244, 257, 258, 260, 262, 263, 265, 266
Neurofilament, 122, 256, 265, 267
Neuron, 21, 24, 27, 31, 33, 35, 37–40, 42, 43, 46, 47, 49, 55, 79, 81, 85, 99–101, 103, 104, 115–122, 124, 125, 140, 148, 160, 161, 175, 176, 184, 185, 187, 189, 192–194, 207–209, 212–216, 226, 242, 244, 251–253, 255–257, 262
Neuroserpin, 257–259, 261, 265, 267
Neurotoxic lesions, 103
Non-amyloid-β component (NAC), 13, 27, 45, 46, 160, 162, 176, 178

O

Oligodendrocyte, 43, 102, 121, 123, 124, 209, 211–213, 216, 234, 235, 238–244, 263
Oxidative stress, 122, 123, 175, 176, 179–181, 183, 184

P

p25α, see Tubulin polymerization promoting protein/p25
p62, 84, 86–88, 100, 257
Paclitaxel, 106, 233
Paired helical filament (PHF), 33, 35, 262
Pale body, 263
Paramagnetic relaxation enhancement, 162, 163
Parkin, 80, 82, 83, 87, 89, 90, 114, 119, 214
Parkinson's disease (PD), 1, 7, 8, 22, 24, 27, 38–43, 45, 46, 80–85, 89, 97–99, 101, 103, 105, 111, 116–120, 124, 159–168, 175, 176, 179, 180, 184–187, 191, 192, 194, 207–210, 212, 214–216, 225, 235, 236, 242–244, 255, 258, 261, 266, 267
  familial, 160
  Drosophila model, 208
PD, *see* Parkinson's disease (PD)
Perinuclear granules, 263
Phospholipase D, 161, 165
Physiological prions, 1, 8
Pick body, 258, 260, 262, 263, 266
Pick's disease, 22, 39, 121, 244, 262
Polyglutamine diseases, *see* Polyglutamine repeat diseases
Polyglutamine repeat diseases, 50, 105, 117, 122
PolyQ diseases, *see* Polyglutamine repeat diseases
Polyubiquitination, 83
  K63-linked, 82, 84, 86–89
PONDR®, 4, 24, 26, 32, 49, 227, 229, 232

Predictors of intrinsic disorder, 26
Pre-molten globule, 25, 34, 36
Pretangle, 118, 243, 244, 262, 263, 266
Prion diseases, 1, 7, 8, 22, 24, 28, 36, 37, 39, 111, 116, 117, 120, 121, 133–137, 139, 140, 145, 148, 150, 151, 253, 255, 258, 259, 261, 265
Prion protein (Prp), 1, 2, 12, 13, 22, 31, 36, 37, 120, 133–151, 255
  metal ion binding, 143
  native (Prpc), 37, 133, 135–146, 148–151, 155
  scrapie (Prpsc), 28, 117, 120, 121, 133, 135–141, 144, 150, 255, 259, 261
  structure, 138
Progressive supranuclear palsy, 22, 38, 121, 213, 244, 258, 260, 262, 266
Proteasome, 3, 77–83, 85–88, 90, 98, 112–116, 135, 136, 178, 183, 214, 233, 236, 239, 240, 257
Protein aggregation, 6, 25, 37, 77, 80, 84, 89, 90, 97, 99, 117, 133, 134, 146, 151, 216, 236
Protein Data Bank (PDB), 2, 28, 29, 32
Protein deposits, 80, 119, 253, 257, 261, 267
  extracellular, 8, 259, 260, 262, 265
  intracellular, 253, 259, 260, 262, 265
  morphological types of, 259
Protein misfolding, 1, 8, 24, 77, 79, 97–106, 123, 134, 135, 148, 150, 151, 189
Protein quality control, 77, 90, 111, 112, 114–116, 118, 119, 122, 124, 125
Proteinopathies, 21, 24, 214, 257, 258, 261–263
Proteolysis, limited, 13, 22, 23, 48, 50, 231
Proteolytic stress, 77, 82, 85, 88, 240
Protofibril, 164, 175, 177–183, 186, 188, 189, 193, 194, 210

R

Rapamycin, 105
Reactive oxygen species (ROS), 116, 122, 176, 182–185, 187, 188, 192, 194, 252
Residual dipolar coupling, 4, 163
Residual secondary structure, 159, 161, 162

S

Scrapie, 8, 22, 36, 120, 133, 134, 137, 138, 146, 151, 255
Scrapie prion isoform, *see* Prion protein (Prp), scrapie (Prpsc)
β-Sheet, 6, 12, 13, 34, 37, 50, 52, 56, 138, 140–142, 145, 147, 150, 162, 175, 177, 178, 183, 186, 188, 213, 227, 228

Small nuclear ribonucleoprotein particles (snRNPs), 55, 56
Solid state NMR, 12, 13, 164
Spinobulbar muscular atrophy, 122, 258
Spinal muscular atrophy (SMA), 24, 55, 56
Spinocerebellar ataxia (SCA), 23, 46, 50, 51, 81, 122, 257, 258
SPT3-TAF9-ADA-GCN5 acetyltransferase (STAGA), 51
Stabilization centers, 228, 229, 232
β-Strand, 4, 8, 12, 13, 133, 134, 138, 163, 164, 232
Substantia nigra, 24, 40–43, 103, 118–120, 160, 161, 175, 176, 181, 184, 194, 209, 242, 261
Superoxide dismutase, 83, 98, 122, 267
Survival of motor neurons protein, 55
Synaptic vesicle, 160, 161, 164, 165, 167, 168, 176, 210, 214, 216, 242, 243
Synphillin-1, 83, 214
Synucleins, 45, 46, 159–161, 163
α-Synuclein, 1, 2, 5–8, 10, 11, 13, 22, 27, 28, 31, 37–46, 83, 85–87, 89, 99, 105, 117–120, 175–194, 207–216, 225, 234–236, 242–244, 254, 255, 257, 259, 261–263, 265–267
  aggregation, 216, 236
  amyloid fibril form, 159
  binding to lipid membranes, 159
  inclusions, 39, 43, 191, 257
  induced cell death, 120, 185, 194
  fibrillization, 175, 177, 178, 180, 181, 183, 185, 186, 189, 190, 192–194
  post-translational modifications, 179
β-Synuclein, 22, 31, 45, 46, 159, 160, 254
γ-Synuclein, 22, 31, 45, 46, 159, 160, 254
Synucleinopathy disorders, *see* Synucleinopathies
Synucleinopathies, 22, 37, 39, 43, 45, 176, 179, 186, 190–192, 194, 207–216, 225, 235, 243, 244, 258, 263
α-Synucleinopathy, *see* Synucleinopathies

T
TAR-DNA-binding protein-43, 214, 256, 258, 259, 261–263, 265–267
TARP, 101
TATA-box-binding protein (TBP), 31, 47, 51, 52
Tau
  hyperphosphorylated, 35, 121, 213, 243, 254, 260, 261, 265, 266
Tauopathies, 22, 34, 35, 39, 97, 106, 117, 121, 213, 244, 258, 262, 263, 265, 266

TDP-43, *see* TAR-DNA-binding protein-43
TFE, *see* Trifluoroethanol (TFE)
Thioflavin T fluorescence, 11, 177, 192
TorsinA, 182, 189, 191, 192, 216
Transmembrane AMPA receptor regulatory protein, *see* TARP
TPPP/p25, *see* Tubulin polymerization promoting protein/p25
Transmissible spongiform encephalopathy (TSE), 5, 36, 39, 120, 134–138
  therapy, 136, 137
Trifluoroethanol (TFE), 141, 146, 148, 151, 227, 232, 235
Trinucleotide repeat, 28, 256, 258, 265
Tuberin (Tsc2), 105
Tuberous sclerosis, 105
Tubulin polymerization promoting protein/p25, 208, 211–213, 225–250, 257, 266
  in brain, 241
  interacting partners of, 212
  occurrence of, 236, 241, 243

U
Ubc13, 78, 84, 86, 89
Ubiquitin
  -activating (E1) enzymes, 78, 79
  -conjugating (E2), 78, 79, 114
  -ligating (E3) enzymes, 78, 79
  immunoreactive inclusions, 262, 266
  modifications, 77, 81–84, 90
  carboxyterminal hydrolase L1, 105
Ubiquitin-binding protein 62/sequestosome-1, 86, 257
Ubiquitin-proteasome pathway (UPP), *see* Ubiquitin-proteasome system (UPS)
Ubiquitin-proteasome system (UPS), 77, 79–81, 86, 112–114, 120, 122, 178, 179, 181–183, 189, 240, 257
Ubiquitination
  non-proteolytic, 82
Unfolded protein response, 111, 112, 115
UPS, *see* Ubiquitin-proteasome system (UPS)

V
Vanishing white matter disorders, 117, 123

W
Wallerian degeneration, 97, 103, 104, 106, 122
  slow (WldS), 104, 105

X
X-linked inhibitor of apoptosis protein (XIAP), 98, 99